MUNI
PRESS

SCIENTIA
EST
POTENTIA

To my daughters Anna and Daniela

MASARYK
UNIVERSITY
MONOGRAPHS
Vol. 2

Jana Šmardová

WHAT TUMORS TEACH US

PARALLELS IN CELL AND HUMAN BEHAVIOR

Illustrated by Jana Koptíková
Translated by Jan Šmarda
Language revision by Benjamin J. Vail

Masaryk University Press
Brno 2023

The publisher would like to thank the Faculty of Science of Masaryk University and the LifeM and Roche companies for their financial support for the book.

This book was supported by the "Scientia est potentia" Masaryk University fund.

Book reviewed by:
Prof. Joseph Lipsick, M.D., Ph.D.
Prof. Jiřina Relichová, Ph.D.
Prof. Jan Žaloudík, M.D., Ph.D.

Painting "As Above, So Below" (the main graphic motif of the book)
by: Pierre Favre
(www.pierredestleonard.ch)

Illustrated by:
Jana Koptíková, Ph.D.
(IBA, Faculty of Medicine, Masaryk University)

ISBN 978-80-280-0376-0
ISBN 978-80-280-0377-7 (online ; pdf)
https://doi.org/10.5817/CZ.MUNI.M280-0377-2023

Table of Contents

Acknowledgements 9
A. Introduction 11
B. Overlaps 13
1A. The healthy multicellular organism 17
1B. Systems 29
2A. The multicellular organism: A system that can create cancer 33
2B. Eleven non-deadly sins 39
3A. Self-sufficiency in production of growth signals: Resistance to signals
for cell cycle arrest 43
3B. Man's race with himself: The truth does not win 51
4A. Damage of apoptosis 63
4B. Death is taboo 73
5A. Unlimited replication potential 83
5B. Reluctance to age 95
6A. Increased genetic instability 105
6B. Non-compliance with the rules 121
7A. Induction of angiogenesis 131
8A. Reprogramming of the energy metabolism 141
7B. Selfish misuse of resources 151
8B. Wasting 151
9A. Formation of metastases 165
9B. Underestimation of the value of relationships and home 181
10A. The ability to evade the immune surveillance 191
11A. Chronic inflammation 199
10B. Circumvention of laws 205
11B. Constant mobilization 213
12A. The tumor microenvironment: Two perspectives of tumor development . . . 217
12B. Pride and introspective individualism 233

13A. The tumor suppressor p53 245
13B. Wisdom and responsibility 263
14A. Multicellularity 271
14B. Social arrangement 285
15. What do tumors teach us? 299
Glossary 305
References 311
Name index 343
Subject index 345

Acknowledgements

The idea for this book was born at the end of 2008. Working on it has been a long, lonely road. For many years I searched for a form to write a book, got lost and returned to the beginning. A painting by the Swiss painter Pierre Favre, which I first saw in 1999 in the office of his French brother-in-law, the biologist Pierre Jurdic, accompanied, inspired and supported me all this time. It was a painting of a girl's figure painted into the nucleus of an osteoclast – a bone cell that Pierre Jurdic had photographed as part of his research. This painting, which I named *As Above, So Below* after acquiring it in 2014, symbolized and reminded me of the idea of general laws that apply analogously at different levels of life. I would like to thank Pierre Favre for the painting and for the kindness with which he allowed me to use it as the logo for "overlaps" and my lectures, as well as for the permission to use it on the cover and as a distinctive graphic element of this book. I would also like to thank Pierre Jurdic for giving me the opportunity to meet the painting and its author.

Around the same time I acquired the painting, in 2014, Dr. Jana Koptíková joined me to help with the book's figures and graphics. I first met Jana in 2002 when we were working together on an oncology textbook. Later, our ability to collaborate in a pleasant and fruitful way was confirmed while working on several joint papers. Perhaps that is why Jana accepted the challenging work on the book, work with a completely uncertain outcome that promised only some adventure and freedom. I thank Jana for years of persistent work, but also for friendship and unfailing support and understanding in moments when courage or faith or both left me. I also thank Jana for much fun and joy we had while working together. And above all, I thank Jana for the results of her work on the book, for the figures and the graphic design, which reflect my ideas better than I could have done myself.

In the summer of 2016, for the first time, I found the courage to show the first chapters of the book in progress to a first-time reader. She was my great colleague, a distinguished geneticist and teacher – Prof. Jiřina Relichová. I would like to thank

Jiřina for her many years of inspiration, support, and encouragement, and especially for her unequivocal support of this book and for her valuable comments.

I would also like to thank the other people who have read excerpts from this book and shared their perspectives and comments with me, encouraging me to continue writing. By name (and in the order in which they read and commented on excerpts of the manuscript), I thank Dr. Filip Trčka, Dr. Jiří Studený, Prof. Peter Tavel, Dr. Klára Maliňáková, and Ing. Libor Teplý. My special thanks go to Prof. Jan Žaloudík, who was my long-planned first reader of the finished manuscript. His assessment, understanding and support meant a lot to me. I also thank Dr. Zdeněk Řehák for providing the positron emission tomography results and their interpretation, and Prof. Petr Hořín for critical reading of the chapters on the immune system.

I would also like to thank the staff of the publishing house Munipress, who assisted me in the final stages of the manuscript processing, for their valuable comments, advice, and help. Namely I thank the director Dr. Alena Mizerová, and her co-workers Mrs. Martina Hovorková and Mrs. Radka Vyskočilová. I also thank to my husband Jan for translation of this book to English, and Dr. Benjamin J. Vail for proofreading the English version.

I also thank the Department of Experimental Biology and the Dean of the Faculty of Science of Masaryk University for important financial support.

And last but not least, I would like to thank my family. Mom and Dad, who, although uneducated themselves, always cultivated respect and a thirst for knowledge in me, supported me in my aspiration and dreams. They rejoiced in my work on the book, but they died before I could finish it. But they always believed in its completion. I thank my wonderful daughters Anna and Daniela, who shared my dream of the book for years and cheered me up, always inspiring and strengthening me in their own way. Finally, it would have been difficult for me to work on such a lengthy and uncertain project as writing this book without the reliable, supportive, and loving background that my husband Jan has given me over the years. Without his support, his patience, and most of all, his love and trust in us and in me, it would have been much more difficult to write this book. Thank you so much for that!

A. Introduction

Ethologist and Nobel Prize winner Konrad Lorenz wrote in his book *Civilized Man's Eight Deadly Sins*: "Far from being an insurmountable obstacle to the analysis of the organic system, a pathological disorder is often the key to understanding it. We know of many cases in the history of physiology where a scientist became aware of an important organic system only after a pathological disturbance had caused its disease" (Lorenz, 1974, p. 5). For tumors – currently one of the most common disorders of the human body – this is the perfect truth. Understanding the rules broken by tumors and followed by healthy cells provides an important insight into the fundamentals of healthy multicellular organisms. It is the tumors that remind us how perfectly organized the healthy body is, and how breathtakingly sophisticated is the scenario that makes all these incredibly different, diverse, yet interconnected cells coexist and work together in harmony.

And that is exactly what tumors teach us. Or they can teach it. They show us clearly and painfully what the violation of basic rules means for coexistence and cooperation in the community of cells that make up the multicellular organism. Perhaps they can also teach us something about the rules of coexistence and cooperation in our human community. Or at least, perchance we can get some insightful and playful suggestions to improve or correct the way humans live together and cooperate. In the pages that follow, I will refer to these free analogies as "overlaps."

The development of a tumor begins inconspicuously, just as a cluster of several proliferating cells. Cells that gradually multiply and, step by step, acquire more and more properties that increasingly distinguish them from healthy cells. The pathological behavior of these cells is reminiscent of the behavior of us – humans. Considering the harmonious perfection of a multicellular organism and the dramatic effects of tumor development, one begins to wonder if cancer is really just a disease and a matter of cells. What if cancer represents a more general principle? A more general failure of complex, multi-layered systems? Perhaps tumors thrive not only in our bodies, but also in our lives and in society as a whole. If so, it might be worth investigating whether the characteristics and behaviors that distinguish tumor cells from healthy cells are not parables or analogs for the characteristics and behaviors of us humans. And do they not then represent such characteristics or behaviors that pose a threat to society as a whole?

It can be argued that simple transfer of knowledge from biological to social systems is foolish, just as the functioning of living systems cannot be explained solely on the basis of understanding physical and chemical processes. This is undoubtedly true.

However, this book does not intend to provide a literal and authoritative transfer of knowledge from biology to the social sciences. It is more like an experiment, a trial, a game. We can use the biological system here as a starting point, as inspiration for analogies and reflections on human behavior. And what is the point? Some do not find one. Some may even see this book as pure nonsense. On the other hand, if only some of the findings about tumors and cancer were more generally applicable, and we were aware of all the limitations and simplifications we are making, the conclusions could be extremely useful to human society. While we already know the consequences of cancer and its effects in cells and multicellular organisms, it is difficult, if not impossible, to assess the behavior of people in society that we might label as "cancerous." We already have a lot of experience with the diagnosis and prognosis of biological tumors. We have developed tools to intervene in their further development and cure them. In contrast, our experience with human "tumor behavior" is very limited. The "overlaps" of biological knowledge with the human world could help us to become more sensitive to the "cancerous behavior" of people, groups, and especially ourselves in our own lives. And awareness of the possible consequences of human cancerous behavior could inspire, stimulate, and motivate us to become less tolerant and supportive of conduct that has bad consequences for ourselves and others. This awareness could help free us from many prejudices and from what we think are the unchangeable conditions of our time.

B. Overlaps

Is it appropriate to think of overlaps?
American writer, theorist, and essayist Susan Sontag would probably disagree. In her book *Illness as a Metaphor: AIDS and its Metaphors*, she writes: "But the modern disease metaphors are all cheap shots. The people who have the real disease are also hardly helped by hearing their disease's name constantly being dropped as the epitome of evil. And the cancer metaphor is particularly crass. It is invariably an encouragement to simplify what is complex and an invitation to self-righteousness, if not fanaticism" (Sontag, 1989, p. 85). Nevertheless, tumors and cancer are used as metaphors. And quite often and in a wide variety of contexts. And this is not a modern phenomenon. Already Publius Ovidius Naso used cancer as a metaphor in his Metamorphoses, written in around 8 AD, in the second book, in a chapter called "Envy and Aglaur":

> *Strenuous she strives to raise her form erect,*
> *But stiffen'd feels her knees; chill coldness spreads*
> *Through all her toes; and, fled the purple stream,*
> *Her veins turn pallid: cruel cancer thus,*
> *Disease incurable, spreads far and wide,*
> *Sound members adding to the parts diseas'd.*
> *So gradual, o'er her breast the chilling frost*
> *Crept deadly, and the gates of life shut close…*
> (Ovidius Naso, 1974)

But let us return once again to Susan Sontag. She writes elsewhere in her book: "To describe society as a kind of body, a well-disciplined body ruled by a 'head', has been a dominant metaphor for the polity since the days of Plato and Aristotle, perhaps because of its usefulness in justifying repression… Rudolf Virchow, the founder of cellular pathology, furnishes one of the rare scientifically significant examples of the reverse procedure, using political metaphors to talk about the body. It was the metaphor of the liberal state that Virchow found useful in advancing his theory of the cell as the fundamental unit of life. However complex their structures, organisms are, first of all, simply 'multicellular' – multicitizened, as it were; the body is a 'republic' or 'unified commonwealth'. Among scientific-rhetoricians Virchow was a maverick, not least because of the politics of his metaphors, which, by mid-nineteenth-century standards, are antiauthoritarian" (Sontag, 1989, pp. 6–7). This raises the question of what is actually the appropriateness of using metaphors. Which ones are acceptable? And when, in what context?

This question has been asked by Bruce H. Lipton, an American biologist and teacher whose research mainly deals with the development of muscle cells. In his book *The Biology of Belief*, he describes his educational experience. "I had been fascinated by the idea that considering cells as 'miniature humans' would make it easier to understand their physiology and behavior," he says. But he is aware of the risks of such a comparison: "Trying to explain the nature of anything not human by relating it to human behavior is called anthropomorphism. 'True' scientists consider anthropomorphism to be something of a deadly sin and ostracize scientists who knowingly employ it in their work" (Lipton, 2005, p. 35). He himself uses the opposite approach in his book, which he calls "cytomorphism" or "subcellularization," and explicitly states that we can learn much from cells. He believes that "cells teach us not only about the mechanisms of life, but also teach us how to live rich, full lives" (Lipton, 2005, p. 27). By conceptualizing his "cytopomorphism," Bruce Lipton fulfills to some degree the ideas and challenges of Carl Richard Woese (1928–2012). Woese was an American microbiologist known for constructing a prokaryotic phylogenetic tree based on sequence comparisons of ribosomal RNA and defining the new kingdom of Archaea. He was involved in introducing the theory of the RNA world and brilliantly interpreted new phenomena in biology throughout his long life. In his extensive essay on the future of biology published in 2004, he wrote: " Biology today is at a crossroads. The molecular paradigm, which so successfully guided the discipline throughout most of the 20th century, is no longer a reliable guide. Biology, therefore, has a choice to make, between the comfortable path of continuing to follow molecular biology's lead or the more invigorating one of seeking of a new and inspiring vision of the living world, one that addresses the major problems in biology that 20th century biology, molecular biology, could not handle and, so avoided. The former course, though highly productive, is certain to turn biology into an engineering discipline. The latter holds the promise of making biology an even more fundamental science, one that, along with physics, probes and defines the nature of reality. This is a choice between a biology that solely does society's bidding and a biology that is society's teacher." He believed that "the main task of biology is to help us understand the world, not to change it. The greatest task of biology is to teach us" (Woese, 2004).

Is it reasonable to think of overlaps?

And is it useful to ask this question? Is it even important to look for an answer to it? Overlaps are not science! And they do not even want to play on it! In this book, the term "overlaps" refers to facts based on the science described in the chapters on tumor biology (Chapters A). Overlaps (Chapters B) are just free analogies, metaphors, ideas, topics to think about, to inspire or to teach. According to Carl Woese, this is the task of the "New" Biology. According to Bruce Lipton, cells have this potential. And perhaps Susan Sontag would accept the overlaps. But who knows? We will not ask her again. She herself died of cancer…

1A. The healthy multicellular organism

The incidence of tumors in humans is not uncommon, nothing rare. It seems that the very basis of the human body, the way it was created and the way it functions, carries the potential for tumor formation.

A healthy multicellular organism is a harmonious community of a large number of cells. Each cell has its function, which it performs at a particular time and place for the maximum benefit of the organism as a whole. The individual cells of the organism are not in competition with each other. On the contrary, they support each other and work together.

The life of every human being begins in the same way: with one cell – a zygote, which is formed by the fusion of two germ cells – sperm and egg. From it, through repeated rounds of cell division and differentiation, gradually develops the embryo, the fetus, the newborn – and the baby then gradually develops and matures into an adult human being (Fig. 1). The body of an adult human is a complicated multicellular system. What do we know about this system?

How many cells are there in the human body?

It is no surprise to anyone that our bodies are made up of a large number of cells. But how many? The bodies of multicellular organisms differ in size and therefore in the number of cells that make them up. From tiny multicellular organisms we can deduce that the number of cells in their adult bodies is not random but on the contrary perfectly regular and accurate. For example, the body of the adult nematode *Caenorhabditis elegans* consists of 959 cells (Potts, Cameron, 2011). Counting the exact number of cells in the body of an adult human is, of course, impossible. In 2013, an Italian–Greek–Spanish research team attempted to make the most serious and rigorous estimate possible. The researchers used a model of an average person – a 30-year-old young adult who weighs 70 kg, is 172 cm tall and has a body surface area of 1.85 m². They admitted that the number they calculated is inherently inaccurate and varies from person to person. Their final estimate of the number of cells in the body of an adult

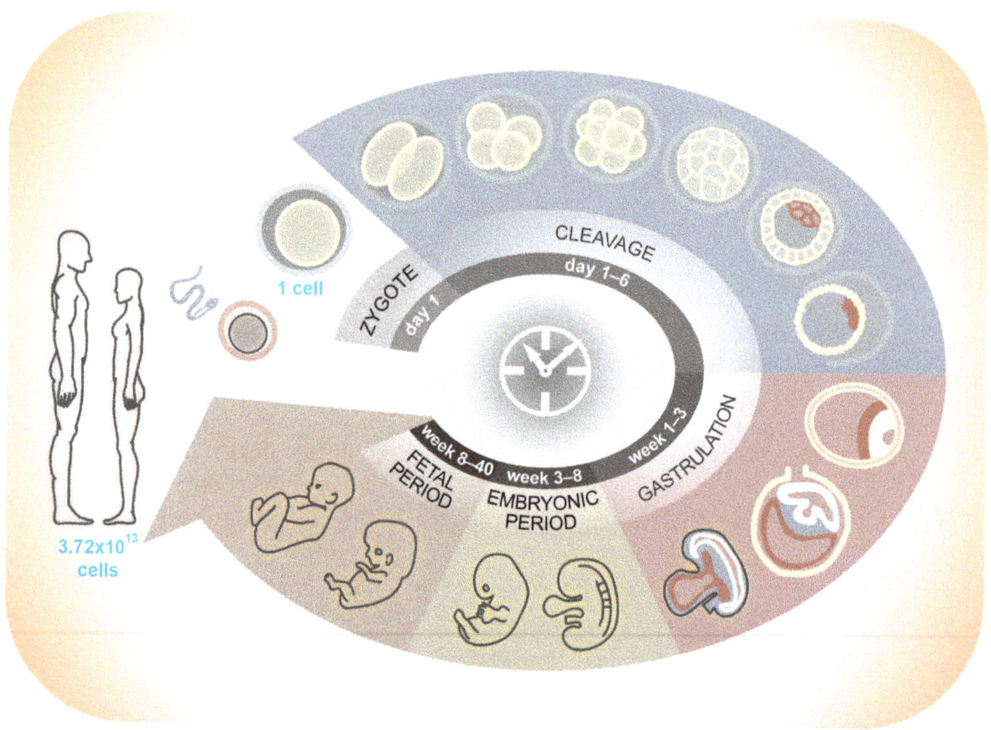

Fig. 1 Development of the human being
The life of a multicellular organism begins with the fusion of egg and sperm into a zygote. It divides again and again, the number of cells increases, the cells gradually differentiate, arrange themselves and form more and more complex structures. The stages of development after fertilization, which last about eight weeks, are called embryogenesis. Around the 56th day of development, when the foundations for all organs have been laid, the human embryo transforms into a fetus and fetogenesis begins. The body of an adult human consists of approximately $3.72 \pm 0.81 \times 10^{13}$ cells, which are differentiated into more than 200 different cell types.

human was $3.72 \pm 0.81 \times 10^{13}$ (Bianconi et al., 2013). This is a staggering number. Just for comparison, there are nearly 8 billion (7.86×10^9) people currently living on our planet. This means that there are 5,000 times more cells in each human body than there are people on Earth.

How many different cell types do we have in our bodies?

A typical feature of multicellular organisms is diversification. Cells differentiate into various specialized forms. We are naturally aware of this fact. We know that there are different cells in our body, such as blood cells (of which there are several types), neurons, muscle cells, epithelial cells covering the external and internal surfaces of organs, liver cells, and many others. But how many different types of cells are there in our bodies? The most common estimate is about 200 to 400 types. For example, the "Cells of the Adult Human Body" catalog published by Garland Science lists 210 clearly distinguishable cell types that can be determined by conventional histological examination techniques: that is, based on microscopic analysis of morphology (shape and structure) and staining. However, this list is not exhaustive as most cell types can be further subdivided into clearly distinguishable subtypes by other methods, e.g. physiological characteristics, degree of differentiation, developmental capacity, and so on. However, even the number 210 is overwhelming and reflects the considerable diversity of cells in our bodies. All these different cells are urgently needed by the body. Each is essential for survival and smooth functioning of the whole. Moreover, each type of cell must be present in a very specific quantity, and even a small deviation from the optimum threatens the viability of the body. Neither a deficiency nor an excess of cell types is tolerated. Deviations from equilibrium in either direction seriously disrupt the harmony of the whole.

A multicellular organism is a highly organized system of different cells

Not even the right amounts of the right types of cells is sufficient for the body to thrive. Also, all cells must be in the right place. Liver cells must not be in the muscles; muscle cells would not serve well in the brain or blood circulation. The nervous system would not be efficient if all the nerve cells were concentrated only in the brain and did not form a network running throughout the body, or if that network was broken somewhere. And the right placement, as well as the right connections – both structural and functional – are much more subtle than the examples given. A closer look at any piece of tissue would show that the order created by the cells in the body is enormous and the tolerance for deviation is low. Every part of the structure must be perfectly placed and arranged.

Considering the large number of cells in the human body, their diversity, and the need for their precise numerical representation and perfect distribution in the organism, two things might interest us. Both are well known, but we are seldom amazed by them. The first is the already mentioned fact that at the beginning of the development of an extensive, highly organized cell community there is always only one fertilized egg (Fig. 1). The nucleus of this cell contains genetic information that largely predetermines the morphology, physiology, and properties of the entire future organism that emerges from it. The second fascinating and also well-known fact is that the individual cells in the body, although so different from each other, all carry almost the same genetic information. This raises extremely interesting questions. How do the individual cell types develop? How do they differentiate? How do they find their place in a complex organism? How does a multicellular organism gradually grow and how is order created? And how is this perfect order maintained throughout the life course? Who or what drives the whole system and its development?

Development of the multicellular organism

Ontogenesis is the process of individual development from the beginning of the embryo until the death of the organism (Fig. 1). The actual beginning of the development of a new individual is fertilization. This is the moment when the germ cells, i.e. the unfertilized egg and the sperm, fuse, resulting in the formation of a fertilized egg or zygote, as mentioned earlier. After fertilization, the egg divides several times. The first division produces two daughter cells, the second produces four, then eight, sixteen, and gradually the number of cells in the developing embryo increases. These first divisions of the zygote are called cleavage. The cells formed at this early stage, the blastomeres, create a structure resembling a mulberry called a morula. A morula is a developmental stage consisting of up to 16 blastomeres. They are in close contact and constantly communicate with each other through a variety of molecular signals. They are similar, function similarly, and send similar signals.

Later, fluid enters the spaces between the blastomeres and the morula develops into the blastocyst. As the number of cells in the embryo increases, the different groups of cells begin to develop differently. Cell division comes under control and the first differentiation takes place. The outer layer of cells, called the trophoblast, surrounds the entire embryo and forms the basis of the future placenta. The embryoblast is an inner cell mass at one pole of the embryo that develops into the new individual being. During the differentiation of the embryoblast, which originally consisted of the same cells, groups of cells are gradually formed that differ from each other and give rise to the so-called germ layers: endoderm, ectoderm and mesoderm. The formation of the germ layers is called gastrulation. A very extensive rearrangement of cells occurs, as the basic orientation plan of the body and the foundations of organs and organ systems are laid (organogenesis). The individual parts of the embryo gradually become more finely specified, and complex tissues are formed, composed of many different types of cells to perform specific functions (histogenesis). While the cells in the morula and blastula still have considerable developmental flexibility and plasticity – they develop according to their position in the embryo – they lose this during gastrulation and acquire a clear and unchanging determination of their fate.

Morphogenesis as a process of formation of body structures has its molecular, cellular and organic levels. At the cellular level, this process includes proliferation, i.e. multiple rounds of cell division; gradual cell differentiation, i.e. diversification and specialization; and also programmed cell death, i.e. termination that accompanies development and occurs at a predictable time and place. At the organ level, cells move and arrange in three dimensions, establishing (and also breaking) mutual physical and functional connections. At the molecular level, these processes correlate with the regulation of gene expression, i.e. the turning of specific genes and gene groups on and off (Vyskot, 1999; Carroll, 2010). Even this brief overview of developmental processes in multicellular organisms raises the question: what drives such a complicated process?

The maintenance of the structure and function of a multicellular organism

The perfection of the multicellular machinery, the interaction and cooperation of cells, does not end with the completion of the organism's development when it reaches adulthood. The multicellular organism has an enormous potential to cope with a number of imbalances and disturbances that may occur during its lifetime due to the external or internal environment. It tends to maintain perfect order. When a part of the body is damaged, the body has a considerable ability to repair or replace the damaged tissue. This repair is usually near perfect, even for injuries that are anything but trivial. We often think of them as trivial because their correction is somehow automatic, without our conscious intervention. Who among us has ever scraped his knee? Even more severely damaged tissue is quickly and completely replaced in a form that is almost indistinguishable from the original tissue. And this is by no means a simply structured tissue. Quite the opposite! It consists of many different cell types that are in the right relationship to each other, correctly aligned and positioned in the tissue and ready to function perfectly.

Cell and tissue regeneration is not just about non-physiological damage. The organism itself is constantly wearing out and consuming some cells. Some of them even very quickly: for example, the cells of the skin surface, intestinal mucosa and more. Other cells wear out more slowly and are replaced only to a very small extent, such as the endothelial cells that make up the lining of blood vessels. Some cells in the body actually function for a lifetime and are almost never replaced. This is the case with the nerve cells, neurons. In any case, in a healthy organism, cells are always replaced at the right time, appropriate rate and in the right places to maintain an optimal structure in the body. So, the question arises again: where and how is all this controlled?

Genome and gene expression profiles

The basic plan and instructions for the development of the whole organism is genetic information stored in chromosomes (Fig. 2), which form the so-called zygote genome. In humans, there are 23 pairs of chromosomes. Twenty-two pairs are the so-called autosomes, and the last pair is the sex chromosomes. In males there are X and Y chromosomes, and in females there are two X chromosomes. Humans have a diploid genome, which means that each gene – with the exception of the genes on the sex chromosomes in males – is present twice in the cell. The human genome contains about 19,000 different genes (Frankish et al., 2019). The zygote, the fertilized egg, contains almost all the genetic information in its genome that is necessary (but not sufficient) for the development of the organism, for its function, and for determining many of its characteristics. Almost all cells, both in the developing and adult organism, contain the same genome, i.e. the same set of identical and equally arranged genes and regulatory sequences.

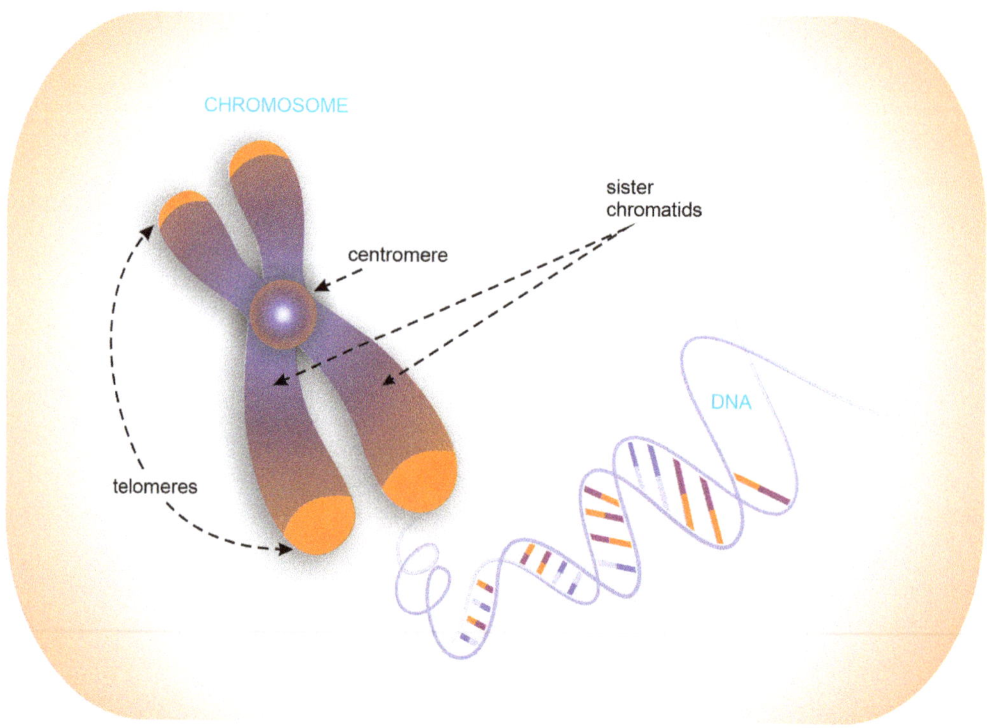

Fig. 2 Chromosome

Chromosomes are special structures located in the cell nucleus. They consist of deoxyribonucleic acid (DNA), which contains genetic information, and proteins. Shown is a condensed, mitotic chromosome with one centromere and two sister chromatids. The ends of the chromosome form telomeres. DNA forms two complementary strands arranged in a double-helical structure that is compact in a mitotic chromosome.

No cell in the body – not even a zygote – actively uses all the genes in its genome. Cells differ significantly in the genes they actively use and the genes that are turned off. An illustrative example is the genes that code for antibodies. Antibodies (immunoglobulins) are glycoproteins that, as part of the immune system of a multicellular organism, are able to recognize and neutralize objects foreign to the organism, known as antigens. Antibodies are produced by a specific type of white blood cells, B lymphocytes. It is clear that the genes that code for antibodies in B lymphocytes are turned on: they are actively expressed to produce their protein products in large quantities. All other cells in the body also have these genes in their genome, but they are turned off. They do not use them. The production of antibodies is the function of B lymphocytes, but not of neurons or muscle cells, for example. On the contrary, the function of a neuron requires a set of proteins that are unnecessary for both lymphocytes and muscle cells. Therefore, a neuron turns on a different set of genes than a lymphocyte or a muscle cell. So, in general, almost all cells in the body have almost the same genome, but they differ significantly in what are called gene expression profiles, i.e. the configuration of genes that are actively used in the cell and those that are turned off. For the sake of completeness, we should add that there are also genes that code for proteins that are necessary for the basic functions of every cell. They are constantly switched on in all cells and are called housekeeping genes.

Cell differentiation, stem cells and tissue structure

Differentiated cells are the basis of the tissues and organs of the adult organism. They perform specialized functions necessary for maintaining the function of a particular tissue and organ. These include the aforementioned lymphocytes and other blood cells, neurons, muscle cells, and many other cell types, totaling about 210 types. In addition, there are always undifferentiated and poorly differentiated cells, often referred to as supply cells. These include stem and progenitor cells. They give rise to the differentiated cells (Fig. 3). By general definition, stem cells are those which, when they divide, give rise to a copy of themselves (they have the capacity for self-renewal) and another more differentiated cell.

If a cell can give rise to all cell types of the embryo and adult, including germ cells (oocytes and spermatozoa) as well as extraembryonic structures, such as the placenta, we call it totipotent. For example, a fertilized egg and early blastomeres are totipotent. Cells that have all of the above characteristics except the ability to form extraembryonic structures are called pluripotent. Embryonic stem cells are pluripotent. All other stem cells found in specialized tissues of the embryo and adult are multipotent, meaning they are capable of forming multiple cell types, but not all, or unipotent, meaning they form only one cell type. More specialized stem-like cells are called progenitor cells. These can be multipotent or unipotent. A prerequisite for "stemness" is the ability to self-renew, i.e. after each division of the stem cell, at least one daughter cell retains the original characteristics of the parent stem cell. The daughter cell that loses this ability becomes a differentiated cell. This specialized cell produces all the proteins required for its specialized function. It may divide several times or not divide at all. Under normal circumstances, a differentiated cell cannot turn back into a progenitor or stem cell; differentiation is a one-way process.

Shortly after fertilization of the egg and the first three cycles of cell division, totipotent cells disappear and are replaced by pluripotent cells of the inner germ layers and multipotent cells of the outer germ layer. In the organs and tissues of the adult organism there is only a small supply of adult stem cells. Normally, they live here peacefully, without profound activity, but they have the potential to divide and differentiate when needed. The range of potential cell types that can arise from a given multipotent stem cell in the adult body is usually limited to those found in tissue. For example, hematopoietic stem cells in bone marrow can give rise to all the cells that make up blood, but not nerve, intestinal, or insulin-producing cells (Fig. 3). Although almost

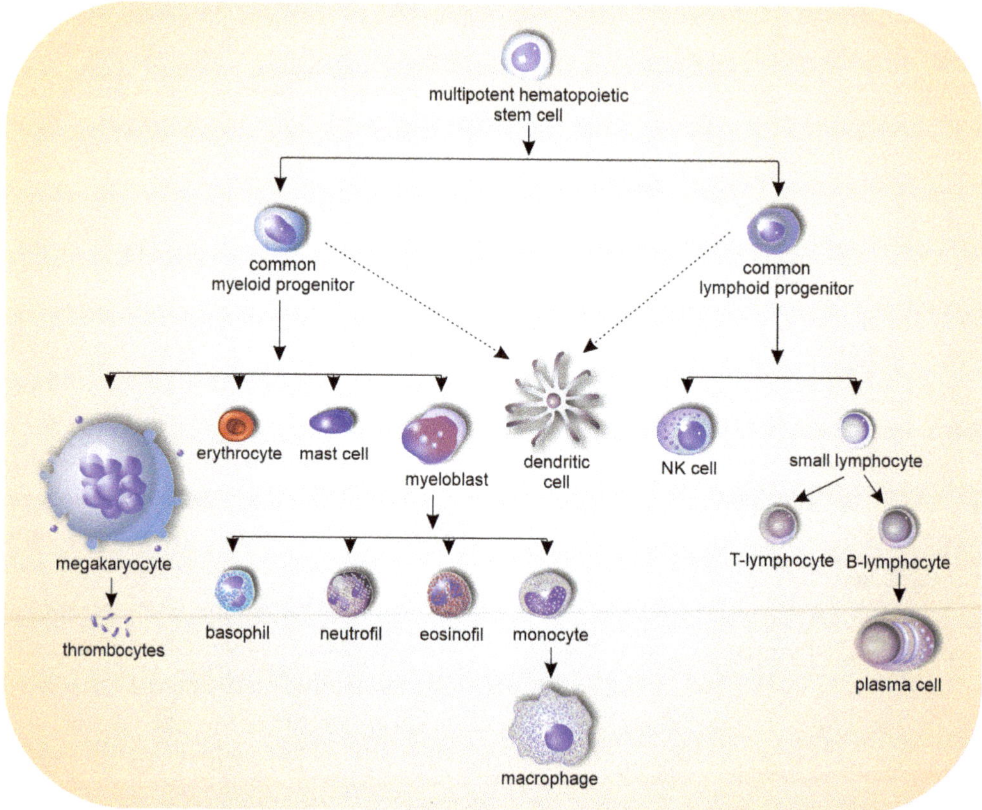

Fig. 3 Cell differentiation

An example of cell differentiation is the process by which immature, unspecialized stem cells gradually give rise to structurally and functionally specialized cells: hematopoiesis, the process of forming mature, fully differentiated blood cells and cells of the immune system. They arise by gradual differentiation from a multipotent hematopoietic stem cell via myeloid and lymphoid progenitor cells.

all cells in the body possess a complete genome, i.e. the genetic information required for the differentiation and development of each specialized cell, most somatic cells are restricted in their development to a certain spectrum of possible phenotypes.

How do cells communicate? Signaling pathways

In a multicellular organism, cells communicate constantly and intensively with each other. Even the blastomeres, which are formed when the fertilized egg undergoes cleavage, immediately begin to communicate. The exchange of signals between cells is necessary for the successful development of the organism. They learn their position and place in the organism, they share tasks and differentiate. But even in a mature organism, intercellular communication is essential for daily physiology. The textbook

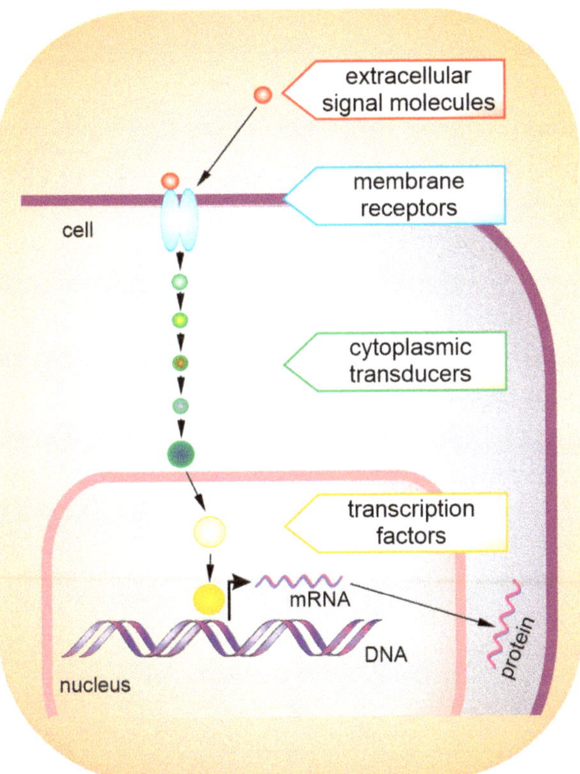

Fig. 4 Common signaling pathways
The receptor located in the cytoplasmic membrane of the cell is activated by the extracellular signal molecule and transmits the signal to the interior of the cell. The signal is then relayed by cytoplasmic carriers to an effector, such as a transcription factor, which can turn its target gene(s) on or off.

Essential Cell Biology states: "As in any busy community, there is a constant hub-hub of communication (in a multicellular organism): neighbors carry on private conversations, public announcements are broadcast to the whole population, urgent messages are delivered from distant sites to individuals, alarms are rung when danger threatens…" (Alberts et al., 1998, p. 481). Cells can communicate with each other directly, when they transmit signals through cellular connections, or indirectly, by releasing extracellular signaling molecules.

What is a signal? And how does the cell respond to it? What is a signaling pathway? In general, we can say that the signaling pathway consists of several points: (1) the extracellular signal molecule, (2) the receptor in the cell membrane, (3) the cytoplasmic signal transducer, and (4) the effector (Fig. 4). The extracellular signal is a molecule – a growth factor, a hormone, a cytokine, an amino acid and similar molecules – located in the extracellular space. Only those cells that are equipped with a suitable receptor for the signal on their surface respond to it. Cells lacking this receptor cannot respond to the signal, even if it is in their immediate vicinity. A receptor is a molecule that is usually located in the cell membrane (a transmembrane receptor). Its extracellular part is directed outward from the cell, while its intracellular part is directed into the cell. The extracellular part of the receptor is responsible for the recognition and binding of the signal molecule.

Where do signals originate, where do they come from? Signaling molecules are produced by cells. A particular signal molecule may be produced directly by the cell, which then responds to it. This is an autocrine type of signaling because it is a type of auto-signaling ("cellular self-talk"). The cell sends a signal and responds to it itself. Paracrine signaling is much more common. This involves signals that originate from a producing cell and affect other cells in its immediate vicinity rather than the cell itself. In endocrine signaling, the signal originates from cells that are very distant from the effector cells. An example of this is the hormone produced by an endocrine gland. The hormone may be secreted and distributed throughout the body or in a large part of the body, stimulating many cells of different types, often very distant from a particular gland. Also in this case, only those cells that have corresponding specific receptors on their surface can respond to the particular hormone.

After the binding of the signal molecule to the extracellular part of the receptor, the receptor changes the structure of its intracellular part. As a result, the signal is transmitted to the interior of the cell. The altered structure enables the receptor to transmit a signal to another molecule inside the cell. This molecule can pass it on to another molecule, and thus the signal can travel through a chain of carriers to finally reach the effector that triggers the cell's response. Signaling pathways can vary in length, branching and cross-talking. Signals can be amplified, attenuated, modulated, and integrated, i.e. combined with other signals (Fig. 5). This means that even if the same signal molecule is recognized by the same receptors in different cells, the response of these cells can be very different, depending on their specific equipment, i.e. their system of transmitters and effectors.

The cell may have a variety of receptors on its surface that decide which signals from the external environment it will respond to. Cells differ from each another in the composition of surface receptors, as well as in the composition of intracellular transporters and, in general, many other molecules. The difference can be small, as in developmentally related, similar cell types (e.g. mature and immature white blood cells), or significant, as in cells that are very far apart (e.g. a blood cell and a neuron). On the other hand, very different cells may carry the same or similar surface receptors and thus respond to the same (e.g. endocrine) signals. Let us not forget that all body cells have almost the same genetic information in their nucleus. The differentiation of cells is related to their development, i.e. to their individual history, to the experiences that caused each cell to use a different set of its genes. One of the possible outcomes of

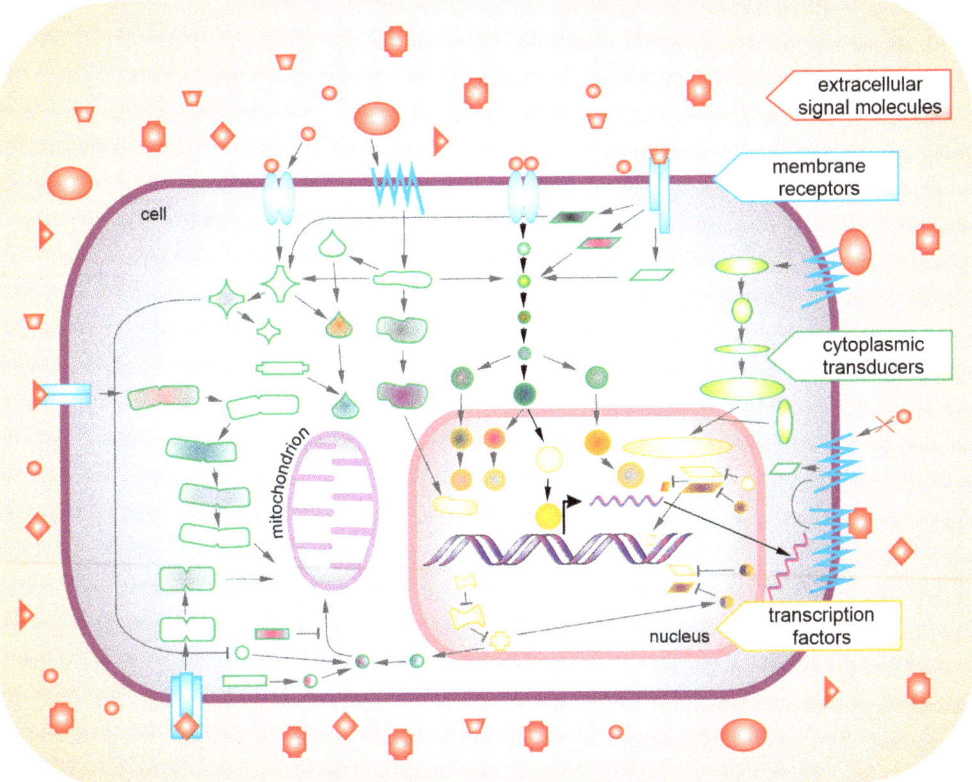

Fig. 5 Signaling cascades
In each cell, there are many different signaling pathways that vary in length, branching, crossing and intertwining. Thus, signals can be amplified, attenuated, modulated, combined and integrated.

signaling events in cells is the turning on or off of a particular gene or group of genes. One could also say that the acquisition of a new experience changes the cell and this change can more or less affect how it will respond to other stimuli, other signals, from then on.

We have described the signaling pathway as a sequence: extracellular signal – transmembrane receptor – cytoplasmic transporter – effector (Fig. 4). This is a very general scheme that has many variants. We have already seen that the spectrum of what can be an extracellular signal is relatively broad. Moreover, the signal need not always be extracellular. The cell is constantly monitoring its own internal environment, its own state, and naturally responds immediately to any imbalance, any deviation from the optimal state. It maintains homeostasis. Signaling pathways are not clearly defined linear pathways, but rather signaling networks: they branch out in different ways, intertwine, talk to each other, and interact with each other (Fig. 5). The end result of the signaling pathway can also be diverse and complex, triggering multiple parallel events. The spectrum of effectors is also large. It includes the aforementioned gene regulation, leading to the expression or silencing of various genes, as well as regulation of metabolism, alteration of the cytoskeleton or other cell structures.

Who controls a multicellular organism?

The structure and function of a multicellular organism is complex. The human body consists of a huge number of many different and perfectly arranged cells. This highly complicated and precise structure develops from a single cell and is extremely stable throughout the lifetime of a human being. Who or what controls all this? Who controls and ensures that the right cells are in the right place to differentiate into the right cell types and perform their proper functions for the benefit of the entire organism? Where and how are these important decisions made in the body?

There are basically only two possibilities. Either there is a "control center," which supervises the entire multicellular organism, assesses its condition, controls and coordinates the individual organs, tissues and cells, and ensures that they work for the benefit of the entire organism, or there is no such control center, and conversely, each cell is responsible for its harmonious integration into the function of the entire body and assumes full responsibility for its own condition, and to a reasonable extent (about 10^{-13} to 10^{-14}), for the condition and fate of the entire organism as well.

Thinking about the extraordinary complexity of the arrangement of the human body and being aware of the low tolerance of any imperfection in this arrangement, one would not even like to believe that the second option is correct! The existence of a control center that gives instructions, advice or commands to individual cells on what to do and how to behave in the body has not been proven. Quite the contrary. Individual cells and groups of cells work largely autonomously. However, they constantly communicate with each other and influence each other (Weinberg, 1998). Each cell constantly monitors its own state and communicates it to surrounding cells. At the same time, it constantly monitors and responds to its environment – the state of neighboring cells and the non-cellular microenvironment. This information is immediately perceived and processed for an appropriate response. This exchange of information takes place constantly. All changes and fluctuations in the state can be continuously reflected and balanced. In this way, equilibrium, homeostasis, is maintained. Without central authoritative control, the cells act quite independently, "from below," and in a coordinated way they create and maintain a perfect multicellular organism.

1B. Systems

What is the system?

A multicellular organism is a highly organized system of differentiated cells. By definition, a system is a complex consisting of parts that interact with each other. Flows of information, matter, and energy can occur between the parts of the system. Thus, a system is not just a set of parts; its quality is not simply the sum of the qualitative contributions of the individual parts. Mutual bonds and relationships develop among the components of the system, and the resulting quality is largely determined and reinforced by the relationships among the individual components of the system, their cross-links, organization, and cooperation.

General systems theory

The foundations of systems theory were laid in the mid-20th century by Ludwig von Bertalanffy, a biologist and philosopher. Systems theory is not concerned with the laws studied by particular scientific disciplines such as physics, biology, economics, sociology, and others, but seeks to understand and explain the principles of phenomena common to these objects. It focuses on the laws common to various living and social systems. It assumes that certain general principles apply to different systems, regardless of their nature. This is a very important idea and a very important concept, because recognizing, naming, and grasping such general principles and laws in one or more systems would allow their application to other systems as well. It would not be necessary to rediscover the same principle repeatedly in different isolated domains. General systems theory thus provides an ideal conceptual framework for unifying the various scientific disciplines and is a tool for transferring principles from one domain to another. Similarly, it provides a conceptual framework for the "overlaps" in this book.

In transferring knowledge from one system to another, it is obviously necessary to avoid superficial, simplistic analogies. Systems are very different, varying in complexity and hierarchies. There are so-called emergent properties, i.e. properties that occur at a certain level of complexity but do not exist at the system level of lower complexity.

Therefore, one cannot automatically assume that what is true for a system with a lower hierarchy is also true for a system with a higher hierarchy. Thus, the findings and conclusions from the biological system cannot simply be applied to the social system. This is not the correct approach. The human community system clearly has a different level of complexity than a system of organisms consisting of cells. Humans have many purely human characteristics and capabilities that are not found in cells. Human society is more complex than a community of cells. Yet Bertalanffy's systems theory is the bridge that connects research from different disciplines. Equipped with "overlaps" we can also proceed and cross this bridge.

Is bottom-up management a common feature of systems? The hive as superorganism

In multicellular organisms, the existence of a control center from which the individual cells in the body receive instructions, advice, or commands about what to do and how to behave has not been established. On the contrary, the individual cells operate largely autonomously and independently and, thanks to constant communication and mutual influence, are able to form a coordinated, well-functioning complex, a system. Is this networked and interconnected but highly autonomous behavior of the individual parts a general feature of living systems?

Phenomenal Bees: Biology of Bee Colony as a Superorganism is the title of Jürgen Tautz's book on the honey bee (Tautz, 2009). In it, he explains that the general tendency of evolution to create increasingly complex structures has produced multicellular organisms that have evolved into a superorganism. Superorganisms (e.g. bee colonies, hives) arise from the union of independent organisms and represent a new level of complexity. As a result, the world of living organisms acquires completely new possibilities. For example, a bee colony as a single biological unit can make decisions that are inaccessible to individual bees. The bee colony as a superorganism is an adaptable complex community of living beings consisting of many thousands of individuals that are constantly active and can adapt to the conditions of their environment and the activities of their colleagues in the nest. The overall behavior of the bee colony is not controlled by a superior authority, but is the result of cooperation and competition among the bees. Tautz gives several concrete examples of the functioning of the bee colony that correspond to this arrangement.

An example is cost optimization according to the clutch supply. In other words: how do bees figure out where to fly and where to collect pollen and nectar so that the distance to the source, and thus the cost, is balanced with the gain, i.e. the quantity and quality of pollen collected? He writes: " No bee from the hive can oversee supply and demand and perform the task of labor distribution. And yet we know from observations and experiments that the bee colony optimally distributes the labor force in space. How is this possible if no one in the colony knows about the overall situation? From a purely technical point of view, the solution lies in a decentralized, self-organizing distribution mechanism. Decentralized means that there is no authority to say 'where the hook is'. Self-organizing means that the pattern of labor distribution used by the superorganism is self-generating, thanks to the many close contacts between individual bees. These contacts are used to exchange information about millions of flowers in the wild. The superorganism stretches its net over more than 100 km², pulling it in where it is profitable, and letting it go where there is nothing to gain. The flying bees, which make up about 5–20% of the bees in the colony, are constantly on the lookout for new food sources and then inform their friends in the nest about their new discoveries" (Tautz, 2009, pp. 73–74).

Other examples of the same principle are honeycomb construction, nest hygiene and air-conditioning. Bees have very effective methods of temperature control. They lower the temperature by bringing in water and creating a draft, and they raise the temperature by producing heat using the pectoral muscle. But how do bees manage to correctly set not only the direction of change (i.e. cooling or heating), but also the exact target temperatures? How does the colony activate just the right number of bees to compensate for unwanted temperature fluctuations? There is a simple but very effective trick based on the fact that different bees react differently to a stimulus. Some bees begin to fan (cool) at even a very small rise in temperature. If this first ventilation effort brings the overheating under control, the problem is solved. However, if it fails to do so and the temperature continues to rise, other bees with the next higher sensitivity threshold respond and also create a draft. And so on, as bees with higher and higher sensitivity thresholds are gradually involved. When the temperature finally begins to drop, the first bees stop cooling (the bees with the highest sensitivity threshold, the last to become active). Gradually, more ventilation commands follow with a lower and lower sensitivity threshold. This strategy is very economical, because only as many controllers are directly activated as corresponds to the intensity of the problem. A prerequisite for such a strategy is the presence of reserves composed of different groups of bees. It is this variety and diversity of bees that enables the superorganism to always respond appropriately to current problems (Tautz, 2009).

Thus, a similar principle applies to the bee superorganism as to the multicellular organism: the decision is not made by a control center, but is the result of communication between different parts of the system – the bees in the hive, and the cells in a multicellular organism. Moreover, the example of temperature control shows how important the diversity, the individuality, of each bee is. If all bees were the same and reacted to the same stimuli in the same way, a gradual, gentle and very precise reaction of the superorganism would not be possible. There is nothing left but to exclaim: the glory of diversity!

Glory to uniqueness!

The human genome contains approximately 19,000 genes (Frankish et al., 2019), of which 6.7% are heterozygous. This means that on average, each human has two different alleles in 1,273 of their genes, while the remaining 93.3% of genes have two identical alleles. During the development of haploid gametes (eggs and sperm), up to 2^{1273} (i.e. 10^{383}) different types of germ cells with a unique combination of gene variants can be produced. This is, of course, many more possibilities than a human can generate and use in subsequent generations, but also a much higher number than all that humans have hitherto used during the entire existence of mankind. This means that of the total number of people living on Earth now, in the past, and that can be expected in the near or far future, no two people (except identical twins) are genetically identical (Relichová, 2009). Each of us carries a completely unique combination of genes in our genome, each of us is a completely unique original that is also the only one capable of exploring this unique part of humanity. With our death, this possibility will be closed forever. And so only each of us can and must explore and unfold our unique potential, in more poetic terms, fulfill the meaning of our own lives.

How to find a place in life? About communication I

How to fulfill the meaning of a unique human life? How do we find the right place in life? How do we find out who we are and what is the right place for us? Perhaps we will look for answers to these questions in this book. Maybe the following chapters about tumor cells and tumors will help us to see more clearly and sharply who we are and who we want to be, and who we do not want to be. But we already know some important information about the circumstances we find ourselves in. Systems operate without any central control. There are no instructions or messages from any center about what we should be and what we should do. We resemble specific cells in their specific places in the body, each in its specific microenvironment, which consists of both specific and variable material equipment and specific and variable signal structure and information. We too live in our specific locations, we are endowed not only with specific and unique genetic information, but also with specific and unique life experiences that arise from our own microenvironment and adaptations in response to environmental variability. In order to be one of these optimally functioning cells, a well-functioning component in a system in a way that is personally fulfilling and at the same time in harmony with the overall complex, we must constantly communicate truthfully: carefully and accurately perceive all signals coming from our internal and external environments, and constantly respond appropriately. This means that we must change ourselves ("turn some of our genes on and off", develop, differentiate) and also contribute to the change of our microenvironment (constantly provide information about ourselves as accurately as possible to others). And in this way – like the cells (and the bees…) – we must proceed truthfully, respecting our uniqueness and not being afraid of differences.

2A. The multicellular organism: A system that can create cancer

What is cancer, tumor, carcinogenesis?

Cancer – or malignant tumor disease – is a diverse group of diseases with a common root. A tumor is a pathological tissue that grows and develops without control. It develops in a multicellular organism, but is not coordinated with it, does not benefit it and does not serve its interests. The process of tumor formation and development is called carcinogenesis.

Basic features of carcinogenesis

In the previous chapter we indicated that there is no control center in the body that monitors the entire system and, based on the information thus obtained, sends instructions to individual cells or groups of cells to direct how they should function and behave. Cells have a high degree of autonomy and individual responsibility. This arrangement brings undeniable advantages to a multicellular organism – if it were not practical and functional, such a method of multicellular control would not have prevailed during evolution – but it also has some weaknesses. When cells are not directed and controlled authoritatively from a center and have a considerable degree of autonomy, some of them can go their own way. This allows such a cell to live and develop, but without supporting the organism as a whole. The perfect representative of such a cell which does not cooperate with the organism and pursues only its own interests, is a tumor cell. Tumor cells are sometimes said to violate essential rules of social behavior, and sometimes they are called selfish cells. A single cell that misbehaves does not pose a serious threat to the organism. A potentially dangerous situation arises when such a cell is able to survive and divide because of a genetic change, giving rise to daughter cells with the same genetic change and the same antisocial behavior. The tissue organization or even the entire organism may be infiltrated by the gradually spreading abnormal cell clone.

There are two basic features of carcinogenesis that are critical to understanding its molecular nature. (1) We refer to tumors as diseases of the genome or genetic diseases because they arise primarily from genetic alterations, or mutations. (2) To transform a normal healthy cell into a completely malignant cell, a single mutation is not sufficient. Carcinogenesis is a multistep process in which multiple mutations gradually accumulate. Each mutation confers a new property, ability, or disability to a cell. As a result, the affected cell is altered compared to other cells of the same type in which no mutation has occurred (Fig. 6).

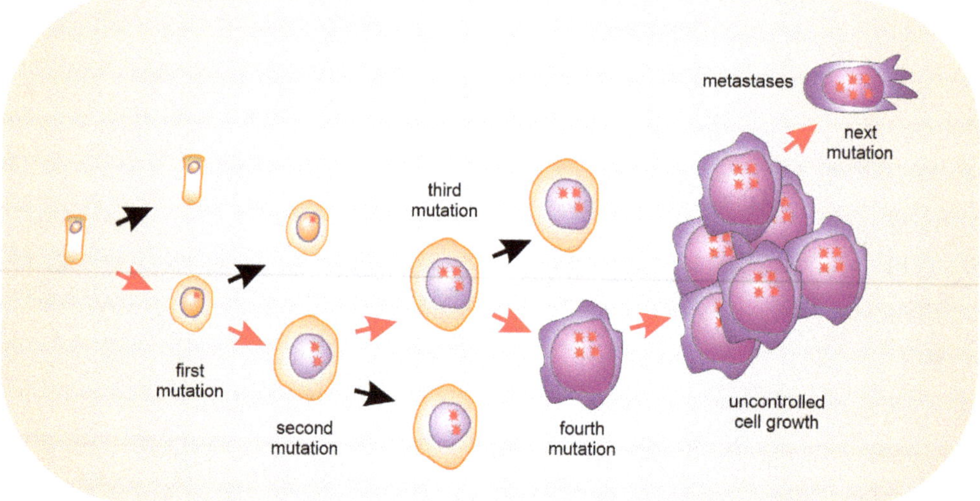

Fig. 6 The formation and development of a tumor cell: A simplified model
The formation and development of a tumor cell is a multistep process marked by a gradual accumulation of mutations. Each mutation confers a new feature to the cell and deepens its tumor character.

The tumor-causing mutations do not harm the affected cells but, on the contrary, provide them with some advantages that the neighboring cells lack. For example, the altered, mutated cells can divide and proliferate even under conditions where unaltered, healthy cells cannot. Or the mutated cells can survive under conditions that are lethal to healthy cells. Each step of cancer development is accompanied by what is called a clonal expansion wave. If a cell happens to acquire an advantageous property that neighboring cells lack, it can expand and overgrow them. At the same time, all of these expanding cells retain the beneficial property that made the expansion wave possible, and they form clones, a mass of genetically identical cells. Hence the name clonal expansion: growth of cells with the same properties (Fig. 7; Weinberg, 2014). The whole process culminates in the emergence of aggressive tumor cells living in a population of healthy "disciplined" somatic cells. Thus, the tumor cells gradually spread,

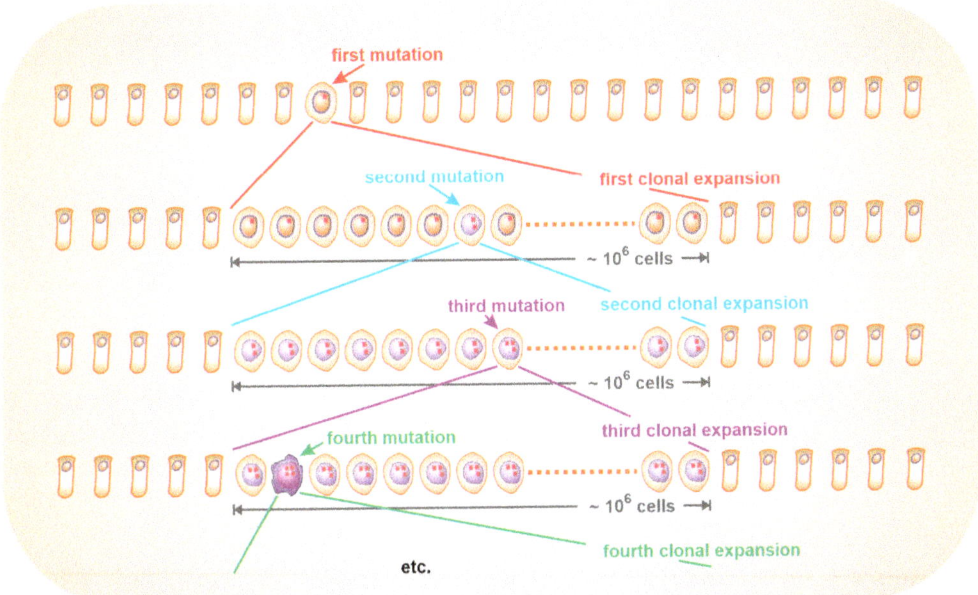

Fig. 7 **The formation and development of a tumor cell: A clonal model**
Each step of the carcinogenesis process is accompanied by a wave of clonal expansion. If a cell acquires a property that favors it over other cells through a random mutation, it can expand and overgrow other cells in its vicinity. All the expanding cells will carry this mutation, so they are genetically identical, constituting a clone. If this clone is large enough (about 10^6 cells), additional mutations may occur in some of its cells (the mutation rate in human cells is about 10^{-6} – see chapter 6A), which gives the cell another advantageous property, and another wave of clonal expansion follows. This process of beneficial mutations and clonal expansions is repeated (adapted from Weinberg, 2014).

suppressing the proper body structures and living at their expense. The advantages resulting from certain mutations in individual cells endanger the health and survival of the entire multicellular organism.

Oncogenes and tumor suppressors

As mentioned earlier, the human genome contains approximately 19,000 different genes (Frankish et al., 2019). In general, mutations in any of these genes can occur randomly. This raises the question of how many genes, once mutated, may be involved in the process of cancer development, and what types of proteins they encode. Mutations associated with carcinogenesis occur mainly in two types of genes, which we call proto-oncogenes and tumor suppressors. Their functions are opposite and have been defined according to their effect on cell division. Proto-oncogenes encode proteins that stimulate cell division. To have a carcinogenic effect, mutations in these genes must hyperactivate or prevent negative regulation of their products, and are therefore dominant. Any mutation affecting even one allele of the proto-oncogenes is sufficient to convert them to oncogenes. Tumor suppressors, in turn, encode proteins that inhibit cell proliferation. The cancer-promoting mutations in tumor suppressors reduce their activity. They are therefore recessive: to be carcinogenic, the mutation in the tumor suppressor must inactivate both alleles of the gene.

Eleven hallmarks of cancer

There are hundreds of specific genes that may be mutated in tumors and contribute to the process of carcinogenesis (Sondka et al., 2018; Tate et al., 2019). Even tumors of the same or very similar histological type can be caused by mutations in different genes. Therefore, the process of carcinogenesis can be highly individualized, and the pathways leading to tumor formation are very diverse. As a result, it has long been difficult to identify and understand them. Fortunately, we do not need to understand the function of each individual gene to understand tumor development. There is a common denominator for tumor-causing mutations.

Douglas Hanahan and Robert A. Weinberg formulated general principles of tumor development and summarized their basic ideas in a paper published in 2000. Almost immediately the broad community of biologists and oncologists accepted their idea that there are generally seven fundamental changes in the tumor cell and its microenvironment that together contribute to the development of a malignant phenotype. These changes are (1) self-sufficiency in growth signal production, (2) insensitivity to cell cycle arrest signals, (3) damaged apoptosis, (4) acquisition of unlimited replication potential, (5) induction of angiogenesis, (6) acquisition of metastatic potential, and (7) increased genetic instability (Hanahan, Weinberg, 2000).

In 2011 Hanahan and Weinberg published a revision and continuation of their already iconic 2000 paper in which they reviewed and confirmed the previously reported seven typical features of tumor cells and expanded them to include (8) reprogramming of energy metabolism, (9) the ability to evade immune surveillance, and (10) the presence of chronic inflammation, and they also highlighted the importance of (11) the tumor microenvironment (Fig. 8; Hanahan, Weinberg, 2011). Their updated review paper has also received widespread support.

In the introduction to this chapter there was a brief definition of what a tumor is, and therefore also cancer. It is more or less accurate, though not very descriptive. It is difficult to deduce from this definition what exactly cancer is, how insidious and dangerous it is, and why it kills so successfully. Better understanding what "tumor" and "cancer" really mean requires a more detailed explanation of the typical features of tumor cells, tumors, and the mechanisms involved. As Konrad Lorenz mentioned in the opening quote in the Introduction about the importance of pathological disorders for understanding organic systems, such an explanation can provide a deeper insight and understanding of the rules that a healthy multicellular organism and its cells follow.

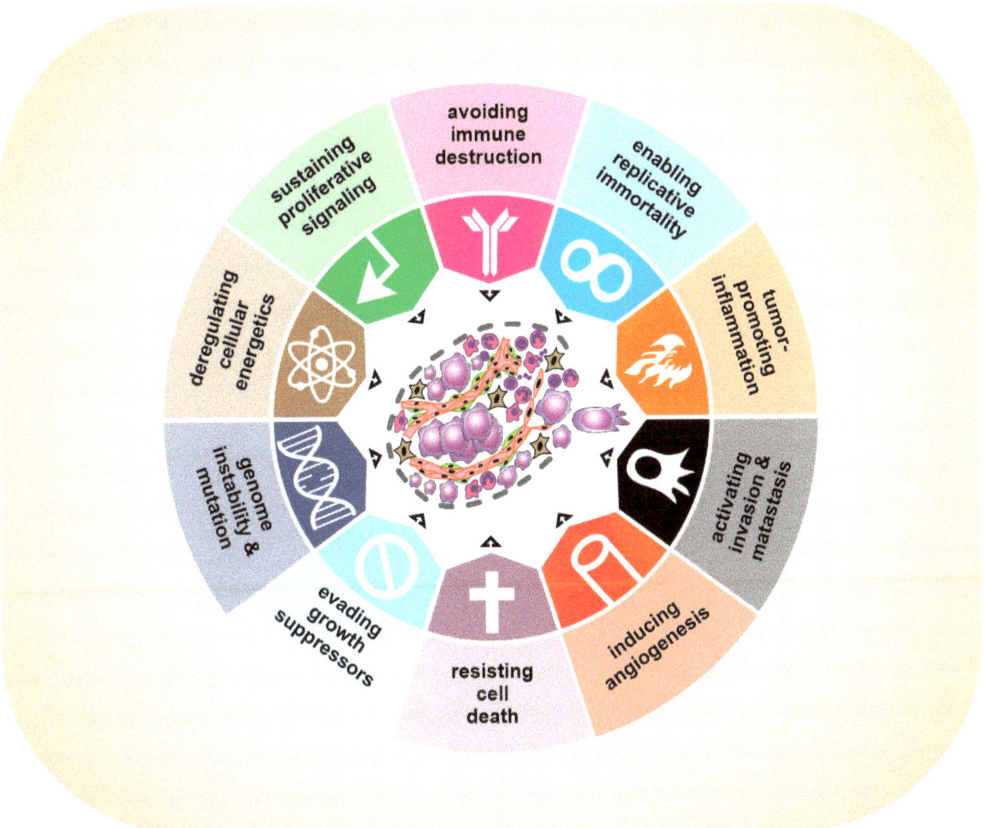

Fig. 8 Eleven typical hallmarks of cancer
Tumors are characterized by eleven features that make them malignant. These include self-sufficiency in production of growth signals, insensitivity to cell cycle arrest signals, failure to undergo apoptosis, acquisition of unlimited replication potential, induction of angiogenesis, acquisition of metastatic potential, increase in genetic instability, reprogramming of energy metabolism, ability to evade immune system surveillance, and presence of chronic inflammation. An altered tumor microenvironment is also implicated in tumor development (adapted from Hanahan, Weinberg, 2011).

Therefore, the following chapters will focus on the individual features that Hanahan and Weinberg described as tumor-specific (Hanahan, Weinberg, 2000; Hanahan, Weinberg, 2011).

2B. Eleven non-deadly sins

Sin

Sin was originally meant as a debt, mistake, error. Error also in the sense of insulting deities. Sin means to miss a target, to take a wrong path. It has also been written about sin: "It is a word, deed, or desire to oppose the eternal law." A plague that spreads like an epidemic… A polluted river that keeps branching off… A bad tree that bears rotten fruit… Sin separates man from society, isolates us, and prevents us from maintaining healthy relationships with others… It kills, because it multiplies, like cancer" (Riedl, 2011, pp. 9–11, 14).

The seven major sins

The seven major sins (also called the seven capital or deadly sins) have been defined by Catholic tradition since the time of Gregory the Great: (1) pride, (2) avarice, (3) envy, (4) wrath, (5) fornication, (6) greed, and (7) sloth. The term "capital sin" does not mean that it is the most serious and grave sin in front of God or the "deadliest" one. The adjective "capital" means that these sins come from the head. They are sins and human weaknesses that produce other sinful actions. The meaning of the descriptor "deadly" in this case is misleading, and therefore the traditional term "seven deadly sins" is not recommended.

The eight deadly sins

The book *Civilized Man's Eight Deadly Sins* by Austrian ethologist and Nobel Prize winner Konrad Lorenz first appeared in the 1970s. Lorenz spent many years studying innate and learned components of animal behavior, patterns of behavior, and their importance for the survival of species. Based on these studies, he tried to understand the essence of the self-destructive behavior that directly threatens human existence.

He described the following eight major transgressions of mankind, which he called – in a nod to the seven major ("deadly") sins – the "eight deadly sins": (1) Overpopulation of the earth, which, because of the superabundance of social contacts, forces each one of us to shut himself off in an essentially "inhuman" way, and which, because of the crowding of many individuals into a small spaces, elicits aggression. (2) Devastation of our natural environment, with destruction not only of our surroundings but also of man's reverential awe for the beauty and greatness of a creation superior to him. (3) Man's race against himself, which pushes the development of technology to an ever faster pace, blinding people to essential values and robbing them of time for the genuinely human activity of reflection. (4) The waning of all strong feelings and emotions, caused by overindulgence. The progress of technology and pharmacology furthers an increasing intolerance of everything inducing the least displeasure. Thus human beings lose the ability to experience a joy that is attainable only through surmounting serious obstacles. The natural waves of joy and sorrow in life ebb away into an imperceptible oscillation of unutterable boredom. (5) Genetic decay. In our modern civilization, apart from an "innate sense of justice" and a few transmitted traditions of right and wrong, there are no factors that exert a selection pressure tending to preserve instinctive norms of social behavior, although, with the growth of society, these are becoming more and more necessary. It is an alarming possibility that the rise of infantilism is turning a certain type of hippie into social parasites. (6) The break from tradition. A critical point is reached at which the younger generation is no longer able to communicate with the older generation, and still less to identify with it. Therefore, youths treat elders like an *alien ethnic group*, confronting it with the equivalent of nationalist hatred. Hence, the continuance of traditions is threatened. The reasons for this disturbance are to be found principally in the lack of contact between parents and children, which even at the earliest stages of infancy can have pathological consequences. (7) The increased indoctrinability of mankind. The increase in the numbers of people within single cultural groups, together with the perfection of technical means, has led to the possibility of maneuvering public opinion into a uniformity unprecedented in the history of mankind. Furthermore, the suggestive effect of an accepted doctrine grows with the number of its supporters, possibly in geometric progression. There are cultures in which an individual who purposely keeps aloof from the influence of the mass media, for example from television, is regarded as pathological. Deindividualizing effects are desired by all those whose intention it is to manipulate large groups of people. Opinion polls, advertising, cleverly directed fads and fashions help the mass producers on the western side of the iron curtain, and the functionaries on the other side to attain what amounts to a similar power over the masses. (8) The arming of mankind with nuclear weapons constitutes a threat easier to avert than the seven other developments described above (Lorenz, 1974, pp. 102–103).

Eleven non-deadly sins

Lorenz wrote *Civilized Man's Eight Deadly Sins* as a warning against human activities that may pose a direct threat to human existence. These eight different processes threaten to destroy not only our civilization, but also humankind as a species. His texts – "sermons," as he calls them – were originally written as transcriptions of radio broadcasts that generated a great response. By the time of the first printed edition in 1973, his pessimism had already begun soften, but Lorenz published the book precisely because he saw each of the eight "deadly sins" as a serious threat to humanity (Lorenz, 1974).

Hanahan and Weinberg described the eleven features that are typical of tumors (Hanahan, Weinberg, 2011). No feature, by itself, is capable of causing the transformation of a healthy cell and tissue into a tumor cell and tissue, and is therefore not lethal *per se* to the affected organism. Carcinogenesis is a multistep process in which there is a gradual accumulation of mutations and consequent progressive development of a fully malignant tumor phenotype. It is the combination and gradual accumulation of all the features that leads to the development of a dangerous and even lethal disease. Thus, none of the traits and characteristics I will attempt to describe in the overlap chapters are "deadly sins" *per se*, either. It is their accumulation and cross-links that can create a dangerous combination with tragic consequences. Each distinct characteristic that describes a tumor can therefore be considered a "non-fatal (i.e. non-deadly) sin," following Konrad Lorenz. Recall, however, that the seven major (i.e. also non-deadly) sins are described as "weaknesses behind other sinful actions." It is amazing how apt the notion remains, even figuratively, that "sins" together can lead to the development of cancer. Just as apt as other definitions of sin: "… a plague that spreads like a disease… A polluted river that keeps branching off… A way of life that is simply pernicious…" (Riedl, 2011, pp. 8–11).

In the following chapters, I will try to deal with and explain the non-deadly sins in social life on the basis of eleven typical signs of tumors. The race of the organism against itself caused by the ability of tumor cells to (1) produce growth signals on their own and (2) be resistant to signals that stop the cell cycle is the first sin. And I will show you that tumor cells are liars – the second non-deadly sin.

3A. Self-sufficiency in production of growth signals: Resistance to signals for cell cycle arrest

What is the cell cycle?

When we think of cell division, we usually first imagine a sequence of events that begins with the visibility of chromosomes in the nucleus (prophase), their arrangement in the equatorial plane, and their connection to the mitotic spindle via the centromere (metaphase). Then the chromosomes split into two chromatids, which are drawn to the opposite poles of the cell (anaphase), and the actual division of the entire cell into two daughter cells follows (telophase). This is mitosis (or M phase), the most visible and impressive part of the cell cycle, but it is only one of the four phases (Fig. 9). The entire cell cycle includes all the steps that the cell goes through from its formation to its division. The actual division – mitosis – is necessarily preceded by a synthetic phase (S phase), in which the cell duplicates (replicates) its genetic material so that it can later be completely passed on to its daughter cells. Even a small amount of missing or excess genetic information can seriously affect the viability or functionality of the daughter cell. The S and M phases are separated by the G2 phase, and the G1 phase separates the M phase from the S phase. Thus, the cell cycle is a cycle with four phases that always proceed in the same order: G1–S–G2–M. The name of the G phases is derived from the word gap, indicating that nothing important happens during these periods. However, the G phases are also full of cell activity. In the G1 phase, the cell grows, RNA and proteins are synthesized, and a supply of nucleotides and enzymes for DNA replication is created. In the G2 phase, organelles duplicate and structures necessary for subsequent cell division are formed, such as the mitotic spindle.

Step by step

Thus, the entire cell cycle is divided into several successive steps. They follow one after another, always in a precise, unambiguous order. The given sequence of events must be

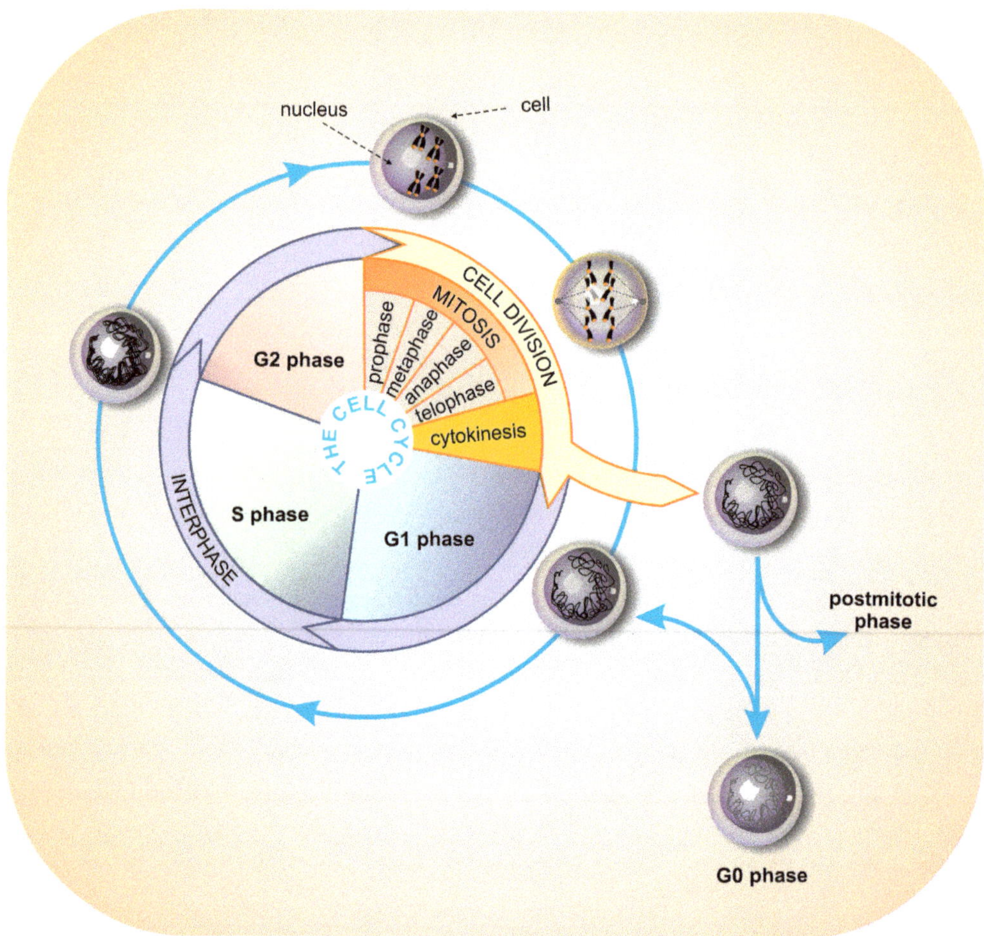

Fig. 9 The cell cycle
The cell cycle is a sequence of four phases that always follow each other in the same order: G1–S–G2–M. The genetic material (DNA) is replicated during the synthetic phase (S phase), divided during mitosis (M phase), and then the cell divides into two daughter cells. G0 represents a long-term resting phase from which the cell can return to the active cell cycle and divide based on specific signals. Mature, differentiated cells may enter a postmitotic phase that is irreversible.

followed, even if one of them takes longer than usual. Then the subsequent activities are also delayed, but the sequence of steps is maintained. How and by whom are these steps and their sequence ensured?

The main players in the cell cycle are two groups of proteins: cyclin-dependent kinases and cyclins. Cyclin-dependent kinases (CDKs) are enzymes that phosphorylate, i.e. add phosphate to, a specific group of target substrates. However, CDKs cannot do this on their own. In order to recognize their target substrates and label them with phosphates, they must be bound to a member of the cyclin protein family. The name "cyclin-dependent" aptly describes a key property of these kinases: their function is completely dependent on the binding of the corresponding cyclin. Cyclin-dependent kinases are constantly present in the cell – throughout the cell cycle – but their function depends on the presence of their partner cyclin. In contrast, cyclins – as their name implies – appear periodically in the cell, always reaching their peak at a very specific phase of the cell cycle (Fig. 10) and disappearing after due to degradation.

Each step of the cell cycle is associated with the activity of a specific pair of cyclin and cyclin-dependent kinase molecules. Together they phosphorylate the target proteins, which then perform the specified activities or transmit this signal to other proteins. In summary, at each phase of the cell cycle, the cell undergoes changes and undertakes a whole range of different activities aimed at fulfilling the tasks and duties of that particular phase and preparing the cell for entry into the next phase. These

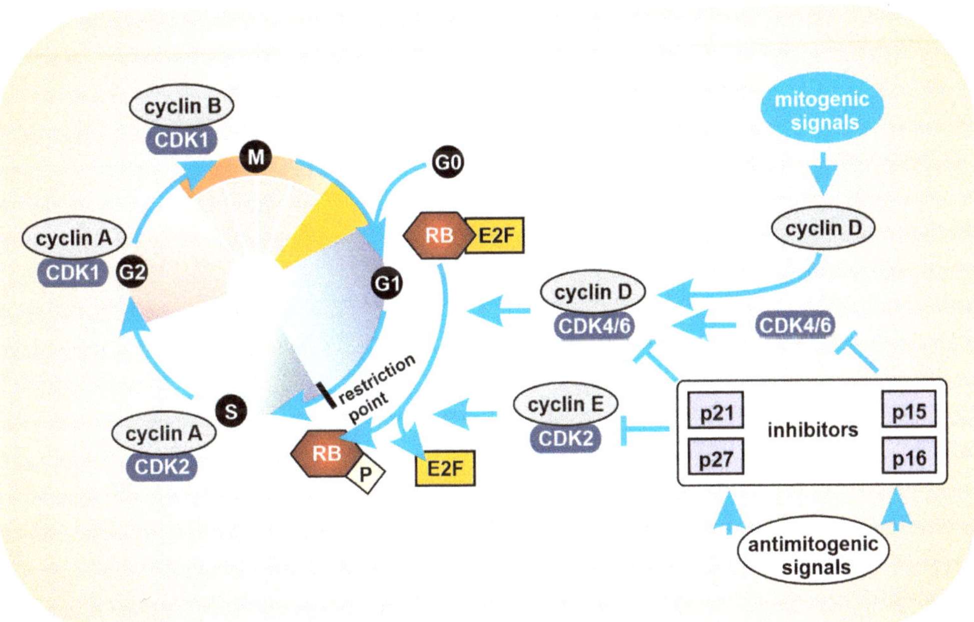

Fig. 10 Cell cycle control
Cyclin and CDK complexes and their inhibitors are key molecules controlling cell cycle phases. Passing the restriction point, which precedes entry into S phase, depends on phosphorylation of the RB protein. The degree of RB phosphorylation determines its ability to bind to E2F transcription factors, which, when released from binding to RB, initiate S phase of the cell cycle. The RB protein is phosphorylated by the complex of cyclin D and CDK4/6 and subsequently by a complex of cyclin E and CDK2, unless prevented by high concentrations of inhibitors. Release of the E2F factors from RB binding leads to entry into S phase and irreversible onset of cell division (adapted from Lundberg, Weinberg, 1999).

activities include the synthesis of cyclin, which, after binding to the appropriate CDK, provides for the initiation and immediate execution of the next step of the cell cycle.

Full completion of the previous phase is necessary to enter the subsequent phase. This is ensured by several checkpoints located at different points in the cell cycle. A checkpoint is a molecular mechanism by which the cell verifies that the current step of the cell cycle has been completed and that nothing can prevent the cell from moving on to the next phase. This ensures that the individual phases of the cell cycle actually follow each other in the correct order, and that no phase is omitted, repeated, overtaken by or overlaps with another phase.

Cyclin-dependent kinases and cyclins are important players in the cell cycle. However, there is a third group of no less important molecules that cooperate with them. These are inhibitors that block the assembly or activity of complexes of cyclins and CDKs, halting the cell cycle just before they can initiate the appropriate phase. Inhibitors are another tool that enables effective cell cycle control.

Restriction point, the beginning of cell division

The sequence of cell cycle phases resembles the domino effect: one falling tile hits the next and activates it to bump the next, and gradually all the dominoes fall in exact order, just as in the cell cycle phases. The crucial question is, what is the first stimulus that triggers this process? This first domino and control over it are key elements in understanding the regulation of cell division. The molecular nature of this first piece has already been largely elucidated. We know that this essential decision point (called the restriction point) of the cell cycle occurs at the end of the G1 phase. Therefore, all the important information about cell division – the signals that need to be taken into account – are directed to this point. If the signals the cell receives from its external and internal environments do not clearly indicate the start of cell division, the cell will remain in G1 phase until the signals change and the cell cycle can continue. The cell may remain in this quiescent phase for a very long time. Even a cell that is not dividing can still perform its functions. This long-term resting phase is called the G0 phase. Its essential feature is that it is reversible. The cell can resume the cell cycle from the G0 phase at any time – depending on appropriate signals. In addition, cells can enter the post-mitotic state. This state is irreversible and prevents the cell from re-entering the cell cycle again (Fig. 10). It can live and perform a number of functions, but it will never divide again. The postmitotic state is common, for example, in terminally differentiated cells.

Restriction point regulation

Division into two daughter cells is a critical, irreversible event that significantly alters the cell. For example, cell division is associated with cell maturation – the progression of cell differentiation. This means that after completion of the cell cycle, not only do two cells emerge from a parent cell, but these daughter cells also reach a more differentiated state. They mature and can therefore function differently from the mother cell. Apart from the importance of cell division for the dividing cell itself, the fact of cell division is also important for the organism as a whole. Where there was once only one cell, now there are two cells. We have already seen that the order of the multicellular organism is high and allows only slight variations. The optimal structure and function of tissues depends on the presence of the correct cell types, their proper number, and the correct spatial arrangement among them. Perfect control of the cell cycle of all the cells that make up a tissue is an absolutely necessary condition for its construction and maintenance throughout the life of the organism. Whereas unicellular organisms (e.g. bacteria, yeasts) are highly autonomous and their division is limited primarily by the availability of nutrients, each cell in the multicellular community of the organism must be attentive to the needs of its neighbors when deciding on cell division and must subordinate its division to the others' needs. A cell in a multicellular organism will divide only if it receives clear instructions from its environment to do so, i.e. if it receives a mitogenic signal or a signal that stimulates mitosis – cell division.

The restriction checkpoint is sometimes called the R point, sometimes the Start. This is because a cell that is allowed to pass this point is destined to go through the entire cell cycle. According to some authors, the restriction checkpoint could also be referred to as the "stop" because it is the only checkpoint at which the cell cycle can be halted for an extended period of time (Alberts et al., 1998). This dual conceptualization of this important checkpoint perhaps reflects the fact that even the decision to enter the process of cell division, i.e. the decision to pass the restriction point, occurs at two levels. Mitogenic signals must be present to initiate the cell cycle, while the signals that arrest the cell cycle, the antimitogenic signals, must be absent. This means that the concentration of cyclin D, which forms the complex with CDK4/6, must be high enough and the concentrations of the inhibitors that could block this complex must be low. The cyclin D – CDK4/6 complex phosphorylates the RB protein, which is a key factor for the cell to pass through the restriction point. RB is sometimes referred to as the guardian of the restriction point or the brake on cell proliferation. When the RB protein is not phosphorylated or is hypophosphorylated, it forms a tight complex with the E2F transcription factors and prevents them from acting as transcriptional activators of the target genes required to initiate S phase. As long as the RB protein "holds" the E2F protein, the cell remains in a quiescent state, i.e. it stays in the G1 or G0 phase. Thus, passing through the restriction point is controlled by the phosphorylation status of the RB protein. RB receives all information from both inside and outside the cell via mechanisms that affect the rate of its phosphorylation. This protein has the power to determine whether or not the cell will enter the next cell division cycle. Once the phosphorylation rate of the RB protein reaches a critical level, the E2F proteins are released and initiate the S phase. The die is cast. The cell divides.

Self-sufficiency in the production of growth signals

A normal, healthy cell divides in a multicellular organism only when it is useful to the organism as a whole. The organism's interest in cell division is expressed by mitogenic signals. A healthy cell needs mitogenic signals from its environment to enter an active, proliferative state. In contrast, it is typical for a tumor cell to divide independently of the interests of surrounding cells, completely ignoring their needs and in the absence of a mitogenic signal. The tumor cell itself can stimulate its own division. There are many ways it can do this. For example, it can produce – inadequately, nonphysiologically – a growth factor by itself, to which it then responds quite "adequately." While most growth factors are produced by cells of one type to stimulate the division of cells of another type in their environment (by paracrine signal transduction), tumor cells often produce the proliferation-stimulating growth factors themselves. The situation in which the cell produces a growth factor to respond to itself is called autocrine signaling. The cell responds to its own stimulation as an objective demand from and requirement of its environment, although in fact it is "talking to itself." Mitogens, similar to other extracellular signals, are transmitted to the cell via transmembrane receptors.

Another cause of permanent activation of cell division (growth factor-independent) may be mutations that alter the structure and function of the receptor so that it remains permanently switched on even in the absence of the appropriate signal. Furthermore, the same effect can be caused by malformation of the receptor itself, such that it is strongly sensitized to otherwise subthreshold amounts of signal molecules, or does not need to be externally activated at all, despite an unchanged structure. The effect is the same: a cell with a high concentration of receptors on its surface receives (and internally fires) a permanent signal to initiate cell division, although this is not in the interest of the organism as a whole. The signals are transmitted from the receptors to the effectors via cytoplasmic transporters. They can also mutate into a permanently active state in the absence of a physiological stimulus. In general, there are several specific ways to achieve sustained activation of mitogenic signaling pathways, all with the same effect that forces a cell to proliferate continuously.

Insensitivity to cell cycle arrest signals

A cell that is self-sufficient in the production of growth signals might be expected to be sufficiently independent in its control of division, sufficiently "unbrakeable" and sufficiently ready to embark on the journey to tumor transformation. Fortunately, this condition alone is not sufficient to completely deregulate the cell cycle. One could even say that a cell self-sufficient in production of growth factors is in a state of "full throttle," but also taking a break. To get a car moving, one must both step on the gas pedal and release the brake. Similarly, the cell must also release the brakes to begin uncontrolled division. Not only must the cancer cell generate its own mitogenic signals, but it must also find a way to ignore the signals that stop the cell cycle. These brake signals can come from the external environment or from inside the cell. An example of an external signal that stops the cell cycle is contact inhibition, which is consistently observed in cell cultures. Healthy cells that reach a critical proximity – when they begin to touch – signal to each other that they are already in close contact and that further division is inappropriate. The same effect can also be triggered by processes emanating from within the cell itself.

Each cell is equipped with mechanisms that constantly monitor its own state. For example, the integrity of DNA, which is critical for the reliable storage of genetic information, is closely monitored. Any damage that could result in altered genetic information being passed on to daughter cells during cell division is grounds for triggering a signal that effectively prevents the cell from dividing. Similarly, a lack of certain nutrients, energy sources, or building blocks necessary for the formation of the two healthy daughter cells is the source of a signal that effectively stops the cell cycle. The key messengers that signal the need to stop cell division are two important tumor suppressors. The aforementioned Rb protein, which controls passage through the restriction point, mainly mediates the action of external antimitogenic signals. Internal signals are preferentially mediated by the p53 protein. We will hear much more about this tumor suppressor later (in chapter 13A). The mechanisms that cause the cell to become insensitive to signals that stop cell division are analogous to those we learned about in the previous chapter on achieving self-sufficiency in growth signals. Instead of activating signals that cause cell division to constantly "hit the gas pedal," these changes are simply inactivated, and the brake on the cell cycle is permanently disabled.

What is wrong?

Cell division is a fundamental process in the life of a cell and has great implications for its future. Therefore, it is carefully controlled and regulated, involving many different proteins with different functions, which together form a large, well-organized and smoothly functioning team that works together according to clearly defined rules. Figure 10 shows a control room of the cell division center, where signals and messages from all collaborators are evaluated.

The control room is located in the cell nucleus, but the team members – the field workers – are scattered all over the cell, including its surface, collecting information and forwarding it to the control room via an established courier chain. Each field worker and courier takes care of each piece of information. A mistake by of any one of them can lead to a wrong decision to start cell division.

However, such a mistake may be made not only by a field worker, but also by a member of the top management – in the control room. With all the fatal consequences, of course. For example, the Rb protein itself can be altered by mutation. This protein decides, based on all the information constantly flowing into the control room, whether to release E2F, which then initiates the synthetic phase of the cell cycle. Even if all the signals – mitogenic and antimitogenic – are correctly delivered to Rb, it can lose control of the E2F factors due to mutation. And then it's all for naught. The restriction point is permanently unlocked and completely free to pass.

3B. Man's race with himself: The truth does not win

Speed, power, growth

Tumor cells lose the ability to control their division and rush forward. Uncontrollable, unstoppable, at their own pace, they override other cells and ignore the needs of the home tissue. And in the race with the surrounding regulated, receptive, communicating cells, they win. However, this victory is illusory and short-lived. Healthy cells never "step on the gas" of their own accord; they begin to divide only when they receive clear instructions and signals from their environment. In contrast, cancer cells can do that! If we attribute human characteristics and our commonly accepted values to them, we might perceive cancer cells as very efficient and intensely alive. They are highly motivated – they are self-sufficient in producing mitogenic signals! And they are persistent and indestructible – they do not perceive any disagreement, doubt or discontent in the cells surrounding them (not even their own); they are insensitive to signals that could hinder them. They do everything their own way, they behave as if they were detached from the chain. They do not pay attention to anything. So they cannot pay attention to the interests and needs of others. And so, they ignore the interests of the organism as a whole and thus their own long-term interest.

Take a look at tumor cells, their inexorably fast multiplication and their permanent brake-less and accelerated state, and you will see the general speed of our modern lifestyle, our restlessness, our relentless striving and our constant hurry. Evidence of and testimony to our obsession with speed, haste and high performance is described in the book *In Praise of Slow* by London journalist Carl Honoré, aptly subtitled *How a Worldwide Movement is Challenging the Cult of Speed*. It has become a world bestseller. In the introduction, the author recalls the words of German economist Klaus Schwab: "We are moving from a world in which big fish eat small fish to a world in which fast fish eat slow fish" (Honoré, 2012, p. 11). The parallel to rapidly proliferating tumor cells that first "eat up" healthy, i.e. regulated and thus slower, cells in their environment and then "eat up" and decompose the entire organism is obvious. Carl Honoré shows with many examples how speed and obsession with it have invaded and affected all areas of life, leaving many people damaged and leading lifestyles that are superficial and often incompatible with health and sometimes even with life itself.

Czech economist Tomáš Sedláček, in his book *Economics of Good and Evil*, takes a similar view of our often pointless, autotelic pursuits when he recalls the importance of observing the Sabbath: "Sabbath observance conveys the message that the purpose of Creation was not creation as such, i.e. as endless activity; as diligence for diligence itself. Creation was created to find rest, completion and joy in it. The meaning, the completion of the creation, is not in the further creation, but in the rest in the midst of the created. Translated into the language of economy: the meaning of profit is not in its constant increase, but in the rest in the existing profit. This dimension has disappeared in today's economy. The economy has no intention of resting. Today we know only growth for the sake of growth, and if the economy is flourishing, that is not a reason to rest, but to keep striving" (Sedláček, 2009, pp. 63–64). Moreover, Tomáš Sedláček points out that we live and act so fast that we often forget to think about the direction and especially the purpose of our course. We do not have time to think about whether the goal of our striving is good for us and for the whole community to which we belong.

Karoshi

One of the characteristic features of our modern "fast and intense" time is *karoshi*, "death by overwork." This phenomenon was first described in Japan in 1987 as occupational sudden mortality due to excessive stress and overlong work. The most common causes of karoshi deaths are heart attacks or strokes. In Japan, karoshi is listed as a distinct category in death statistics, and there is ongoing debate about whether suicide deaths due to depression and anxiety also fall into this category. One-fifth of Japanese thought to be are at risk of karoshi, and the number of these deaths is steadily increasing. It is spreading among young people and beyond Japan's borders. Karoshi threatens everyone who works so hard that they do not even realize they are working overtime. The problem is increasing to such an extent that one can speak of a karoshi epidemic.

One of the eight deadly sins by Konrad Lorenz

Konrad Lorenz calls this feature of contemporary life a race of man against himself, and describes it as one of the eight deadly sins. He quotes his teacher Oscar Heinroth, who is reported to have said that "after the wings of the Argus cock (ornamental feathers that favor courtship but hinder flight), the working pace of modern man is the stupidest product of intraspecific selection." And he asks, what damages the mind of modern man more: his blinding greed for money or his enervating haste (Lorenz, 1974, pp. 25–27)? Konrad Lorenz was an ethologist, an expert on animal behavior and the relationship between behavior and the ability to survive. Armed with this knowledge, he writes: "One of the worst effects of haste, or of the fear engendered by it, is the apparent inability of modern man to spend even the shortest time alone. He anxiously avoids every possibility of self-communion or meditation, as though he feared that reflection might present him with a ghastly self-portrait, such as that of Dorian Gray… A being unaware of the existence of its own self cannot possibly develop conceptual thought, word language, conscience, and responsible morality. A being that ceases to reflect is in danger of losing all these specifically human attributes" (Lorenz, 1974, p. 28).

The connection between haste, escape from self-reflection, and loss of thought, morality, and responsibility is truly a warning. And the parallels of this connection – also a warning – can be traced back to the process of cancer development, as we shall see. Konrad Lorenz points to the relentless haste of modern man as one of the deadly sins. Today's concept of the process of carcinogenesis views each typical feature of tumors as only one element in a relatively long chain of factors that only together pose a lethal threat. No feature is lethal by itself (in the sense that it could cause a lethal disease – cancer). On the other hand, loss of cell division control has two features typical of tumors: self-sufficiency in growth signals and loss of sensitivity to signals that stop the cell cycle. In addition, it is the rapid proliferation that is associated with the induction of some other signs of tumors. In general, then, none of the tumor signs alone can trigger a tumor, but the loss of the ability to regulate cell division – a permanent frenzy and a high, literally self-destructive, pace – represent "the sin and (human) weakness that underlies other sinful actions" (Riedl, 2011, p. 11).

What speed is appropriate?

In his book *In Praise of Slowness*, Carl Honoré presents a philosophy of slowness that calls us to slow down and calm down in many areas of our lives. But it is not a simple call for general deceleration. The author literally says that his philosophy of slowness understands the words "fast" and "slow" as more than just descriptions of changes in speed. They are symbolic of ways of being or philosophies of life. He urges, "Be fast when it makes sense to be fast, and slow when the situation calls for slowness. Try to find just the right pace." Being slow, then, means "controlling the rhythm of one's life and setting one's own pace" (Honoré, 2012, pp. 20–21). That is, there is no fixed, standardized, uniform optimal pace of life. "Slow" – which in the terminology of cell biology means "physiological" – is a pace that is appropriate, adequate, effective (not purposeful), meaningful, balanced, and consistent. It ensures consistency with other cells, the environment, circumstances, conditions, and opportunities. As we see with healthy cells in the body. While for the epithelial cells of the intestinal mucosa such a proliferation rate is appropriate, which ensures complete regeneration of the mucosa every four days, the appropriate proliferation rate of the endothelial cells, which form the vascular lining, is incomparably slower. Only one in a hundred thousand endothelial cells is in the active phase of cell division. And only rarely, in the event of injury or damage, does the proliferation rate of the endothelium increase dramatically but transiently. But even this dramatic increase in proliferation is consistent and physiological. It is an appropriate, balanced response to the immediate, real state of the organism – the injury. But for both cell types – intestinal epithelium and vascular endothelium – we know their tumor analogues, intestinal cancer cells and angioma cells that proliferate precipitously. To be exact: faster than desired, appropriate, effective. And this "stampeding" – although not fatal in itself – can be the beginning of a process whose outcome is fatal.

But if there is no standard, optimal, uniform – i.e. "correct" – rate of division, how does the cell know how fast it must divide to be "consistent"? How does it know when to enter cell division and, conversely, when to stop and rest before the restriction point? In multicellular organisms, there is no centralized control of the organism, so no one tells the cell anything. There are no commands. The cell has to decide for itself. On what basis? It has to make a decision depending on the current situation. It must listen carefully to all the signals – from the external environment and from its own interior – and evaluate them accurately. It must communicate properly.

Communication, communication noise and lies

Communication is defined as a particular type of social interaction in which one individual consciously conveys a message to another individual. In other words, it is a two-way process of transmitting information in social contact through various means.

Communication consists of five basic elements: information source, sender (encoder), channel, receiver (decoder), and destination. Sometimes a sixth element is referred to as noise, which affects the five aforementioned elements and can lead to a breakdown in communication. Communication noise is generally defined as an interfering signal that changes during communication and corrupts the transmitted message so that it is distorted in various ways and, as a result, may be received incorrectly. At the human level, there are a number of causes of communication noise, ranging from technical or mechanical issues to physiological, psychological, verbal and nonverbal, semantic, and other problems. In the cellular environment, communication noise can result from a completely natural fluctuation of the individual molecules involved in signal transduction, which can affect this process and its impact. The effect on the message also depends on the relationship between the signal strength and the noise level. Under normal physiological conditions, such natural fluctuation of individual molecules does not affect the meaning of the transmitted messages and represents only a kind of subtle information "hum" and modulation in the subtext of the main message. However, tumor cells do not "hum," but fundamentally change the "main message." A "yes" becomes a "no" and vice versa. This is a lie!

Cell lies

An example of the false cell communication that occurs in tumor cells is autocrine stimulation, a kind of cellular soliloquy. Actually, this is not true communication. By definition, communication is the passing of information between individuals, while the autocrine cell talks to itself. It constantly stimulates itself to divide, ignoring the needs and "opinions" of other cells. But this autocrine signal is false, unphysiological, and does not reflect the real needs of the tissue and the organism. Tumor cells lie to themselves about how important they are. Another example of damage causing false cellular communication is the alteration of receptor structure and function. The damaged receptor remains permanently active even in the absence of an activating signal molecule. Cells with such a damaged receptor do not have to care about what is happening around them, and harmony with other cells is disrupted. Furthermore, false stimuli can arise at other levels of the signaling pathway, as cytoplasmic transporters or various effectors. Similarly, there are many ways in which we, as human beings, lie to and deceive ourselves. And equally diverse are the pathways by which we reject information. Signaling pathways that receive and transmit antimitogenic signals, i.e. signals that prevent a cell from dividing, are permanently "turned off" in tumor cells. As a result, they do not care, they do not see, they do not hear, they permanently ignore the bans. And nothing can stop them.

Truth and lie

We encounter the terms truth and lie all the time. We use and understand them quite naturally. However, they deserve a closer analysis, and then it may turn out that their definitions are not so simple. It may seem that truth simply expresses agreement with reality: what agrees with reality is true; what does not agree with it is a lie. But a statement that does not agree with the facts may result from ignorance or an error, and in that case we do not usually call it a lie. So, there are two conditions that are crucial for a lie. First, the liar knows the truth and has the choice to tell it or to lie, and second, the purpose of the lie is to deceive and harm someone or to gain an unfair advantage or escape punishment. The key is the good or bad intention, the effort to benefit or harm someone with his statement (Stránský, 2012; Sokol, 2014).

If we consider truth and lies further, it turns out that everything is much more complicated than we might think at first. For example, if I tell the truth about someone (i.e. something that really happened or is happening to him), it may be insulting. By revealing a negative, embarrassing, intimate, or threatening fact, we can use the "truth" in bad faith to harm another. On the other hand, we know the so-called merciful lies, for example when we try to protect someone or not to burden them with bad news. Merciful lies we often do not even perceive as lies, because they are not intended to harm anyone. Then we sometimes talk about so-called prosocial or "white" lies. This is the case when we want to do a favor, avoid a conflict, appease someone, or just be polite. These phenomena are ordinary, "physiological," socially accepted "noises" that are usually not essential information, do not distort our perception of reality, and do not interfere with our communication.

However, the relativization of truth can be a major problem. It can expand far beyond the innocent and acceptable limits of politeness and consideration and open the door to a real lie. The postmodern belief in the value of heterogeneity, plurality, and equivalence of views, and in the idea that there is virtually no single truth, probably helps to keep this door open. In many areas of life, we are gradually getting used to and admitting that we do not live in a world of facts but in a world of mere interpretations. Sometimes this leads us to feel that the world is complicated, that we cannot know all the connections, and we simply resign ourselves to not knowing the truth. Sometimes we excuse the obvious lie quite pragmatically due to its usefulness and utility. After all, we know lies also from the world of animals. Animals also deceive and camouflage themselves by using all kinds of mimicry. From this, some evolutionary biologists, i.e. scientists and experts, deduce that morality and fairness do not play a role in nature and that lying can have an evolutionary benefit. Many non-experts then readily agree…

The human lie

We can probably find human equivalents to all forms of cellular lying in cancer cells. Autocrine signaling sometimes resembles self-glorifying, self-celebrating, self-promoting media campaigns, bombastic press conferences, and newspaper articles – how good we are, what we have accomplished, and what we will accomplish in the future. To achieve greater impact and effectiveness, facts are presented in a somewhat simplified and selective manner, and merits are often emphasized through exaggeration. For many insiders, this evokes an indulgent smile; for the uninitiated, possibly incomprehension and disgust; for the inexperienced, admiration and often unrealistic expectations. And for most – in today's media landscape saturated with bombastic messages – perhaps not much of a reaction at all. The principals or authors of media campaigns, however, are often reinforced in their egocentric attitudes, lack of judgment, and insensitivity to the real needs of their fellow human beings.

Autocrine stimulation need not be bad *per se*. On the contrary, there are situations in which it is very important. For example, autocrine stimulation seems to apply to stem cells and is a component of the signals that maintain their capacity for self-renewal, that is, to keep their phenotype unchanged during cell division, whereas the phenotype of most cells changes irrevocably when they differentiate during division. This ability of stem cells not to change during cell division is a necessary condition for their "stemness" (see chapter 1A). Therefore, a press conference organized to convey important information that needs to be published and shared immediately may be quite appropriate and useful. And it can also help the organizer reflect on himself, become aware of his responsibilities, and network with others. It is the question of meaning and purpose that determines what is right, what is coherent, and in the interest of the system, and what might jeopardize it. In general, advertising and other products of PR managers promoting companies, products, services, brands or ideas can draw serious attention to news and new opportunities, but they can also be just tools of manipulation, artificial demand, and illusory needs. Self-promotion can have a meaning – to satisfy a real need – but it can also have a purpose – to increase sales and profits. Lying can also contribute to our diminishing ability to distinguish meaning from purpose. And then the lie can spread even faster…

Our post-factual, post-truth age

In 2016, the German Language Society chose the term "postfactual" as its word of the year. The same term was chosen by the Oxford Dictionary as the most prestigious English word of 2016. At that time, two major political changes were shaking the world. The so-called Brexit, i.e. the decision of the United Kingdom to leave the European Union, and the election of Donald Trump as President of the United States. Both successful campaigns relied on obvious falsehoods, distortions of reality, and alarmism. Although their critics pointed out the contradictions to reality, voters did not hear this or did not care. Facts were not important for decision making; emotion-based stories were much more important.

Events comparable to Brexit and the election of Donald Trump are occurring with increasing frequency around the world. Why? What is going on in public opinion? And will this continue to be the case in the future?

Arguably, politics has never been entirely free of lies. The exploitation of misinformation (misleading information, fake news, rumors) as a means of propaganda and manipulation has a long history, possibly dating back to the dawn of civilization. Any invention that makes the dissemination of information more efficient (clay tablets – printing press – rotary printing – radio – television – Internet) is immediately used as a tool for more effective and sophisticated dissemination of misinformation.

At present, the Internet in particular has completely changed the means and scope of dissemination of information and misinformation. It has opened up entirely new ways of disseminating and manipulating news. The "traditional" media had their editors (so-called gatekeepers) who decided what was published and who paid attention to the truthfulness and verifiability of the reports. "With the advent of the Internet, many things have changed. Among them is the speed at which information reaches people. In today's media world, speed is the most important factor! Whoever gets the news first wins. The second one is not talked about. Truthfulness takes a back seat…," say the authors of *The Best Book about Fake News* (Gregor, Vejvodová, 2018, p. 75).

The world of the Internet is flooded with lies! Unfortunately, not only because some journalists do not take the time to check the truthfulness of the news they spread, but mainly because there are enough of those who lie quite deliberately, with a clear intention and for a clear purpose. With the Internet and the resulting much easier access to information, for example, classic information control – censorship – is no longer possible. Deliberately flooding the Internet with untrue or confusing information is thus a modern form of censorship. In the huge flood of information, it is much more difficult to find true information that is not infiltrated by lies. An excess of information is the new censorship! And lies are of course an effective means of manipulation and control.

Why do we believe in lies?

Where does our great tolerance of lies come from? Why do we give them so much acceptance? Why do they have such power over us?

With the Internet, the availability and amount of news that floods us has increased tremendously. The human brain is not prepared to process such a large amount of information, so it creates shortcuts: it generalizes, simplifies, creates stereotypes. Above all, people do not seem to look for the truth. Once we form an opinion based on information, we are very reluctant to change it. This is probably related to what is called cognitive convenience. It forces us to avoid facts that would strain our brains. And changing one's mind requires effort. Therefore, we prefer to expose ourselves only to information that supports our opinion. Reports that prove us right even trigger dopamine releases, "happiness molecules." Contrarily, the so-called "backward effect" has been described. When we are confronted with facts that would disprove an already formed opinion, we not only do not change our mind, but reinforce commitment to our original belief.

So the point is, how do we form opinions? True information and misinformation have the same chances of success with open-minded people. The way a piece of information or misinformation is presented can be opinion-forming. And it turns out that opinion formation is strongly based on emotions. In the decision-making process, emotions play a more important role than logic. That is why any shocking news with a touch of sensation, tragedy, passion, personal suffering, or public humiliation has such a great impact. Appealing to the need for solidarity and fairness and emphasizing the majority opinion also works well because it allows us to have a good opinion of ourselves. And an absolutely reliable "seller" of information is fear and evoking a sense of threat. Therefore, it is not difficult to influence people's opinions. For those who seek to manipulate opinions, it looks worth it to be shocking and not tell the truth, because if we have already formed an opinion based on information, it is very difficult to change it, even under the weight of a factual argument. When information associated with feelings is accepted, it is much more difficult to refute the resulting opinions. Facts are displaced by feelings. We are the post-factual people… (Gregor, Vejvodová, 2018).

The role of social networks

The role of social networks in spreading misinformation and lies today has already been mentioned. Treacherous websites can hide the truth through a flood of lies. In addition, social networks use algorithms that tailor search results to users by exploiting what the system knows about their browsing preferences, making it harder to access balanced information. They show like-minded people similar posts, creating opinion bubbles in which similar-thinking groups reinforce each other's beliefs and reject opposing views. This is referred to as homophilic sorting. Like-minded people meet much more easily in social networks than in the real world, and belonging to one opinion group makes it difficult – often completely unconsciously – to encounter a different opinion or conflicting information, making a change of mind less probable.

Moreover, an overwhelming excess of news can lead to apathy. One gives up searching for the truth, because "everyone is lying anyway." Such an attitude is actually a victory for disinformation campaigns and propaganda. Creating an atmosphere of hopelessness, mistrust, anxiety, and fear generally destroys our confidence in the functioning of the human community and democracy, and questions or obscures the meaning of human life in general (Horký, 2016; Neumajer, 2017).

Thou shalt not bear false witness against thy neighbor

The fact that we can speak and express ourselves, think and imagine, inevitably brings with it the possibility of pretending, dissimulating and lying. On the other hand, language and communication would be utterly meaningless if we lied in ways that aroused suspicion and created lasting doubt. The deep significance and immense importance of truthfulness and truthful communication is underscored by the fact that refusing to bear false witness is one of God's ten basic commandments. The Christian Ten – besides the requirements "Thou shalt not kill!" "You shall not steal!" and "You shall not commit adultery!" – includes the eighth commandment forbids lying, cheating, and giving incomplete or misleading information. Moreover, this commandment is in line with the other commandments, although we do not seem to have such a big problem with lying. False testimony, slander, false accusations, misinformation and rumors are almost a regular part of our lives today. Everyone tells lies! So why do the Ten, a set of basic and central rules for human behavior and society that have stood the test of centuries and generations, place so much emphasis on living in truth? Why do the Ten guide us not to deceive our neighbors and, on the contrary, to take responsibility for conveying correct knowledge (Chvála, 2014; Klimeš, 2014)?

Theory of gossip

A possible explanation is offered by the so-called theory of gossip. Gossip is usually condemned as something inappropriate. Talking about someone behind their back is considered improper, or at the least indecent, and can even lead to exclusion from decent society. But the gossip theory comes up with a revolutionary idea. It says that gossip plays an important role in maintaining social order: it is not work, but gossip that humanizes the monkeys! Our language evolved mainly to gossip! Until primates lived in small groups, simple sounds were enough for them to inform each other about dangers, their moods, and the state of their relationships. Then, in small groups, members managed – and still manage – to personally find out who is who, who can be trusted, who can be teamed up with, who should be listened to and followed, and who, on the contrary, should be kept away from and guarded.

Man is a social creature. The key to the survival and especially to the great evolutionary success of *Homo sapiens* was and is cooperation, even in large groups where not everyone can know everyone else personally. At the same time, a prerequisite for the functioning of both large and small groups is good knowledge of who is who. It is possible, therefore, that the development of language and the exchange of detailed and accurate information about community members, i.e. gossip, can replace the immediate personal knowledge of each individual, thus enabling the smooth and long-term functioning of large groups. Gossip, according to one definition "the informal communication between two or more people concerned with the actions or behavior of persons not present at the time," reportedly accounts for 65 to 90% of all human communication. In the life of a community, gossip serves as a means of social control and maintenance of standards and norms of behavior. Violations of norms, evasions and non-compliance with community standards are sooner or later exposed and pointed out, and the other members of the group are warned about "sinners." Of course, gossip can also become an instrument of lying and manipulation or punishment and social exclusion, including punishment and social exclusion of liars. False testimony about the neighbor, deliberate distortion of reality, i.e. lying, undermines the community, is devastating and toxic to the system like chemical poisons or radiation. Genuine communication is a basic requirement for successful community life, for the functioning of the system (Chvála, 2014; Soukalová, 2015; Harari, 2014). And tumor cells only confirm and underline this experience. Lying is pervasive tumor behavior, it is typical and an integral part of the catastrophic process of tumor formation and development.

Truth and speed

What is the connection between truth and speed? The one has already been mentioned above. Because of the enormous worship of speed and competition to see who can deliver a new message first, verification of truth content has taken a back seat. Considering how the human psyche works and how difficult it is to undo a belief once it has been accepted, it becomes clear that excessive speed contributes not only to the creation but also to the spread and maintenance of lies (Gregor, Vejvodová, 2018). According to popular English saying "A lie is halfway round the world before the truth has got its boots on." Moreover, on social networks we are constantly overwhelmed with many messages and stimuli, which can significantly weaken our concentration and remove context. Our capacity for pattern recognition – to perceive news in context, to consider the factors of time and variability, and to perceive the repetitiveness of events – can be completely undermined by the information overload. This can significantly weaken our ability to think about, compare, and evaluate the messages that reach us and synthesize them into a more complete understanding of what is happening around us, where our own place is, and who we ourselves are (Lipold, 2019).

Loss of specialization

The loss of self-awareness and knowledge of one's own place in the system due to "excessive speed" is also observed in the cells of the multicellular body. Cell division is closely related to cell differentiation. There is an indirect relationship between them. As the cell becomes more specialized, its proliferation potential, i.e. its willingness to divide, usually decreases and vice versa. Once the cell begins to divide recklessly, it ceases to differentiate. Damage or loss of differentiation, i.e. loss of specialization, is a typical feature of tumor cells. Tumor cells seem to have forgotten their original function, purpose and mission within the system. They proliferate vigorously but cease to develop, mature, and differentiate – or differentiate inadequately. The less differentiated the tumor cells are, the more aggressive the tumor they form. And so the question arises: what about us, full human beings? Is it possible that the enormous speed at which we live, produce and accumulate makes us forget our own development and personal growth? We value speed, action, flexibility and versatility so much that many people tend to do everything, and often everything at the same time. Is not the downside of this attitude the loss of differentiation? The lack of specialization, of maturity, of really deep knowledge, of concentration skills and focus? And also the lack of individuality, uniqueness, authenticity? And as a consequence – within the whole social system – the loss of variety, diversity, differences?

Loss of specialization and victory of the fastest

A prime biological example of the loss of specialization and the victory of the fastest is cancer of the immune system – lymphoma. Lymphocytes in a healthy immune system are concerned with their diversity, the proverbial uniqueness. There are hundreds of millions of different viruses, bacteria, animal cells and other foreign substances, as well as altered cellular components, that are potentially dangerous. The job of a healthy immune system is to recognize these and respond appropriately. To do this, the immune system needs a large number of lymphocytes that differ in their ability to recognize specific targets – it relies on numerous and highly specialized cells, the experts. A lymphoma, however, is a tumor composed of a single type of immature, undifferentiated lymphocyte – the winner of the undeclared competition to be the fastest.

Some lymphomas (e.g. Burkitt's lymphoma and mantle cell lymphoma) develop from a cell with dysregulated division. Such a cell multiplies with enormous speed, grows and thus "defeats" all other lymphocytes, which have to make way for the winner. Is it possible that these defeated experts and their specific specializations are then missing in the body? Very likely yes! And is it possible that this victorious "skilled" lymphocyte has not had time to learn, mature and differentiate? Yes, indeed! But everything seems to be fine – the fastest cell won and defeated all other competitors! And if speed is the criterion we use to judge, then this cell is also the "best"! And who would not admire, celebrate, support and bet everything on the winner? What to do with the slow and unsuccessful? Who cares that we may need them sometime in the future? What if a threat suddenly appears in the body in the form of an antigen that is not part of the winner's "portfolio"? But at this speed, who has time to look that far ahead?

4A. Damage of apoptosis

Programmed cell death

Just as cells are born by cell division, they also die by death. They may die by accident, irreversible mechanical or chemical damage, or due to loss of their ability to maintain their integrity and viability. Such non-physiological, pathological cell death is called necrosis. It is a passive process in which the cell usually enlarges, its ion balance is disturbed, the cell membranes lose their function, and the cell disintegrates. The cell contents are released into the environment, triggering an inflammatory response around the necrotic cells (Fig. 11).

However, cell death does not necessarily have to be the result of an accident, an unintentional, accidental injury. It is also an integral and physiological part of a multicellular living organism. For example, billions of cells in the bone marrow and intestine of a healthy adult human die every hour (Alberts, 1998). The same type of programmed cell death is essential for the development of the organism.

The term "programmed cell death" was coined to describe the phenomenon of a cell dying at a specific place and time. The most important form of programmed cell death is apoptosis. This is a sequence of biochemical processes that lead to degradation of the cytoskeleton, changes in the shape of the cytoplasmic membrane, shrinkage of the cell, and fragmentation of the nucleus and chromosomes (Fig. 11). This is followed by gentle degradation and removal of cell debris by neighboring cells and cells of the immune system, especially macrophages, without disturbing neighboring cells and triggering an inflammatory response. Thus, apoptosis is fundamentally different from necrosis. It is an active, perfect self-destruction cascade that involves a whole complex of interconnected steps that occur in precise sequence.

The nematode: An ideal model for studying cell death

Fundamental insights into the molecular mechanisms and developmental relationships of programmed cell death, or apoptosis, have been gained by studying the nematode *Caenorhabditis elegans* mentioned earlier in Chapter 1A. The average life span of a

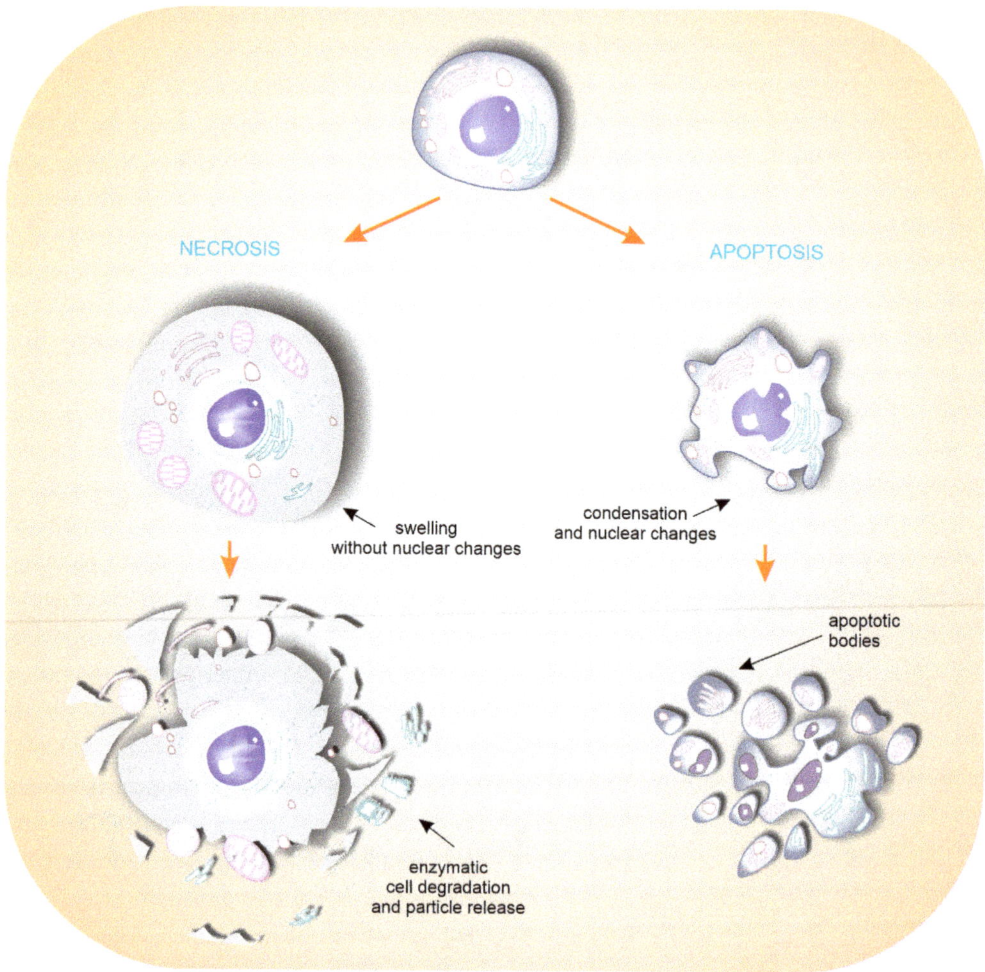

Fig. 11 Necrosis and apoptosis

In necrosis, the cell enlarges, the ion balance is disturbed, and the cell membranes break down. The cell bursts and the cell contents spill out into the surrounding area. This triggers an inflammatory response in the tissue surrounding the necrotic cells. In apoptosis, the cytoskeleton is degraded, the cell shrinks, the nucleus and chromosomes are fragmented, and the cytoplasmic membrane is preserved. The cell disintegrates into apoptotic bodies, which are subsequently removed by neighboring cells or cells of the immune system, especially macrophages.

nematode is about 20 days. Its body is about 1 mm long, transparent, and has 959 cells at adulthood. Its accessibility, size, and relative ease of observation make the nematode body such a clear and tangible terrain that it has become an ideal model for studying cell death. And not just cell death. The development of the nematode has been mapped with such precision and detail that we know the individual path and fate of each of its cells. We know exactly the pedigrees of the entire so-called cell lineages: we know when and where each cell will divide, how it will change, how it will work, and also when it will disappear. It has been accurately mapped that 131 out of 1,090 somatic cells die during nematode development (Potts, Cameron, 2011). And it is known exactly when and which cells undergo apoptosis. Thus, it is quite clear that this type of cell death is not due to any accidents, mistakes, or errors. It is a legitimate, inevitable and necessary process – the interruption of which, on the contrary, leads to failure of nematode development.

Mechanism of cell death

The study of apoptosis in nematodes is useful not only because it reveals developmental patterns and thus the physiological role of apoptosis in multicellular organisms. An equally important benefit is the insight into the molecular mechanism of apoptosis. All four key proteins responsible for apoptosis have been described in the nematode, along with the genes encoding them. All of these proteins have their homologs in human cells. However, the molecular mechanism of apoptosis of human cells is much more complicated than that of the nematode and its study is more difficult. Knowledge of the key regulators of nematode apoptosis has facilitated the discovery of the mechanism of human cell apoptosis and thus our understanding of the association of apoptosis disorders with tumor development.

Caspases are the most important molecules in the process of apoptosis of human cells. They are proteases, that is, enzymes that degrade other proteins and activate themselves by proteolytic cleavage (Fig. 12). In human cells, twelve different caspases have been described. Some of these are called initiation caspases. They are proteolytically cleaved in response to apoptosis-inducing signals and function at the beginning of the entire activation cascade. The initiation caspases activated in this way cleave and activate effector caspases. Effector caspases, as the name suggests, are responsible for execution of the apoptosis process. They gradually degrade key structures and enzymes of a cell undergoing apoptosis. In addition, they also proteolytically cleave other effector caspases that are part of the apoptotic cascade, thereby activating them, resulting in amplification of the proapoptotic signal. Once a cell enters the process of apoptosis, this process accelerates and the cell dies very efficiently and rapidly (Fig. 13).

What activates programmed cell death

There can be many reasons to trigger apoptosis in a cell. The death of certain cells may be a natural part of development. Programmed cell death is a tool for maintaining tissue homeostasis and eliminating damaged cells. Therefore, at the molecular level, there are a variety of triggers of programmed cell death activators. These include the death signals that the cell receives from its environment, to which it then responds with the immediate onset of apoptosis. Similarly, the absence of pro-survival signals from the cell's environment can trigger apoptosis. Once the supply of these pro-survival

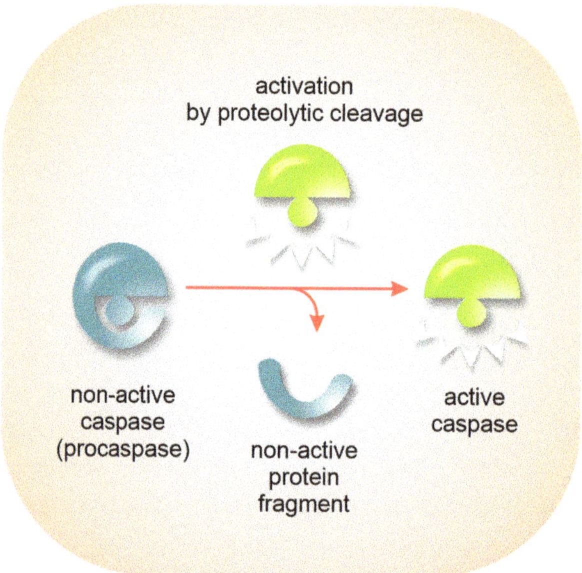

Fig. 12 Activation of caspases
Caspases are proteases, i.e., enzymes that cleave other proteins. An inactive caspase, a so-called pro-caspase, is activated by proteolytic cleavage, often by another, active caspase.

signals fails, the cell dies. This dependence on pro-survival signals from other cells ensures that cells survive only in the right places at the right time. Anoikis (from Greek – homelessness) has a similar function, and means apoptosis which is caused by the loss of cell contact with the surrounding extracellular matrix. Programmed cell death can also be induced by a lack of oxygen (hypoxia), the presence of toxins, viruses or bacteria. Apoptosis can also be caused by extensive irreparable damage to cellular DNA or unphysiologically high activity of some oncogenes and other stimuli.

Intrinsic and extrinsic apoptotic pathways

Obviously, the activators of cell death can originate from the external environment of the cell as well from within the cell itself. The molecular mechanisms controlling this process correspond to the origin of the stimulus. Apoptosis can be triggered by either the extrinsic or the intrinsic pathway (Fig. 14). In the extrinsic pathway, specific transmembrane death receptors are activated by binding of the death ligand. The receptor mediates the transmission of the signal to the interior of the cell, whereupon the initiation caspase (usually caspase 8) is activated and initiates the entire proteolytic cascade.

The intrinsic pathway represents the cell's response to stimuli originating from within the cell itself and is associated with mitochondria. Mitochondria are sometimes referred to as cellular power plants because they are the major organelles of cellular

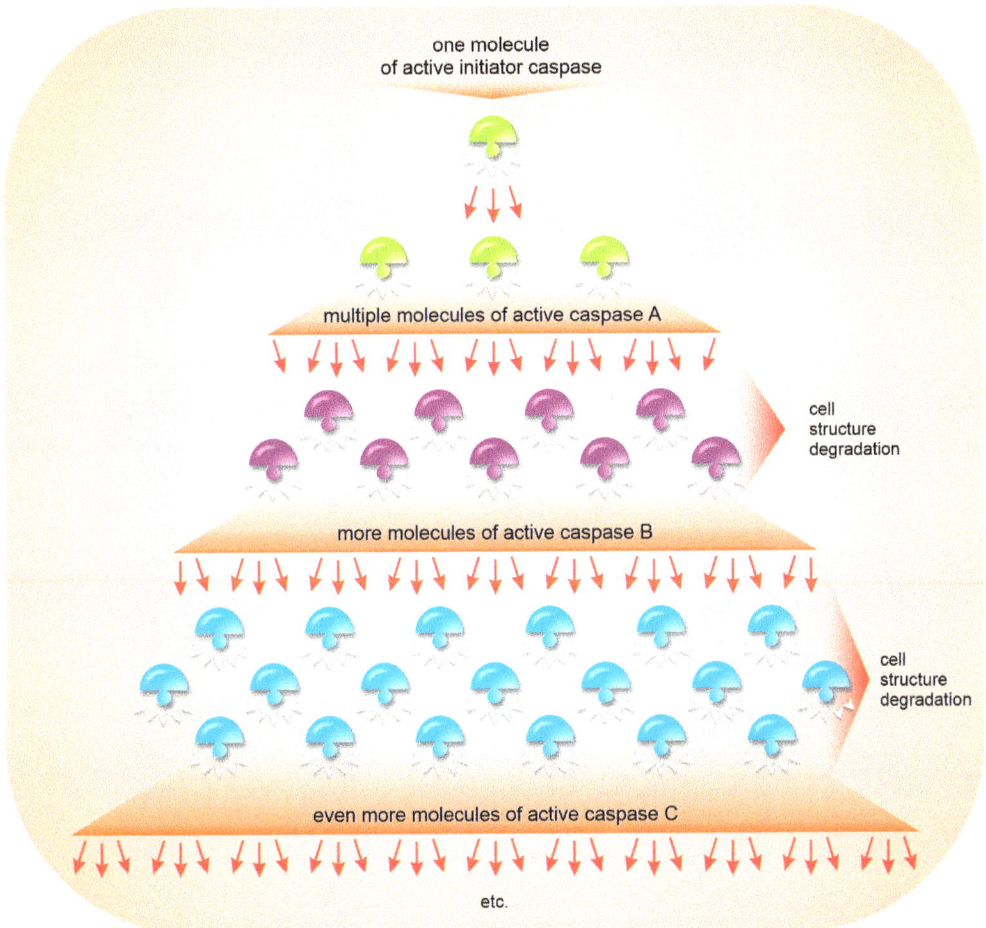

Fig. 13 Caspase cascades
Caspases are activated by proteolytic cleavage by other caspases. Initiation caspases are activated in response to apoptosis-triggering signals and subsequently activate other caspases called effector caspases. Each active caspase can activate many other subsequent caspases. In this way, a cascade is created – an amplifying chain reaction that enables very effective and rapid degradation of the cell (adapted from Alberts et al., 1998, modified).

energy metabolism, where cellular respiration occurs. Various apoptotic signals can hit the mitochondria. They can both stimulate and inhibit apoptosis, and their integration occurs in this organelle. When the balance of signals tips toward apoptosis, cytochrome c (in addition to other proteins) is released from the mitochondria. Cytochrome c is located inside the mitochondria, more specifically on the outside of the inner mitochondrial membrane, and is an essential component of the respiratory chain. After its release into the cytoplasm, it becomes part of the apoptosome, the assembly of which leads to the activation of initiation caspase 9 and thus subsequent effector caspases.

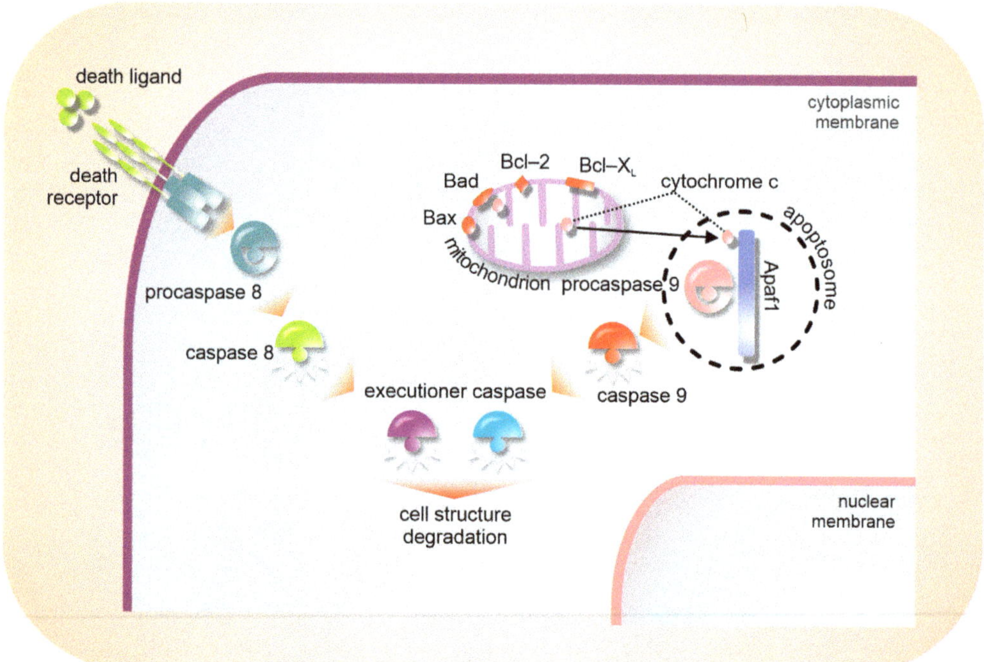

Fig. 14 Extrinsic and intrinsic pathways of apoptosis induction
In the extrinsic pathway, specific transmembrane death receptors are activated by binding of the death ligand. Through them, the signal is transmitted to the interior of the cell, where it activates caspase 8 and subsequently a cascade of other effector caspases that lead to efficient cell degradation. The intrinsic pathway is associated with mitochondria and their key molecule cytochrome c. The release of cytochrome c from mitochondria is controlled by Bcl-2 family proteins. After successful release, cytochrome c participates in the formation of the apoptosome, followed by activation of initiation caspase 9 and then a cascade of effector caspases, just as in the case of the extrinsic pathway.

Extrinsic and intrinsic apoptotic signaling pathways merge at the moment of initiation caspase activation. The actual execution phase is already common to both pathways (Fig. 14; Thornberry, Lazebnik, 1998; Igney, Krammer, 2002).

Regulation of apoptosis

If the apoptotic apparatus is so powerful that it irresistibly leads the cell to death and effective decomposition, it is clear that both the initiation of apoptosis and its entire course must be very strictly and precisely regulated. The apoptosis process always begins with the reversible signaling phase, during which the entire process can still be stopped. Therefore, in the signaling phase, a very careful evaluation of all signals must occur, culminating in a competent decision to initiate an already irreversible execution phase of apoptosis. An example and the personification of this "weighing of all pros and cons" is the integration of pro-apoptotic and anti-apoptotic signals in mitochondria. Multiple related proteins of the large Bcl-2 protein family are involved in this process of carefully weighing pro- and anti-apoptotic signals in the mitochondria of human cells. Some proteins of this family (e.g. Bax, Bim, Bik) are pro-apoptotic, whereas others (e.g. Bcl-2, Bcl-XL, Mcl1) are anti-apoptotic. The balance between them determines the final state of the mitochondrial membrane. An intact mitochondrial membrane impermeable for cytochrome c is an effective barrier to the intrinsic pathway of apoptosis. Conversely, release of cytochrome c from mitochondria and activation of the initiation caspase terminates the apoptosis signaling phase and activates the caspase cascade (Cory, Adams, 2002; Letai, 2008).

However, even at this point in the apoptotic pathway, the cell still has a number of tools with which it can very effectively affect the course of apoptosis. For example, the IAP (inhibitor-of-apoptosis) proteins can bind to both procaspases and active caspases and block their activation or activity. Thus, the IAP proteins are potent inhibitors of apoptosis. On the other hand, there are regulatory proteins – such as Diablo/Smac – that interact with the IAP proteins and prevent them from binding to caspases. Thus, they act as inhibitors of apoptosis inhibitors. Their existence shows how the process of cell death is precisely tuned in human (and generally in mammalian) cells (Fig. 15; De Laurenzi, Melino, 2000; Goyal, 2001).

A cell death program is present in each cell

Cell death is called programmed because it occurs at a predetermined place and time during development, as has been convincingly and repeatedly shown in nematodes. However, activation of the cell death program can also be a response to certain risky or accidental situations, such as irreversible or excessively dangerous cell damage. The term "programmed" is also an expression of the fact that the entire process is a genetically determined sequence of functionally interrelated steps that follow each other in a certain order. This genetic program of cell death – a detailed step-by-step guide to highly organized self-disposal – is inherent in all cells in the body. And not only is this program part of every cell's genetic makeup, it is always ready to go. Caspases, the most important enzymes of apoptosis, are constantly present in the cytoplasm of all cells. They are in an inactive state in the form of procaspases. They are activated only by appropriate signals. This constant readiness of all cells in the body to self destruct demonstrates the great importance of this process. Further evidence of the importance of programmed cell death is also seen in the fact that, although for a time it appeared to scientists that programmed cell death and apoptosis were synonymous, there are in fact several types and forms of programmed cell death that are more or less related, share more or less common features, and may overlap. These include, for example, aponecrosis, necrosis, paraptosis, autophagy, entosis, and programmed lysosome-mediated cell death (Kreuzaler, Watson, 2012; Radogna et al., 2015).

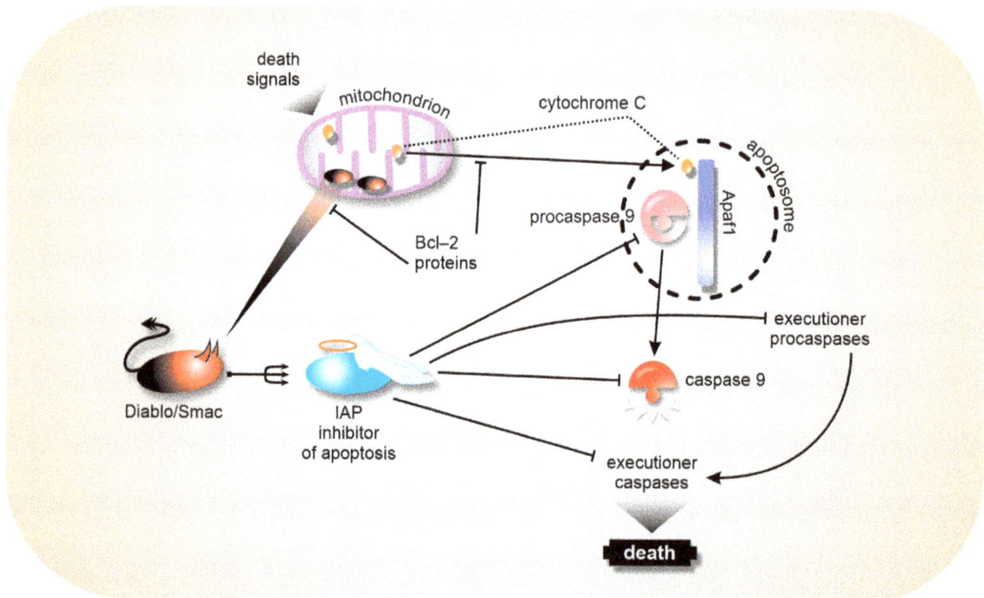

Fig. 15 Fine-tuning cell death
The cell is equipped with a number of tools that very effectively influence and thus fine tune the course of apoptosis. The IAP proteins can block the activity of caspases, effectively stopping apoptosis. The Diablo/Smac regulatory proteins block the activity of the IAP proteins and, conversely effectively stimulate cell death (adapted from De Laurenzi, Melino, 2000, modified).

Autophagy, which does not necessarily lead to cell death, occupies a special place among them. Autophagy – literally "self-eating" – is the process by which the cell breaks down its own structures (proteins, nucleic acids, and some organelles) into simple building blocks (e.g. amino acids, nucleotides) that can then be recycled. This restores macromolecule synthesis and energy metabolism. This is an adaptive mechanism necessary for survival under adverse conditions, and the cell can use it to respond to certain forms of stress, e.g. nutrient and oxygen deprivation, oxidative stress, and the like (White, DiPaola, 2009; Zhou et al., 2012).

Failure of apoptosis and human diseases

In the nematode, a relatively simple animal with a small number of somatic cells and with a limited number of cells undergoing apoptosis during development, perturbations of the apoptotic process result in significant developmental defects. Misdirected programmed cell death can be expected to have severe consequences in humans as well. And so it is. Both excessively high and excessively low rates of cell death have serious effects. Excessive apoptosis, i.e. an unphysiologically high loss of some cells, is most likely the cause of a number of neurodegenerative diseases such as Alzheimer's, Huntington's and Parkinson's, and is also associated with other diseases such as multiple sclerosis and AIDS. However, insufficient apoptosis also has serious consequences and is associated, for example, with some autoimmune diseases and clearly also with the development of tumors.

In chapter 3A on the cell cycle, we saw that loss of control over cell division significantly disrupts cellular homeostasis and can lead to excessive proliferation and thus excessive cell mass formation. Insufficient cell death can have a similar effect. Homeostasis – the maintenance of a stable, physiological cell number – is generally provided by the balance between the rate of cell division and the rate of cell death. An unphysiological increase in cell mass can result from both excessive cell proliferation and insufficient cell death (Fig. 16).

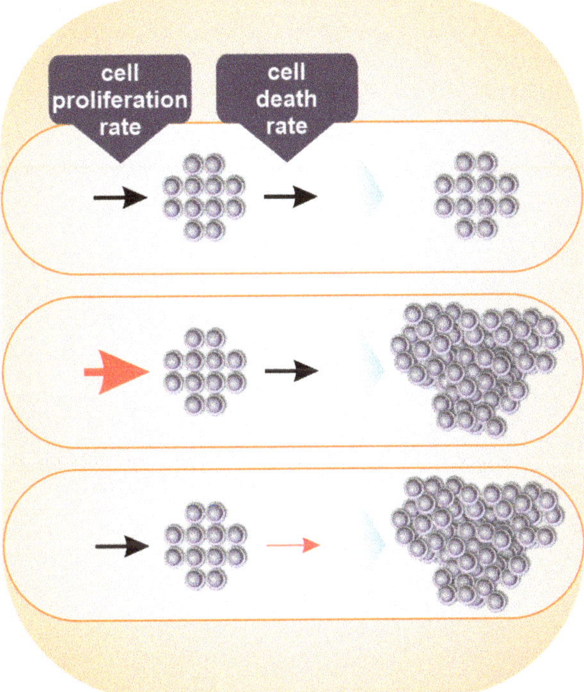

Fig. 16 Balance of cell division and cell death
Cellular homeostasis – a stable, physiological number of cells – is determined by the balance between the rate of cell division and the rate of cell death. A pathological increase in cell mass can be caused by both excessive cell proliferation and insufficient cell death.

The cell death program, and in particular the constant readiness of the cells of the multicellular organism to undergo it, is one of the mechanisms that protect the organism from tumor transformation processes. An integrated, well-functioning programmed death process protects the cell from transformation by several mechanisms. One of them is based on the fact that cell death is an effective cellular response to a high proliferation rate. Apoptosis can be triggered by a high level of activity of some oncogenes (i.e. excessive mitogenic stimulation), oxygen starvation, or the lack of certain building blocks – events that typically occur in rapidly proliferating cells. In addition, well-functioning cell death programs are a good protection against metastasis formation because they are triggered when a cell is in the wrong place in the body and does not receive appropriate signals from its environment (cell survival and death signals; anoikis). Conversely, apoptosis disorders and a general increased resistance to programmed cell death contribute significantly to the tumor phenotype. And in this context, it should be noted that one of the key proteins that can very effectively and extensively stimulate the cell to undergo cell death processes is the tumor suppressor p53. We will hear a lot about this tumor suppressor later (chapter 13A).

Once again lymphomas

In the previous chapter we showed how disturbed cell division is related to the development of some lymphomas. However, the same effect – an excessively increased number of cells – can also be achieved by the opposite way, namely by an insufficient rate of cell death. During their relatively complicated development from progenitor cells, maturing lymphocytes go through a certain phase, a kind of quality control, which targets those lymphocytes that are not fully functional or capable of interacting with endogenous antigens due to autoreactivity. Such lymphocytes are immediately and very effectively removed by apoptosis. In mice, for example, it is estimated that up to 70% of daily B-cell production dies by apoptosis (Krejsek, Kopecký, 2004), and according to some authors the death can be as high as 95% of all B-cells (Peelengaris, Khan, 2006). This means that lymphocytes at this stage of maturation are very sensitive to the induction of apoptosis. In fact, there is a high proliferation rate on one side and a high apoptosis rate on the other side, and both processes balance each other. And there is a type of B lymphoma (follicular lymphoma) in which oncogenic transformation does not begin with a disruption of cell division as in Burkitt's lymphoma or mantle cell lymphoma, but conversely with a disruption of apoptosis. Follicular lymphoma cells arise from a lymphocyte that has acquired resistance to cell death.

4B. Death is taboo

"To die is an evil: for the gods have thus decided. For otherwise they would be dying."
(Sappho)

"Think how great is a blessing is a timely death, how many have been injured by living longer than they ought." (Seneca, "On Consolation to Marcia")

What is our attitude toward death? Balanced? Controlled and managed? Do we perceive death as an inseparable and integral part of life, which it undoubtedly is, and do we understand its significance? Do we accept death serenely, with dignity and humility? Or is the concept of death omitted from our vocabulary, banished from our lives, and we want nothing to do with it? Is death an evil that even "the gods recognize by their immortality," or does it also bring something good and its absence harms us? Cell death and the role it plays in the development of tumors can help us find answers to these questions.

What is the meaning of cell death: An evolutionary viewpoint

Nick Lane is a British evolutionary biochemist and author who studies the origin of life and evolution, specializing in bioenergetics and the processes by which cells gain energy. In his book *Power, Sex, Suicide* (Lane, 2005), he addresses the multifaceted role of mitochondria in the life of cells and in the evolution of multicellular organisms. He considers the formation of eukaryotic cells capable of efficient energy production by mitochondria as the most important milestone in the evolution of life. He proves that the formation of a eukaryotic cell was a prerequisite for the development of multicellularity and occurred only once in evolution (Lane, 2005). In contrast, the mechanism of apoptosis, which is inextricably linked to mitochondria, has evolved several times independently, i.e. repeatedly. This also suggests that this mechanism was and is necessary for the functioning of multicellular organisms as a result of strong selection pressure – the need to remove damaged cells that could endanger the whole multicellular organism. The ability of cells to undergo apoptosis is essential for maintaining the integrity of multicellular organisms. Nick Lane writes: "The multicellular individual is made up of cells that collaborate together for the greater good. Nonetheless, this cooperation is not a cellular love fest – it is enforced by the death penalty for any cells that try to abscond and return to their ancestral way of life." True devotion to a multicellular way of life requires the ultimate sacrifice – death for the good of the whole. He adds, "Occasionally, selfish cells escape detection and evade the death penalty, and when they do the result is cancer. Cancer cells replicate furiously, in their own interests rather than those of body, undermining the integrity of the body. Ultimately, having evaded death temporarily themselves, they bring about the death of their erstwhile master, and with it their own" (Lane, 2005, p. 230).

Death as the absence of life

Nick Lane notes that, from a historical perspective, there has long been a reluctance among biologists to give apoptosis any real meaning. "Biology is, after all, the study of life and there is a sense in which death, the absence of life, is beyond the remit of biology" (Lane, 2005, p. 218). Even scientists who have studied the nematode and the fate of its individual cells acknowledge that while it has long been clear that the fate of individual cells is determined as a normal part of development, the fact that cell death is an integral part of that development initially seemed inconceivable. Death was thought to be a deviation from normality, and it was not clear why cells should be created whose only obvious function is to die. The authors literally described the involvement of cell death in the development of the organism as "intellectually unsatisfying" (Potts, Cameron, 2011). Not only is this a very honest and frank admission (especially when considered as part of a typically scientific paper written in a rigorous and strict style), but it also reflects and probably quite accurately captures our problem with accepting death as a natural, "normal" part of life in general.

After all, quite similar is the statement – albeit from a very different field – of Ira Byock, an American physician who specializes in palliative care. In his book *Dying Well, The Prospect for Growth at the End of Life*, he explains, among other things, the importance of palliative care and why it is still difficult not only to understand but also to really support dying people – both financially and humanly. He explains why death seems like a problem, "and that the current health care system is dominated by a culture of curative medicine that is determined to save life at all costs – literally, it aims to prevent death at all costs." Byock even suspects that current medical education acts like a filter through which people see only options for further life-prolonging treatments. He believes this reflects the fact that "Western culture has no vision or direction regarding the end of life" (Byock, 2013, pp. 274–278).

Czech journalist Petr Třešňák came to the same conclusions in his article *The Prize for a Good Death*. Třešňák: "Western civilization suffers from a strong fear of death. It perceives it as an unwelcome tragedy without comprehensible value, which deserves to be rejected by all available medical means" (Třešňák, 2013). Bernard Jakoby, who studies the end of life, also believes that "death is accepted as the destroyer of all hopes, as the worst thing that can happen to us in life" (Jakoby, 2005, p. 11). Czech sociologist Jiřina Šiklová has described death as "expelled" from contemporary society, although, as she writes, "the awareness of death, i.e. the awareness of the limitation of our life, is a fundamental sign of man" (Šiklová, 2013, p. 11).

Is apoptosis a cellular "suicide" indeed?

But back to cells and cell death. What can be deduced about our general relationship to death from the way we think about cell death? Apoptosis is a very common term for cell suicide. Perhaps to emphasize the differences between necrosis and apoptosis. Nick Lane describes necrosis as a "violent, unexpected, swift demise, in which the carpet is left stained with blood and gore and triggers an unruly inflammatory response equivalent to an incendiary police investigation," while apoptosis reminds him of "the silent, premeditated swallow of a cyanide pill in which all evidence of the deed is spirited away" (Lane, 2005, p. 218). Suicide is defined as the act of taking one's own life; an act in which a person intentionally causes his or her own death. It seems then, that the link between apoptosis and murder lies precisely in the "intention," in a voluntary decision that comes from within the cell (and from within the person). But is this really so?

The processes of apoptosis are "programmed" and carried out by the machinery present in the dying cells themselves. Actually, it is not a voluntary (and certainly not an arbitrary) death. Apoptosis is the inevitable consequence of one or more specific and serious causes. There are several phenomena that trigger apoptosis, such as the inability of the cell to survive and function properly due to, for example, severe cellular damage or loss of basic functional equipment. Apoptosis may also be a relevant response of a cell to severe external conditions (oxygen deficiency, pH extremes, etc.) to which it is unable to adapt. Finally, apoptosis can also be triggered naturally as part of the development and remodeling of the organism – as exemplified in the nematode.

From the cell's point of view, then, it is an inevitable but involuntary response to a particular stimulus – a command emanating from the external environment or from internal structures. It is obvious that the cell would not go through this process if it were not inevitable. Certainly, the process itself, apoptosis, is programmed, internally controlled, ordered, follows a fixed plan and rules, and requires considerable active participation by the cells themselves. In the final stages, the participation and cooperation of other cells is required, but it causes no inflammation, no confusion, no need for a dramatic reaction of the microenvironment. It is as if the cell accepts the inevitability of its demise with humility, discipline and without resistance. On the contrary, it actively cooperates and participates in this process. It is – in contrast to the wild, uncontrolled necrosis that causes inflammation and threatens the microenvironment – a "well dying."

A well dying that Ira Byock describes in his book of the same name: "I consciously reject the term a well death. The term 'well dying' is probably more appropriate to describe the experience people want to have at the end of their lives" (Byock, 2013, p. 51). Byock refuses to confuse death with dying and leaves open the evaluation of death as such.

A terminology full of violence

Let us return to the question of why we call apoptosis cell suicide. Some authors try to weaken the word suicide, even when it is used in its original meaning. For example, the Czech psychologist Pavel Říčan, in a book on child bullying, intentionally uses the Czech term "self-killing" instead of "self-murder" for "suicidium" because "the latter has a persuasive accent that is not appropriate for some suicides" (Říčan, 1995, p. 18). Similarly, the Austrian philosopher and essayist Jean Améry applies the concept of voluntary death to suicide, adding, among other things, "I know that this act often occurs in a situation of great pressure" (Améry, 2010, p. 13). As if the word suicide (self-murder) contained too much violence and arbitrariness in a context involving children or understanding the voluntary ending of life. The frequent use of the term self-murder for apoptosis is therefore surprising.

Even from the brief quote from Nick Lane mentioned earlier, describing the relationships and fates of cells in a multicellular organism, it is clear that calling apoptosis cellular suicide does not end the terminology of violence: "This collaboration is not a cellular love fest – it is enforced by the death penalty for any cells that try to abscond and return to their ancestral way of life. True commitment to the multicellular way of life demands the ultimate sacrifice – death for the greater good…" (Lane, 2005, p. 230).

If such terminology were used only in a text aimed at generating interest and publicizing a biological phenomenon to the widest possible audience, it might be understood as literary exaggeration. Yet, the same style is also used to describe apoptosis in a textbook of cell biology: "The suicide machinery is regulated by signals from other cells. Some act as killer signals, activating the suicide machinery of the cell." The very next paragraph states, "For a multicellular organism, programmed cell death is a common-place, normal, and generally benign event" (Alberts, 1998, p. 587), which clearly demonstrates that we do not understand programmed cell death as something negative but, on the contrary, as a natural part of the development and maintenance of cellular homeostasis. At the same time, instead of the term "killer signal" we can use the more neutral "death signal." Instead of "suicide" we can easily use the neutral and completely appropriate term "self-destruction," etc. So why do we use such expressive, emotionally tinged terminology that has a "judgmental accent" (Říčan, 1995, p. 18)? Is this just a poetic license? Or just a habit that became common after the word suicide was first used as an exaggeration? Or does this "judgmental" terminology unconsciously reflect our unacknowledged fear and rejection, condemnation of death?

A contribution to the debate supporting the last option could be an illustration attached to an article in the prestigious scientific journal *Nature*. The article, entitled "The Little Devil of Death," begins with the words, "Cell suicide is an essential, everyday event in the developing and adult multicellular organism. But cells must not die without good reason, and 'roadblocks' at several points in the cell-death signaling pathway prevent wanton cell suicide." The article was a commentary on two original scientific articles reporting the discovery of a "roadblock" – a Diablo/Smac inhibitor that blocks IAP apoptosis inhibitors. While the IAP proteins that block apoptosis and thus "save" the cell from death are depicted in the scheme as little angels, their inhibitor, which on the contrary promotes cell death, takes the form of a little devil (Fig. 15; De Laurenzi, Melino, 2000). Even if we admit that these funny symbols are used with exaggeration, they contain a clear evaluation. They suggest a (relatively general?) attitude toward (cellular?) death and survival. The absurdity of such an evaluation – "death is bad, survival is good" – becomes clear (even at the cellular level) the moment we realize that the development of tumors is clearly related to a reduced readiness of cells to enter the processes of programmed cell death. Thus, insufficient activity of the "devil" Diablo/Smac and, conversely, too much zeal on the part of the "angel" IAP proteins contribute to the development of tumors. So who is the angel and who is the devil in reality?

Cell suicide is anthropomorphism!

We have already mentioned anthropomorphism, the effort to ascribe human traits and characteristics to a non-human reality or phenomenon. "True scientists consider anthropomorphism to be a something of a deadly sin and ostracize scientists who knowingly employ it in their work" (Lipton, 2005, p. 35). According to suicidologists – experts who study suicides – "suicide is specifically and universally human. No animal obviously takes its own life, and even young children do not do so. Suicide is a human privilege." There has yet to be a single known suicide "in which the destruction of consciousness goes so far as to completely obliterate human nature" (Améry, 2010, p. 60). And can something without consciousness be intentional? Jiřina Šiklová says that "the consciousness of one's own death, of one's own finitude, distinguishes humans from other living beings" (Šiklová, 2013, p. 13). It seems beyond doubt, then, for several reasons, that suicide is a specifically human matter. And is it not then equally clear that calling apoptosis "cellular suicide" is the ultimate anthropomorphism? And surprisingly, this anthropomorphism is widely accepted and unchallenged by the entire community of molecular and cellular biologists! What does it say? What does it say about us?

Good dying: Apoptosis or necrosis?

If tumor cells have a reduced ability to enter the processes of programmed cell death (and this is almost always the case), it does not mean that they cannot die. They are much more resistant to death and survive where healthy cells would not. They can withstand much greater external stress as well as a much greater degree of internal damage than healthy cells. But when the pressure of external conditions or the level of internal damage reaches a threshold, tumor cells will eventually die anyway. Never by apoptosis – they cannot – but by necrosis. As mentioned earlier, necrosis is a wild, disorderly cell death. The necrotic cell breaks down, and its contents are released into the environment. The impact of necrosis on the environment surrounding the dying cell is significant. Inflammation occurs, causing confusion, chaos, and general danger. In this death, the "carpet is left stained with blood and gore" (Lane, 2005, p. 218). From the perspective of the cell itself, the difference between apoptosis and necrosis does not seem to be great. After all, the result is the same: death of the cell. But the path to death is different, the dying is different. And death is not to be confused with dying, as the above quote from Ira Byock (Byock, 2013) already suggests. Dying is still one of the stages of life. It is a process, and only at the end of this process is death. The Czech poet Jiří Wolker wrote that he was not afraid of death, but he was afraid of dying (Walker, 1984). And he apparently expressed the feelings of many people, not only of his own generation. Apoptosis, in contrast to necrosis, stands for an orderly, organized, peaceful course of dying. Perhaps it represents the good dying that most of us wish for at the end of our lives. How are the themes of death and dying related? And are they connected at all? And can the cells teach us anything?

Is it possible that our relationship with death affects the way we die? Czech journalists Zdeňka Trachtová and Jan Jiřička, in their article "Death in Czech: We Pretend It Does Not Exist and End Up in Undignified Circumstances," see this connection. They believe that "the tabooing of issues related to death means that dying often takes place in unpleasant, stressful conditions instead of an ending in dignity" (Trachtová, Jiřička, 2014). Similarly, Ira Byock uses specific stories from his rich experience to show how non-acceptance of death can make the dying process extremely complicated (Byock, 2013). Bernard Jakoby, who has long been involved in research on the end of life, agrees: "In a life oriented almost exclusively to the values of this world, there is no place for individual or collective preparation for death" (Jakoby, 2005, p. 12). And he goes even further when he states, "The technical level of modern medicine prolongs not only life, but also dying itself, which unfortunately is often undignified as a result… Because doctors increasingly intervene artificially in the dying process, the natural course of the process is blocked" (Jakoby, 2005, pp. 13, 47). And although Jakoby explains this by saying that "medical science perceives death as its arch enemy, as evil" (Jakoby, 2005, p. 13), the problem, of course, is not just on the medical side of science. It reflects a social order and the attitude of society as a whole.

If death is generally considered a "loss" (Trachtová, Jiřička, 2014), then we are all, without exception, doomed to lose from the beginning. And is it even possible to live with the knowledge that the only possible outcome of our lives is defeat? Is it not then a fact that "we have in fact isolated and pushed aside death, which is an essential part of life, as if it did not exist" (Jakoby, 2005, p. 13)? Is it not logical that we repress death because it is the only way to live, to strive, and to try, regardless of what inevitably awaits us? Is it not ultimately logical that death – understood as evil and loss – must be tabooed, because otherwise the awareness of death would deprive us of the motivation and will to live? However, we pay a high price for this in the form of dying badly. Just like a cell, which, if it loses its ability to go into an orderly, calm and smooth process of programmed cell death – apoptosis – under appropriate circumstances, will sooner or later die by necrosis – in a dramatic, chaotic, disorderly way.

Death is taboo. The tabooing of death, its non-acceptance and rejection, opens the door to an unpleasant, uncomfortable dying. And not only that. Tabooing death is one of the eleven non-deadly sins. What about this? Where is the path to a more reasonable and harmonious relationship with death? Is there anything positive about death at all?

Death as a creative tool: A creative combination of YES and NO

Johann Wolfgang Goethe wrote: "To create is to leave out" (*Schöpfen heißt weglaßen*). Nick Lane puts it similarly, "Just as the sculptor chips away at a block of marble to create a work of art, so too the sculpting of the body is achieved by subtraction rather than addition" (Lane, 2005, p. 218). In this statement, he is referring, among other things, to the detailed study of nematode development, which showed that while cell death may seem "intellectually unsatisfying" (Potts, Cameron, 2011), it is a normal, undeniable, and most importantly, a completely indispensable and irreplaceable part of the successful functional development of a multicellular organism. Damage to programmed cell death leads to significant developmental defects. (And as mentioned earlier, it is also involved in the development of tumors!) Without apoptosis, no healthy, functioning body can be formed. Thus, the basis of the development of a multicellular organism is a combination and interplay of two processes: Cell division and apoptosis. A combination of birth and death, emergence and decay, YES and NO.

Biological models provide many more vivid examples of how not only birth, creation, addition, activation, and initiation, but also death, extinction, withdrawal, inhibition, and termination are creative and essential to the development of new entities.

In the previous chapter 3A on the cell cycle, for example, we saw how important cyclins are for the regulation and smooth functioning of cell division. Their effects are based on their regular occurrence in the cell, always at a very specific phase of the cell cycle. And this cyclic recurrence– so specific to their activity and importance that it is reflected in their name – is achieved not only by precise regulation of their synthesis but also, and equally important, by precise regulation of their efficient degradation. Failure on either side would lead to failure in the control of cell division.

Another example is cell differentiation, mentioned earlier. All cells in the body have almost the same genome, but they differ significantly in the genes they actively use and those they have turned off. Successful cell differentiation is accompanied by the gradual, tightly regulated activation of new genes. The protein products of these genes are involved in the function and modification of the developing cell. And it has been repeatedly shown that cell differentiation is equally fundamentally dependent on the gradual and tightly regulated silencing of genes whose protein products, if left in the cell, would block the process of cell differentiation.

A creative combination of YES and NO: Steve Jobs

Steve Jobs and his life story probably need no further introduction. In 1976 he founded Apple with Steve Wozniak, a company that developed some of the first successful personal computers. Apple was a progressive and very successful company based on its explicitly formulated marketing philosophy. This included empathy (an intuitive connection to the customer's feelings) and investment (an awareness that design, not just function, sells the product), as well as what's known as Focus of Interest: "In order to do a good job of those things that we decide to do, we must eliminate all of the unimportant opportunities" (Isaacson, 2011, p. 78). The philosophy that a company's success depends not only on what it does, but in large part on what it doesn't deal with, was demonstrated again by Jobs in 1997 when he returned to the bankrupt Apple after a hiatus of several years, and saved the company by ceasing production of about 70% of its products. He later commented, "Deciding what *not* to do is as important as deciding what to do" (Isaacson, 2011, p. 336). In the context of the "Apple" project, Steve Jobs brilliantly mastered the principle of NO. In his "Own Life" project, he faced similar challenges. After returning to Apple, he was full-time Chief Executive Officer, but continued to work full-time at his Pixar film studio, which he'd founded when he left Apple, and was the father of three young children. Looking back, he rated that time as "rough, really rough, the worst time in my life. I had a young family. I had Pixar. I would go to work at 7 a.m. and I'd get back at 9 at night, and the kids would be in bed. And I couldn't speak, I literally couldn't, I was so exhausted. I couldn't speak to Laurene. All I could do was watch a half hour of TV and vegetate. It got close to killing me…" (Isaacson, 2011, p. 334).

In 2003, Steve Jobs was diagnosed with pancreatic cancer. He kept this information secret for some time. He spoke about his illness for the first time in 2005, when he accepted Stanford University's invitation to give a talk at the opening of the academic year. He said, among other things: "Remembering that I'll be dead soon is the most important tool I've ever encountered to help me make the big choices in life. Because almost everything – all external expectations, all pride, all fear of embarrassment or failure – these things just fall away in the face of death, leaving what is truly important. Remembering that you are going to die is the best way I know to avoid the trap of thinking you have something to lose. You are already naked. There is no reason not to follow your heart" (Isaacson, 2011, p. 553).

Death as the measure of life

The story of Steve Jobs opens the final reflection on death. The last and perhaps the most disturbing and inspiring. What would life be like if it had no end? Would we live it better if the awareness of its end did not frighten and burden us? Those who study death and dying agree that people often do not realize what really matters in life until the very end. It is only with the realization that their own death is something obvious and near that they begin to appreciate familiar things (Šiklová, 2013), begin to ask questions about the meaning of their own lives (Jakoby, 2005), or, like Steve Jobs, find the courage to live a true, authentic life (Isaacson, 2011). In this context, it is worth recalling the legacy of Elisabeth Kübler-Ross, a Swiss physician who devoted her entire life to caring for the dying and their loved ones, researched the dying process, made several groundbreaking discoveries about it, and wrote several books about it.

At the end of her life, together with her colleague David Kessler, she published a profound, touching book called *Life Lessons*, which is about lessons of authenticity, love, relationships, fear, etc., that we need to master in order to live a good, quality life (Kübler-Rossová, Kessler, 2013). Reviewers repeatedly emphasize that the book is not about dying, but about living. It is really about life lessons, about how to live, and not about how to die, as the reader might perhaps expect from the authors. This book in particular clearly shows that living, dying and death are inseparable. And even more than that. This book (as well as many others) is a testimony to the fact that death and dying are conditions for quality of life. The authors make no secret of the fact that it is precisely their rich experiences and lessons of dying that have clarified for them what is truly important in life. The knowledge described in Life Lessons is obtained from the dying, who themselves are already departing from life. Thus, death and dying are resources for understanding life, they are its measure, and awareness of them is a way to experience life better. And perhaps the relationship between life and death is entirely reciprocal: just as dying and death teach us something about life, the way we fulfill, use, and experience life prepares us for dying and affects our ability and willingness to embrace death. "We fear death in part because we have used the years of our lives so badly," wrote American archbishop Fulton Sheen (Sheen, 1969, p. 222). This quote is a reminder, but also a clear inspiration, of how and why to face one of the non-deadly sins. Death does not have to be taboo.

5A. Unlimited replication potential

Tumor development is a long process leading to the gradual accumulation of many genetic alterations in tumor cells. It proceeds in waves of clonal expansions based on the proliferation of cells that happen to have acquired a favorable property providing them with the ability to overgrow other cells, expand, and form rapidly growing clones (see chapter 2A – Fig. 7). For the organism, whose satisfactory overall condition depends on the responsible behavior of each of its cells, waves of expansive cell growth that ignore the needs of the organism are, naturally, very dangerous. One might think that the cells that have overcome the protective barriers meant to ensure cellular homeostasis that we have described in previous chapters (i.e. the control of cell division and cell death) can divide and multiply indefinitely. For a time, this may indeed happen, but sooner or later the cells run up against one or other of the protective barriers that have evolved in multicellular organisms. One of these barriers is the limitation of replication potential.

To understand what limited replication potential means, it may be helpful to recall how it was discovered. It was first described in the context of growing human cells in tissue culture. Human cell cultures are made by exposing a piece of human tissue to trypsin, a digestive enzyme from the pancreas that dissolves the intercellular connective tissue holding the cells together. The tissue is thus broken down into individual, isolated cells. These are transferred to culture dishes containing a nutrient medium with all the necessary components for cell life and stored under appropriate conditions, such as the correct temperature and oxygen concentration. The cells sink to the bottom of the culture dish and adhere there. Under these conditions, they can survive and begin to divide. The division process continues until the cells cover the entire bottom of the culture dish. Normal, healthy cells then stop dividing in response to contact inhibition signals. Now, the cells can be detached from the bottom of the culture dish (e.g. again using trypsin), transferred to a fresh medium, and divided into several culture dishes, where they grow again until they cover the entire bottom. This process of cell dilution and culture is called passaging. Cells can be repeatedly passaged to obtain a relatively large amount of cells for different purposes.

However, when human cells are grown in tissue culture, they have been repeatedly shown to each divide about 50 to 70 times, and after reaching this limit (called the Hayflick limit after one of the discoverers, Leonard Hayflick), they stop dividing further. They are still alive, but change their phenotype and lose their ability to divide. This phenomenon is called cell senescence, aging, or the M1 point (Fig. 17). Thus, an internal setting, an internal cell division counter that limits the total number of divisions, has been repeatedly demonstrated in human cells. This system for limiting replication potential is based on internal factors – a kind of cell counter. Understanding this counter that limits replication potential has been linked to the discovery of telomeres and their function (Hayflick, 1997; Mathon, Lloyd, 2001; Shay, Wright, 2005).

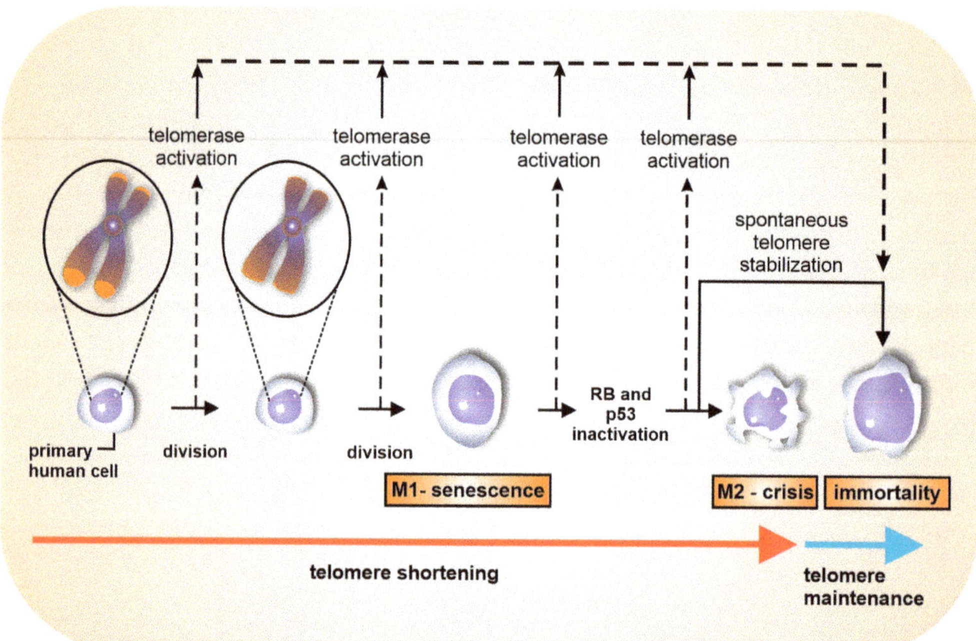

Fig. 17 Telomere hypothesis: The shortening of telomeres determines the replication potential of human cells

Cells grown in tissue culture are capable of dividing approximately 50 to 70 times. When this number of cell cycles is reached, shortened telomeres lead to senescence (M1 point): cells adopt a senescent phenotype and stop dividing. When tumor suppressor RB and p53 signaling pathways are damaged, cells continue to divide, but after reaching the point of critical telomere shortening, a crisis occurs (M2 point). In rare cases, cells manage to avoid the crisis by spontaneously activating telomerase. This stabilizes telomere length and renders the cells immortal (adapted from Mathon, Lloyd, 2001; Shay, Wright, 2005, modified).

Telomeres

Genetic information written in DNA is stored in the nuclei of eukaryotic cells. Long strands of DNA, together with bound specific proteins, form structures called chromosomes. Telomeres, located at the ends of chromosomes are also composed of DNA and proteins (Fig. 2). However, the structure of telomeric DNA and proteins is different from the other parts of the chromosome. In human chromosomes, telomeric DNA consists of repeating six nucleotide sequences TTAGGG. This repetitive sequence is not part of genes; it has no coding function. But there is a specific protein complex bound to them called telosome or shelterin. The main function of the telosomes is to protect the ends of chromosomes. As we will see later (in chapter 6A), cells constantly monitor the state of their DNA very closely, and when damage is detected, they immediately begin to repair it.

One serious type of DNA damage is a double-strand break. Once a DNA double-strand break occurs, the cell sets in motion a complex cascade of events that lead to perfect repair. The ends of linear chromosomes would be perceived by cellular mechanisms as DNA double-strand breaks if they lacked the "shelter" – the telosome. In an attempt to repair them, the chromosomes could become damaged or fused together (Fig. 18). This would subsequently complicate or directly prevent some functions of the chromosomes. Therefore, these specific protein complexes are formed at the telomeres to protect the ends of the chromosomes and prevent them from being perceived as DNA breaks and triggering unwanted repair signals (Wright and Shay, 2005; Xin et al., 2008; Frias et al., 2012).

Length of telomeres

The length of telomeres varies in different cells. For example, the telomeres of chromosomes of embryonic cells are very long and consist of up to 15 thousand nucleotides. In contrast, the telomeres of chromosomes of old or aging cells are quite short. How is this possible? In each cell cycle, genetic information is duplicated or replicated in the synthetic phase (S). DNA replication gives both daughter cells a complete and identical copy of the genetic information of the dividing parental cell. For this purpose, the cell is equipped with a very complex replication apparatus that duplicates DNA. This mechanism replicates the original DNA sequence almost perfectly except for a "tiny" difference. It cannot replicate the linear DNA to the very end. Thus, the last 50 to 100 nucleotides at the end of the chromosome remain unreplicated (Deng, Chang, 2007).

Although daughter chromosomes are almost identical to the original maternal chromosome, they are these 50 to 100 nucleotides shorter. They have shorter telomeres. This does not mean much for the daughter cell, because the telomeric DNA is long, it has no coding function, and thus the daughter cell is not missing any genes or other

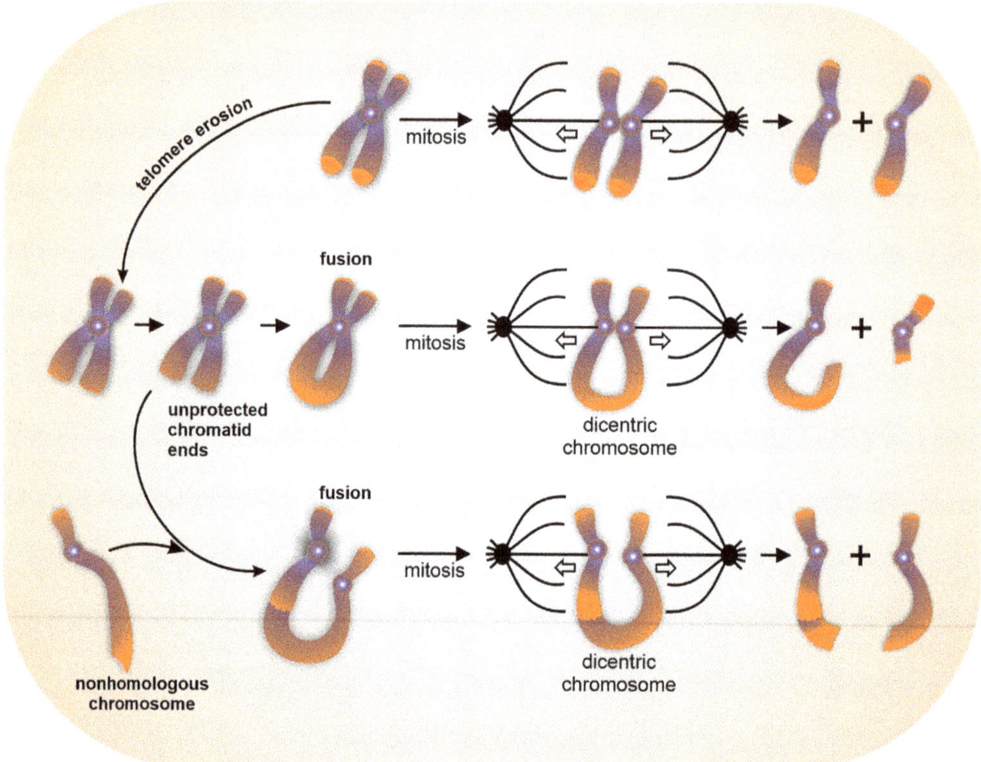

Chromosome ends that are not protected by telosomes due to short telomeres can fuse. This leads to the formation of dicentric chromosomes that cause severe chromosome aberrations in subsequent mitoses (adapted from Weinberg, 2007, modified).

important pieces of genetic information. With each subsequent cell division, the telomere shortens by another 50 to 100 nucleotides.

The problem arises only when telomeres become so short that they can no longer perform their function. When the telomere is too short, the telosome cannot properly form around it. And the insufficiently hidden end of the chromosome is perceived by the cell as a double-strand break of the DNA. This must not happen! And so, at the moment when the telomeres are still long enough to allow the successful formation of the protective complex, but are already so short that their further shortening could be dangerous during the next cell divisions, a signal is generated in the cell that initiates the state of senescence, cell aging.

A cell can change its morphology and metabolism, but the top priority is to stop the cell cycle, to block it. It must reliably ensure that it does not enter the next round of cell division. Telomere shortening thus acts as a mitotic counter, i.e. a cell division counter. Telomere length determines the replication capacity of the cell. In other words, it determines how many times a cell can safely divide (Shay, Wright, 2005). Poetically, telomeres are compared to candles that gradually burn down and measure the lifespan of the cell. When the candle – the telomeres – burns out, the cell grows old and later dies (Bártek, 2007). Less poetic are the events associated with the "burning down of the candle" called the replicative senescence checkpoint (Harley, 2002).

Limited replication capacity as a protective barrier against tumors

Carcinogenesis is a multistep process of gradual accumulation of mutations in tumor cells. It progresses step by step, mutation by mutation, to a profound malignant phenotype. Each step – mutation – is accompanied by a wave of clonal expansion, i.e. expansive, repeated proliferation of cells with a new mutation (Fig. 7). The waves of clonal expansion repeat with each acquired mutation until the fully malignant phenotype is reached. This would repeat indefinitely if the cell was not held back by limited replication capacity. During the repetitive waves of clonal expansion, the cell proliferates intensively and divides repeatedly, the telomeres shortening accordingly. And at some point, the telomeres are shortened to such an extent that a signal initiates senescence and prevents the cell from proceeding to the next division. This mechanism also stops the process of carcinogenesis before it can endanger the entire organism. Thus, limited replication capacity – i.e. the predetermined, limited number of cell divisions that each cell can undergo – should protect the multicellular organism from the development of tumors (Shay, Wright, 2005).

The disadvantage of this protective barrier is that it limits the overall life expectancy of cell lines, the possibility of their development and differentiation, and thus the potential of all tissues to regenerate, heal, and renew themselves, which contributes to the overall aging of the organism.

Telomerase

Nevertheless, tumors do exist, often reaching a very malignant stage, and the tumor cells from which they arise carry many mutations. It is obvious that during development they somehow manage to overcome the protective barrier of limited replication potential. This is illustrated by the long-term cultivation of tumor cells in tissue cultures. While normal somatic cells (derived from embryonic or adult tissues) have been repeatedly shown to have a limited lifespan in tissue culture, tumor cells can often be cultured indefinitely. For example, the classic tumor cell line HeLa (derived from Henrietta Lacks' cervical carcinoma) has been continuously cultured and used for experiments in laboratories around the world since 1952 (Skloot, 2011). Tumor cells must therefore be able to bypass or break through the barrier of limited replication potential and thus achieve immortality. How do they manage to do that? What trick do they use?

Most body cells have limited replication potential. However, there are physiological, healthy cell types that have unlimited replication potential. This means that the progressive shortening of telomeres does not occur during their division. These include, for example, germ cells. These are the cells that give rise to a whole new individual after fertilization, consisting of $3.72 \pm 0.81 \times 10^{13}$ cells. Clearly, germ cells could not perform this task if they had limited replication potential. Similarly, limited replication potential would substantially impair the function of stem cells, which are the starting cells for entire lineages of cell types. The ability to maintain a stable telomere length is also shared by some other body cells that divide repeatedly, such as cells in the proliferative zone of the intestinal crypts or in the basal layer of the skin.

Some somatic cells may acquire the ability to temporarily maintain a stable telomere length during cell division. This is the case of antigen-stimulated lymphocytes, cells of the immune system that respond to acute immunogenic stimuli and for which it is desirable that they proliferate massively. Similarly, the lobular endothelial cells of the mammary gland are stimulated to divide intensively during pregnancy, and here, too, a restriction in replication potential would mean that the performance of a desirable, useful task would be limited or prevented. Thus, there must be a physiological mechanism that allows cells to maintain, either permanently or temporarily, an unlimited replication capacity. This mechanism is based on an enzyme complex called telomerase. When telomerase is active in the cell, it participates in DNA replication, ensuring that chromosomes are fully replicated and that their length remains stable regardless of the number of completed cell division cycles. Selective activation of telomerase ensures the full replication potential of selected cells or cell groups when desirable for the functioning of the organism. Conversely, silencing telomerase limits the replicative capacity of most body cells to a degree that minimizes the potentially deleterious effects of a cell entering the pathway of carcinogenesis (Harley, 2002; Granger et al., 2002).

For the sake of completeness, and to understand the broader context, recall that the gene encoding telomerase (or two genes, since telomerase consists of two different subunits) is present in all cells of the body – both cells in which telomerase is active and gives them unlimited replicative capacity and cells in which telomerase is not active. Even cells lacking active telomerase, i.e. with ever-shortening telomeres, have the genes that code for telomerase. They are just silenced, switched off.

Telomeres, telomerase and tumors

When cells go down the path of tumor transformation, they proliferate intensively but also soon exhaust the capacity of their telomeres. Once telomeres reach a critical shortening, they generate a signal that halts cell division and ends the advanced process of cell tumor transformation. More specifically, this is what would happen if the replication senescence checkpoint always functioned properly. However, even during the first steps of oncogenic transformation, the first mutations occur in the developing tumor cells, leading to initial changes in some cellular functions. These often include damage to the regulation of cell division and to the mechanism that triggers apoptosis. Thus, these cells can no longer respond appropriately to critically short telomeres. The cell cycle may continue even though the telomeres are clearly too short, and induction of senescence may fail. The chances of entering programmed cell death may also already be significantly reduced in mutant cells.

Translated into the language of molecular mechanisms: we know that for successful induction of cellular senescence, the tumor suppressor RB (discussed in chapter 3A) and the p53 signaling pathways, which we describe later (chapter 13A), must function properly. We know from tissue culture experiments that a cell with a damaged replication senescence checkpoint can indeed continue to divide and undergo another 15 to 20 cycles of cell division. But then it reaches a point that we call the crisis or M2 point (Fig. 17). At this point, the structure of the chromosomes is so disrupted, for example by fusions of chromosomes with unprotected ends, that the viability of the cell is dimaged, and it usually perishes (Fig. 18). In rare cases (with a frequency of about 10^{-6}), a cell can exit this "death cycle," escape the crisis, and become immortal. By far the most common mechanism for inducing immortality is activation of telomerase. Once telomerase is activated, cells proliferate without further telomere shortening (Fig. 17) (Shay, Wright, 2005).

The ability to maintain a stable telomere length is characteristic of almost all malignant tumor cells. In most tumors (85–90%), this ability is associated with telomerase activation. Typically, tumor cells with active telomerase have relatively short telomeres, but their length is stable and does not decrease. The remaining 10–15% of tumor cells with inactive telomerase use an alternative mechanism, known as the ALT mechanism, to maintain telomere length. It is based on a completely different principle that is based on DNA recombination. In contrast to the mechanism that uses telomerase activity, this ALT mechanism often leads to the formation of long, sometimes extremely long, telomeres (Henson et al., 2002; O'Sullivan, Almouzni, 2014).

Cellular senescence

Cellular senescence is defined as an irreversible loss of the cell's ability to divide. Senescent cells differ from other non-dividing cells, such as quiescent cells (which can re-enter cell division after an appropriate stimulus) or terminally differentiated cells. Senescent cells have their own specific characteristics. In tissue cultures, they are usually enlarged, flattened, have multiple vacuoles, and sometimes multiple nuclei. When they are part of the physiological tissue of the organism, they usually retain their normal morphology, which is determined by the architecture of the tissue. However, all senescent cells share some common characteristics. Typically, they produce proteins (primarily the p16 and p53 proteins) that induce a senescent phenotype in the cell and block cell division. They do not produce proteins typical of dividing cells, but they produce and secrete certain extracellular factors, including, for example, proteases, extracellular matrix components and, most importantly, many different cytokines that stimulate surrounding cells. The intensity and composition of factors that senescent cells secrete to their environment is so characteristic that they have earned a special name: SASP – *senescence-associated secretory phenotype*. The released factors trigger an inflammatory response and stimulate macrophages, cells of the immune system, to remove senescent cells by phagocytosis, and may also trigger the senescence of other cells in the immediate environment. Part of the SASP effect is also the production of matrix metalloproteinases, enzymes involved in the remodeling or degradation of the extracellular matrix. Senescence thus represents an overall process, a sequence of several events. The first step is (1) the stimulation of cellular senescence itself, i.e. cell cycle arrest and induction of SASP. Signaling molecules released by senescent cells then stimulate the next steps of the entire process: (2) elimination of senescent cells by macrophages and (3) completion of tissue remodeling and regeneration (Muños-Espín, Serrano, 2014). The goal of senescence is not only to stop proliferation and remove damaged cells. Aging cells still significantly influence their microenvironment before they die and are thus involved in decisions about their future fate.

Causes of cellular senescence

Critically short telomeres are not the only cause of cellular senescence. Other triggers of cellular senescence may include DNA damage, high levels of oxygen radicals, nonphysiological activation of some oncogenes, or conversely, inhibition of some tumor suppressors. Sometimes senescence induced by critical telomere shortening is referred to as replicative senescence or replicative aging, whereas induction of senescence by other factors is referred to as premature senescence. Thus, premature senescence is not related to depletion of the replicative capacity of cells, but is the result of a different form of cellular stress (Mathon, Lloyd, 2001). Moreover, like apoptosis, cellular senescence is a necessary to embryonic development and is developmentally programmed. It appears that senescent cells, through SASP, i.e. the specific signaling molecules they produce, play an important role during development by giving "instructions" to surrounding cells and thereby participating in the formation of tissues and organs. In the adult organism, cellular senescence is involved in wound healing and in the resolution of pathological conditions, such as fibrosis. Accomplishment of such a task requires not only senescence of cells and cessation of their division, but also remodeling of tissues by molecules secreted by senescent cells and the subsequent effective elimination of senescent cells (van Deursen, 2014; Burton, Krizhanovsky, 2014).

So, the mechanism of senescence is useful for the organism. It eliminates unwanted and damaged cells, contributes to the healing process and tissue regeneration, assists in embryonic development and tissue remodeling, and last but not least is one of the effective protective barriers against tumors. On the other hand, senescence can also pose a risk to the organism. The positive effect of senescence seems to depend on the effectiveness and completeness of the whole sequence of events of "senescence-removal-regeneration," i.e. on the only short-term appearance of senescent cells in tissues. If all stages of the senescence process are not completed effectively enough and senescent cells remain in tissues for a long time, this may harm the organism, contribute to its aging and even promote the development of tumors. At the same time, the risk of failure of the senescence mechanism increases with time, especially in the aging organism (Muños-Espín, Serrano, 2014; Burton, Krizhanovsky, 2014).

Senescence and aging of the organism: Antagonistic pleiotropy

As the organism ages, the likelihood increases that the "senescence-removal-regeneration" chain will not be completed efficiently and the accumulation of senescent cells will be tolerated. These can then exacerbate tissue damage, contributing to the aging of the organism, rather than preventing aging by removing damaged cells. SASP and the release of pro-inflammatory substances are the main contributors to aging. Cell senescence itself then becomes part of the problem, causing damage and aging, rather than a solution. A mechanism that helps maintain fitness and youth symptoms in a young organism fails in an aging organism; indeed, it accelerates aging (Fig. 19).

Because of this duality, senescence is considered an example of so-called antagonistic pleiotropy. According to the antagonistic pleiotropy hypothesis, genes or processes that have evolved to promote the health or fitness of a young organism may have a detrimental effect on an older organism and contribute to its aging. Thus, senescence is sometimes referred to as an antagonistic feature of aging (Campisi, 2003; Muños-Espín, Serrano, 2014; Burton, Krizhanovsky, 2014).

Aging of the organism

The aging of the organism is manifested by gradual deterioration of bodily functions, accumulation of various defects and diminishing ability to repair them. Aging is accompanied by increased susceptibility to diseases such as diabetes, cardiovascular diseases, neurodegenerative diseases and also tumors.

Progressive shortening of telomeres and cellular senescence contribute to aging, but they are not the only causes and symptoms of this process. It is accompanied by other features. These include, for example, the accumulation of DNA damage, inaccurate organization of chromatin, and the declining ability to maintain proper protein structure (known as proteostasis). Metabolism changes with age, and the ability to regulate and provide the organism with proper nutrition decreases. This is related to a change in the levels of some hormones. Another symptom, and at the same time one of the causes of aging, is increasing damage of mitochondria. In addition, stem and progenitor cells in an aging organism gradually diminish in number and perish. This leads to a decreased regenerative capacity of tissues, including weakened hematopoiesis and immune system

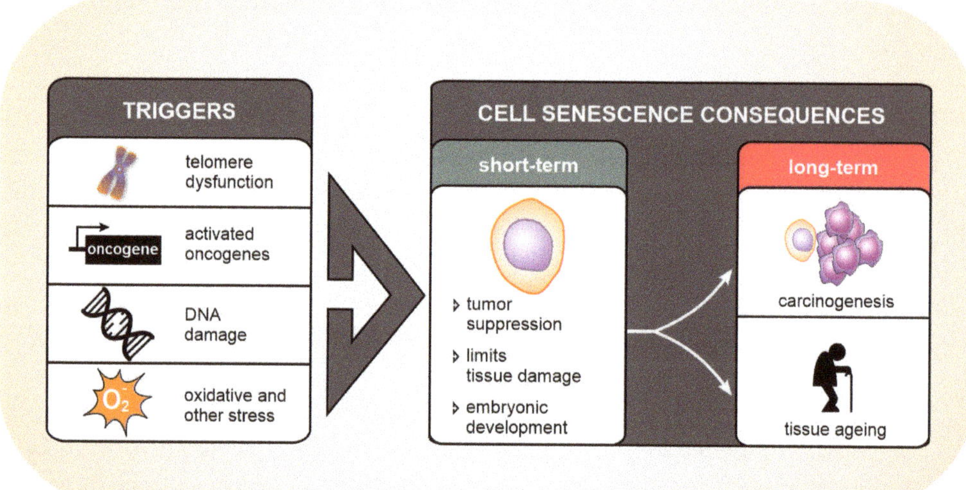

Fig. 19 Biological consequences of cellular senescence
Cellular senescence can be triggered by various causes, such as short telomeres, high activity of certain oncogenes, DNA damage, or oxidative stress. Short-term senescence has a protective function. It protects against tumor formation, is involved in wound healing, and affects embryonic development. In contrast, the long-term persistence of senescent cells in tissues can promote tumorigenesis and favor the development of diseases of aging (adapted from Burton, Krizhanovsky, 2014, modified).

function. In addition to changes that directly affect the cells themselves, intercellular communication also deteriorates with age. This is related to changes in hormones and the nervous and immune systems. Perhaps the most characteristic feature of the altered intercellular communication of the aging organism is a constant slight activation (the so-called smoldering) of pro-inflammatory processes.

Of course, biological aging of the organism is a very complex process. It includes many different types of changes and damages, and all signs of aging are also interconnected and concatenated, influencing and potentiating each other (López-Otín et al., 2013). Similarly complex are the influences and links that exist between aging or its causes and the process of cancer development.

Aging and tumors

Tumors are a disease of older age. Of all the factors that influence the incidence of tumors, age is the most risky. After reaching about half one's life expectancy, the incidence of tumors increases exponentially, and this is probably not unique to humans. The maximum life expectancy of humans (according to the longest-lived individuals) is estimated to be 100 to 120 years, and most tumors occur in humans after the age of 50 to 60. In mice with a life expectancy of 3 to 4 years, tumors begin to occur most frequently at the age of 18 to 24 months (Fig. 20) (Campisi, 2003; De Magalhães, 2013).

Nevertheless, the relationship between age and tumor incidence is not simply linear. For example, tumor incidence stabilizes at a late age and does not increase further. The relationship between age and tumor incidence also depends on the type of tumor – some tumors (e.g. osteosarcomas, testicular tumors) typically occur in younger individuals. Even so, there is considerable overlap between the process of cancer development and the process of aging. According to epidemiological data, fewer tumors occur in families with predispositions to longevity. Experimental manipulations (e.g. genetic or dietary) that slow the aging process also reduce the incidence of tumors (De Magalhães, 2013). The exact relationship between aging and tumor biology is not yet clear, but several cellular mechanisms have been found to be involved in both processes. These include the limited replication capacity of most somatic cells and cellular senescence.

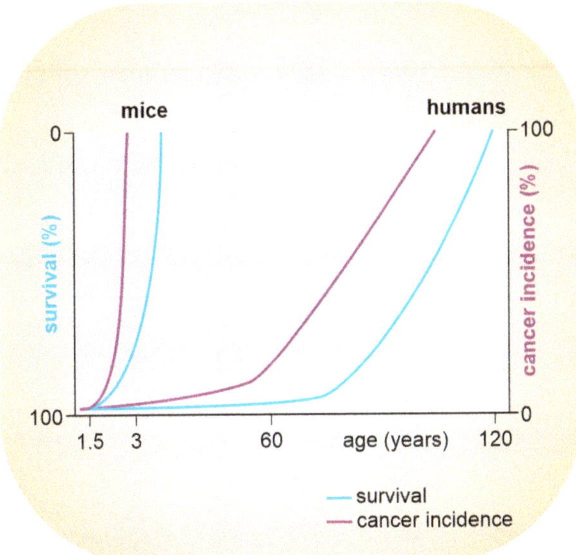

Fig. 20 Tumor incidence increases with age
The incidence of tumors increases with age, growing exponentially from about half the maximum age reached. The maximum life expectancy in humans is estimated to be 100 to 120 years, and most tumors in humans occur after age 50 to 60. In mice, which live three to four years, most tumors occur 18 to 24 months after birth (adapted from Campisi, 2003, modified).

Telomere shortening, cellular senescence, and tumors: Antagonistic pleiotropy

Telomere shortening and cellular senescence – both replicative and premature – can be both beneficial and detrimental to the organism. Transient, short-term induction of senescence (sometimes referred to as acute senescence), followed by tissue remodeling and elimination of senescent cells, is beneficial because it ensures removal of unwanted and damaged cells, tissue regeneration, and is one of the barriers that protects against tumor development. Permanent, chronic senescence, characterized by the inability to effectively eliminate senescent cells, is harmful to the organism and can promote the development of tumors – for example, through the permanent induction of inflammation. Senescence is thus not only an antagonistic sign of aging, but also fulfills the definition of antagonistic pleiotropy in relation to tumor development. It is a protective mechanism against tumors at reproductive age but contributes to tumor development later in life. Similarly, telomere shortening itself fully satisfies the meaning of antagonistic pleiotropy. Limited replication capacity initially protects against tumorigenesis but may later become a mechanism that contributes significantly to tumorigenesis. This is related to both the induction of replicative senescence and the increase in genetic instability, a mechanism discussed in more detail in chapter 6A (Pereira, Ferreira, 2013; Muños-Espín, Serrano, 2014).

Antagonistic pleiotropy: Something for something

Before leaving the chapter on cellular senescence, let us recall the key role that two genes, p16 and p53, play in triggering cellular senescence. They encode two important tumor suppressors, proteins that protect the organism from tumor development. At the same time, these two genes are also involved in a fundamental way in the aging process of the organism, linking these processes at the molecular level (López-Otín et al., 2013). As far as cellular senescence, carcinogenesis, and aging are concerned, that which is referred to in technical language antagonistic pleiotropy could be described in layman's terms as "something for something" (a trade-off). "Something for something" is certainly a general principle that describes the resolution of a situation in which one must choose between two options, both of which have positive and negative consequences. It is a situation where one accepts a small sacrifice, but in the end the positive outweighs the negative. In this chapter, we can use this trade-off principle to accept that aging of the organism is an acceptable price to pay for protection against the development of tumors.

5B. Reluctance to age

"It is only in old age that the success of our lives is decided". (Grün, 2009, p. 39)

Cancer cells refuse to age!

At the cellular level, senescence – aging – is an effective protective barrier for the organism against tumor development. Cells have a fixed lifespan, linked to a specific telomere length, and they live only as long as the progressive shortening of their telomeres allows. When this replication capacity is exhausted, they become old and are removed from the body. In human cells, replication capacity is set to significantly limit the successful completion of tumor transformation. Conversely, this means that successful tumor transformation is necessarily associated with the acquisition of unlimited replicative capacity of the cells, i.e. immortality. Cancer cells do not age. Thus, aversion to aging is another of the non-deadly sins.

We dislike growing old, we adore youth, we dream of immortality

The desire for immortality and eternal youth has probably accompanied mankind since ancient times. For many, Elina Makropulos, alias Emilie Marty, in a story by the Czech writer Karel Čapek (1956) a hundred years ago, is synonymous with this desire. Since then, we have evolved: the present is obsessed with the desire for constant youth. Allegedly, three quarters of Czechs see aging as something bad and something to fight against. Society views older people as inefficient, sick and unfit. "I fear every wrinkle, every gray hair, every extra pound. It puts me in the category of old, i.e. destined to be replaced" (Vavroň, 2015). "The trend requires us not to complain, not to grumble, not to ail. Instead, one is supposed to undergo some cosmetic surgery and intensive physical training, and pretend to be young" (Prudičová, 2015).

This is how Czech journalists summarize their findings on this topic. And, of course, this does not just apply to Czechs. Anselm Grün, the German philosopher and theologian, believes that "we are currently experiencing a crazy desire to be young, and even old people are falling for this madness" (Grün, 2010, pp. 17–19). Elisabeth Kübler-

Ross, an American physician with Swiss roots who was familiar with the dying, the aging, and the elderly, also concluded that "we don't value age in Western culture, we don't see wrinkles as part of life; they're something to be prevented, hidden, removed" (Kübler-Rossová, Kessler, 2013, p. 138).

It seems that, in short, we are a society that "believes in the cult of youth, strength, freshness and newness" (Vavroň, 2015). Indeed, anything bearing the label "anti-age" is a reliable guarantee of widespread interest and commercial success. And so, alongside the never-ending flood of anti-age cosmetics, anti-age body care, and anti-age lifestyle, a whole new field of anti-age medicine has even emerged, the aim of which is to slow down the aging process and achieve the best possible human health.

Can aging be stopped?

In Čapek's story, Elina Makropulos lived for 337 years thanks to her father's miracle recipe (Čapek, 1956). What about the actual possible maximum length of human life? Or at least with our current understanding of it?

Before we address the length of human life, perhaps we can use our knowledge of the end of life of the nematode *Caenorhabditis elegans*. As we mentioned earlier, programmed cell death in this nematode has been studied in detail, providing the starting point for studying the much more complicated apoptosis of human cells. Death and dying of both nematode and human cells have been well described. Surprisingly, however, we know little about death and dying or the mechanisms of death and dying of the organism as a whole. In the case of *Caenorhabditis elegans*, whose average lifespan is about 20 days, the death of the organism has already been explored and we have preliminary knowledge about it. Death of the nematode body has been described as a progressive cascade of cell death mediated from cell to cell by a blue fluorescent substance – anthranilic acid derivatives (anthranilic acid glucosyl esters). The death of the organism takes place as a blue wave spreads throughout the body and gradually initiates the death of all cells and thus the overall death of the organism. The discovery of the "blue death wave" and its key mediator provided scientists with a tool for a possible intervention to block nematode death. They found that inhibiting blue waves can prevent death caused by, for example, massive infection or poisoning, but not age-related death. These results support the idea that the aging of organisms, including human aging, is a complex process that causes death for multiple parallel reasons. Therefore, delaying death will not be easy, nor is eliminating such serious causes of death as cardiovascular disease and tumors in humans like to significantly increase life expectancy (Coburn et al., 2013).

Maximum and average human life expectancy. And what about this?

The average or mean life expectancy of a person is a statistically determined value and reflects the most probable length of life of a person. It is influenced, among other factors, by infant mortality, injury rates, ongoing wars, famine, and also by living and health care standards. In 2019, men in the Czech Republic lived to an average age of 76, and women to 82. In the most developed countries, they reach an average age of up to 90, while in some African countries it is only 40. In general, women reach a higher average age than men, and in Europe this difference is about five years of life.

The maximum human life expectancy is determined on the basis of the longest-living individuals. It is estimated to be about 120 years and does not appear to change significantly over time. In the past, people also lived to be over a hundred years old, but the number of people reaching such an age is increasing today. The potential to reach such an old age is the same for all races and cultures. The difference between a person's expected and maximum age is probably a "window" for "anti-aging medicine." Experiments show that life can indeed be extended to the limit of the maximum possible age. However, this does not mean that we will die healthy if we delay aging. Rather, "only" the phase of human life with increased occurrence of diseases, some of which lead to death, shifts to older age. The onset of age-related diseases cannot be stopped, but only delayed (Gems, 2014a, 2015).

How to determine the maximum duration of human life?

What about the maximum life expectancy of man? Is it inviolable? Are these 120 years sufficient? Is that our evolutionary upper limit?

Humans live much longer than other large primates. Orangutans live 58.7 years, gorillas 54 years, bonobo chimpanzees 50 years, and chimpanzees 53.4 years. Thus, the maximum lifespan of humans is significantly longer than other primates (Gems, 2014b). Is there an evolutionary advantage for humans to live to older ages?

In seeking an answer to this question, it is important to remember that evolutionary fitness is related to reproductive success. Thus, it does not appear that longevity *per se* is an evolutionary adaptation, but rather the question should be asked whether and how longevity is related to reproductive success. The discrepancy between reproductive age and attainable age is particularly striking in women. This is because, similar to other primate females, the reproductive capacity of women ends in the fifth decade of life. After reaching menopause, females no longer contribute directly to reproduction. So why do they – unlike other primates – live to such an old age?

The grandmother theory offers one possible explanation. According to this theory, post-menopausal women significantly increase the reproductive success of their daughters by helping to care for their grandchildren, thus enabling their daughters to have more children. Women's seemingly illogical longevity, which far exceeds their reproductive capacity, appears to be an ingenious adaptation in light of this theory (Hawkes, 2003; Gems, 2014b).

An alternative explanation is based on the assumption that human longevity is related to the gradual extension of male reproductive capacity into old age. According to the patriarch hypothesis, increasing intelligence and the ability to accumulate resources enable successful older men to attract additional, younger women and to reproduce in old age. The parallel increase in the age of women explains this hypothesis in terms of the absence of genes for longevity on the Y chromosome and therefore they are manifested in women. Women are simply "travelers" with men (Marlow, 2000; Gems, 2014b).

The question is how neoteny fits into considerations of the evolutionary context of longevity. Neoteny is a phenomenon in which an organism attains sexual maturity while retaining its juvenile characteristics. In other words, it involves retaining some typical features of a young, immature organism and extending them to the later stages of maturity. This mechanism prolongs or restores the developmental flexibility and plasticity of the young organism, thereby extending its optimal life phase for learning and assimilating information. It is generally a widespread tool of evolution and was probably of critical importance in the development of humans. Compared to other animals and also compared to his own ancestors, modern humans have qualitatively new capabilities at the cultural and symbolic levels, e.g. conceptual thinking and language, which gives him the opportunity to pass on individually acquired knowledge and create a culture that has the potential to evolve rapidly, which gives humans tremendous adaptability (Lorenz, 1974; Veverka, 2013).

On the other hand, the chronically immature human infant requires parental care over a long period of time. The long-term, exhaustive burden on parents of caring for each young infant limits the finite number of infants parents can nurture and raise. The assistance of grandparents could really be a great advantage, supporting the grandmother theory. The grandmother theory is also supported from another, unexpected side. Humans are the only primates to enter menopause, but not the only ones among all animals. A long post-reproductive life phase is extremely rare in nature, but marine biologists have already identified several cetacean species (e.g. killer whales, belugas) that also enter menopause and in which it appears to have evolved independently. Scientists have wondered what evolutionary sense this makes. They believe menopause is related to the social organization of these species. In killer whales, for example, grandmothers teach their daughters' young to seek out and remember areas with abundant food. Thus, the purpose or adaptive advantage of longevity well beyond the reproductive stage is likely to be the intergenerational transmission of knowledge and experience (Ellis et al., 2018a, 2018b).

Aging is not easy

The question arises whether our aversion to aging is related to the fact that we may expect to age and die "prematurely," i.e. not fully realize our life potential since our average life expectancy is well below the 120-year mark, or whether – for other reasons – we do not want to age at all. Just like tumor cells…

According to the common definition of biologists and physicians, aging is a series of disorders and defects that increase with age. Most of them are endogenous in origin, although they can also be influenced by environmental factors, and gradually lead to death (Gems, 2015). This definition views aging primarily on a physical, biological level. And increasing physical infirmity is also what we are most likely to associate with aging.

But aging is undoubtedly a much more complex process that affects other levels of human life as well. In his book *On Aging: Revolt and Resignation*, the Austrian philosopher and essayist Jean Améry offers a very personal and existential view of this stage of life. He looks at (his own) aging from the physical, social and cultural points of view. In addition, he looks at the association of aging with the passing of time and impending death. In the chapter "Not to Understand the World Anymore," he describes how an aging person loses the ability to recognize the latest signs of the times, to understand the pace of technological development and the culture. According to Améry, one has only two choices. Either he remains regressively anchored in his own time and cultural system, or he accepts the new, but at the risk of losing his own individual structure and authenticity (Améry, 2008). Améry's merciless, painful introspection is also confirmed by much younger people, for example, the Czech film critic Kamil Fila, only 36 years old, who admits that it is a "difficult exercise in empathy" for him (even at his age!) to understand a young film audience. He foresees that the layers of society differing in pace of life will become even more distant from each other in the future and communication between them will become basically impossible, as they will speak in different codes (Vlasák, 2016).

A society divided by generations: The sixth deadly sin of Konrad Lorenz

Konrad Lorenz expressed a similar opinion: "The rate of development forced upon today's culture by technology has the result that a considerable part of the traditions still observed by one generation is regarded, rightly, by critical youth as obsolete" (Lorenz, 1974, p. 70). Among other things, this has led him to formulate the sixth deadly sin, which he calls a break with tradition: "A critical point is reached at which the younger generation is no longer able to communicate with the older one, still less to identify with it. Therefore, the younger treats the older like an *alien ethnic group*, confronting it with the equivalent of national hatred" (Lorenz, 1974, p. 102). In the preface to the first edition of the 1973 book, he softened the formulation of this sin, expressing his belief that even "rebellious youths do not reject everything their parents value" (Lorenz, 1974, p. 6).

Nevertheless, the reality of a society deeply divided by generations is reflected in the term ageism, which emerged in the 1960s. Ageism means discrimination against a certain group of people because of their age, based on prejudice against another generation. This ideology stems from a shared belief in the qualitative inequality of the different stages of human life. The term was first used in reference to older people and then gradually expanded to include all age groups, but is still often associated with the exclusion of older people from many aspects of daily life. Sometimes it is also referred to as geriatric or senior ageism, which places a low value on older people, pointing to their redundancy, uselessness, "post-productivity" status, the economic burden they represent to society, their inability to learn and adapt to new circumstances, etc.

When sociologist Jiřina Šiklová had to make a forecast for the development of our society for the period from 2010 to 2030, she assumed a significant aging of society and expressed the conviction that "the pace of changes and discoveries will continue and the gap between generations will widen. Ageism will increase. Younger against older and vice versa" (Šiklová, 2010). Anselm Grün also notes that "today we are more tempted than in the past, to 'devour' and sacrifice our elders. A very common complaint today often has an aggressive undertone toward an ageing society. We are separating the elderly from the community of the younger" (Grün, 2009, p. 5).

And so, on the one hand, we long for longevity, we long to reach old age, but on the other hand, we do not want to be old, we do not want to grow old, and we do not concede much value to old age. We even see old people as a threat. Is there really nothing positive, nothing meaningful about old age and growing old? Or does old age offer special challenges and tasks? Anselm Grün says: "Today we need a new sense of wisdom and age in our society. Appreciating old age will enable us to perceive our own aging positively. Every human being ages. So thinking about old age is not only the task of the elderly; it is important for everyone. Aging is a basic human experience. At the same time, thinking about old age always means thinking about the mysteries of human existence. Man naturally grows old. But whether his aging succeeds depends on himself. To grow old properly is a great art. Aging is a creative process" (Grün, 2010, pp. 6–7). What are aging and old age good for? Can cells and cellular senescence help us find answers to these questions?

Premature cellular senescence: Learning to accept limits

Anselm Grün, who has written several books on aging (Grün, 2001, 2009, 2010), considers accepting our own limits as one of the tasks that old age imposes (Grün, 2009). Ignoring our limitations, going beyond them, pretending that we have more options than we actually do, characterizes our reluctance to age. Good, wise aging accepts and considers these limits in living with newly emerging – and disappearing – circumstances.

Cellular senescence can be triggered by critically short telomeres, but also by other risk factors such as DNA damage. This can cause a cell to age prematurely before its replication potential is exhausted. Similarly, not only age but other life circumstances – physical, psychological, cultural, social – can affect and limit our lives. So our reluctance to age and accept aging may be part of a much broader problem: the problem we have with accepting the limits of life in general. Get over yourself. Insignificant goals make insignificant people! Challenges that describe our times so well. What do they mean? How can we understand them? And how can we respond to them? Are boundaries and limitations a challenge for us to overcome in the sense of rejecting them, ignoring them, or in the sense of accepting them and responding to them with an appropriate change of life?

Nowadays we place great emphasis on "overcoming ourselves," we tend to admire and celebrate the rejecter who is willing to struggle and face difficult challenges to overcome limitations. Humility, accepting circumstances and adapting to them, staying within the given possibilities, are not highly valued or recognized. Accepting limits and facing the situation determined by them is today almost considered cowardice, resignation, defeat. We do not want to give up anything. This is true regardless of the nature of the limitations – physical disabilities or certain life situations.

Indeed, disabilities and loss of strength limit our lives, they hinder some of our activities, they take something away from us. But maybe they also give us something! For example, they can help us to better orient ourselves in our own lives. They can lead us to make a choice, to make a decision: what should we pay attention to? What should we focus on? What should be given up? Who or what should we leave behind? Where should our remaining and limited energies be invested? Limitations and boundaries, narrowing the spectrum of what we are capable of, focus our attention more precisely on what we want to achieve. Acknowledging one's limitations, accepting and respecting them, can mean finding a way to fulfill the purpose and meaning of one's life. And perhaps it even represents a form of Masaryk's "heroism of small work." Tomáš Garrigue Masaryk, the first president of Czechoslovakia, understood dealing with small daily tasks as the antithesis of "titanism," the dazzling overcoming of "great" challenges, revolutions (Čapek, 1990; Petrusek, 2000). Do we not need real courage, inner strength and true heroism to give up fame, admiration and even "just" once-ordinary abilities and powers?

SASP: Tasks of old age

At the cellular level, senescence and apoptosis have something in common. Both processes serve as protective mechanisms that remove unwanted and damaged cells from the multicellular organism. Apoptosis is a program that leads a cell to effective self-destruction, followed by the removal of its remains by macrophages. This process does not affect neighboring cells and does not trigger an inflammatory response. Senescence is a process that also leads to the death of the cell and involves its elimination by macrophages. However, this is preceded by the induction of a senescent state, which includes SASP – a secretory phenotype associated with senescence. A senescent cell produces and secretes substances that alter the surrounding extracellular matrix and affect neighboring cells. They stimulate the phagocytic function of macrophages, affect other cells in the vicinity and trigger an inflammatory response. Thus, the role of senescence is not only to eliminate damaged cells. Senescent cells significantly influence their microenvironment and thus actively help determine its future fate. It is as if an aging cell provides a message about its life experience before its demise and gives instructions to its relatives on how to function in the period immediately after its death. At the same time, the positive effect of senescence is due only to the short-term appearance of senescent cells in tissues and their timely removal. Their longer-term persistence, on the contrary, harms the organism, contributes to its aging and even promotes the development of tumors.

The example of senescent cells shows that old people should share their experiences with others, especially the youth. They should pass on the results of their life and their knowledge, influence their environment. Konrad Lorenz confirms that "extreme conservativism in retaining what has once already been successfully tried is a vital property." And although he does not question the importance of young people's critical examination of parental values, he believes that "the retention of previous experience is even more important than additional acquisition of new ones" (Lorenz, 1974, p. 63).

Similarly, Anselm Grün believes that the task of old age is to "become wise, to see life in context" (2009, p. 13), and he affirms, "We never learn the art of aging just for ourselves, but always for others" (2010, p. 6). Old people should not only share their experiences, but also sense the time when they have to leave. In humans, this "timely passing" is not perfectly analogous to the elimination of senescent cells by phagocytosis, i.e. physical destruction, but primarily with the willingness to give up one's power. "Many people have narcissistic tendencies in old age. There is the experience that people who were very successful in the past tend to overestimate themselves in old age. Some politicians, for example, have destroyed their work records results because they could not get off the scene in time," writes Anselm Grün (2009, p. 3). And he adds that "cooperation without the right to make decisions always meant an intellectual challenge for the older generation" – in addition to facilitating coexistence with the younger generation, which today is generally – and symptomatically – regarded as difficult (Grün, 2009, p. 73).

Aging and death

In Čapek's play, Elina Makropulos lives for 337 years. Although she is tired of life, she seeks a recipe to prepare the elixir of youth to prolong her life at any cost. This is not out of a desire to live forever. The reason is fear of death! Only later she realizes that an unnaturally long life is worse than death, and agrees to destroy the recipe. In his preface to *The Makropulos Secret*, Karel Čapek expresses his expectation that he will be considered a hopeless pessimist because of his "undesirable" attitude toward longevity. But he considers his conviction that an average long life (about 60 years at that time) is good and reasonable to be a sign of optimism (Čapek, 1956, pp. 185–186).

So one of the reasons we dislike aging is the fear of death. As we grow older, we actually come closer to death, and reluctance to age means reluctance to realize that death is near. Aging inevitably brings us closer to death, and everything that comes with aging – every wrinkle, every gray hair, loss of strength – reminds us of approaching death. So it's "logical" to not want to age and to reject everything that warns us about aging.

But aging seems to be the way to get rid of the fear of death step by step. Aging is a process in which we learn, develop, change and thus prepare ourselves for the final phase of life: dying and death. By gradually accepting the changes and associated limitations of old age, we can cultivate and strengthen the very qualities that help us to accept death – which is inevitable anyway. Anselm Grün summarizes this process succinctly: "First we have to give up our own will, then activity, then our own ego, and finally life" (2009, p. 141).

Aging tests our life values

One of the conditions for a happy old age is a life we have lived well and intensely. In old age, people usually regret what they missed and did not fulfill in their lives. "Those who did not live a fruitful life in their youth cannot do so in old age. The deprivation of some experiences cannot be recovered," writes Anselm Grün (2009, pp. 11–12, 44). Elisabeth Kübler-Ross holds a similar view: "For many people, the dreams of youth become what they then regret in old age, and the reason for their regret is not that life is behind them, but that they did not really live through it" (Kübler-Rossová, Kessler, 2013, p. 138). So the anticipation of aging should encourage us to leave procrastination, fears, and excuses behind and enjoy life to the fullest. Here and now! Immediately!

Part of the aging process is increasing limitations: growing powerlessness, lack of independence, helplessness, dependence on the help of others. Aging raises questions. What age have we prepared ourselves for? Where do we stand now? In the midst of a solid, reliable community of kind and loving people? Or are we balancing on a victorious but lonely and uncertain summit, surrounded by defeated people and enemies? Old age tests the values we profess in life. The inevitability of aging and old age can be a constant reminder and encouragement to live wisely and set values right.

Age is the measure and mirror of our life. Aging can positively influence and shape our lives only if we are aware at a young age of its impending arrival, when we are still at the peak of our powers. And therein exactly lies the hidden value and meaning of all the signs of advancing age, all the wrinkles and gray hairs. A realistic self-image plays an irreplaceable role in life, is a compass. It helps to orient oneself in life to what is meaningful and lasting. And it puts into perspective, or at least questions, the sense of the many efforts of the cosmetic, fashion, pharmaceutical and medical "anti-age" industries. Or, it exposes their true purpose, which – besides the clear profit motive – is to serve immaturity and fear.

"Just as there are many young people who are basically panic-stricken about life (but at the same time very much long for it), there are probably even more aging people who are afraid of death. It is the young people who are afraid of life who later suffer just as much from the fear of death. We all feel sorry for the young man who remains in the childhood stage in his thirties – and yet our society admires people who look and act so youthful" (Grün, 2001, pp. 72–73).

Aging and society

The relationship between aging and death is strangely intertwined and ambiguous. On the one hand, aging is a process that can help us deal with and prepare for death. On the other hand, we do not want to grow old precisely because aging reminds us of death and even suggests to us that it is approaching. So we tend to avoid it. We do everything we can to avoid growing old (which we cannot succeed at anyway), and to avoid any expressions of aging (which we succeed at – unfortunately – quite well). Growing old is really difficult in our society.

At the same time, our attitudes toward aging seem to influence the way we actually age. Negative attitudes about aging seem to deepen and worsen the signs of aging. A number of preconceptions about aging, old age, and the elderly are based in part on the negative cultural stereotypes created and perpetuated in society. "The possibility of successful aging depends not only on individual activity, but also on society's attitude toward old age. It must create spaces in which it is possible to grow old well" (Grün, 2009, pp. 11–14). The price for rejecting aging is high: we are not prepared for death and thus not prepared for life: "Only those who are ready to die really live through the second half of life," says Anselm Grün (2009, p. 44).

It seems, then, that unless we accept aging as a natural and even necessary part of life – which would manifest itself, among other things, in ceasing to be prejudiced against older people and beginning to accept and appreciate old age – we will not beat our fear of death and dying. And we will continue to live in the fourth non-deadly sin.

6A. Increased genetic instability

Carcinogenesis is a multistep process

The idea that the transformation of a normal healthy cell into a tumor cell does not occur in a single step was first described in 1953. Architect Carl O. Nordling studied the incidence of death from various cancers in people aged 25 to 74 and found that the probability of dying from cancer increases with the sixth power of age. From this observation, he deduced that at least six genetic changes are necessary for tumor cells to develop (Nordling, 1953). Further studies deepened and refined this original idea, but the model of carcinogenesis as a gradual, multistep process remains valid today and continues to be confirmed. How many mutations are actually required for tumorigenesis remains an open question. It seems that we will never get an exact, definitive answer. It depends on many circumstances, such as the type of tumor, the site of its primary occurrence, and also the type of gene involved. To activate an oncogene, one mutation is sufficient; to inactivate the tumor suppressor, interventions in both alleles are required. In addition, mutations occur randomly in DNA, so tumor cells carry many more mutations than just those that directly contribute to tumor transformation and drive the process of carcinogenesis. We refer to such mutations as driver mutations, as opposed to passenger mutations, which are not associated with carcinogenesis (Beckman, Loeb, 2006; Sjoeblom et al., 2006; Wood et al., 2007).

Mutations accumulate in cellular DNA gradually, usually over many years, which is why tumors are typically a disease of older age (Fig. 20). A relatively long time is required for the accumulation of a sufficiently large number of mutations (de Magalhães, 2013). This is because under normal physiological conditions, the probability that new mutations will arise is relatively low. Based on mathematical models, at one time it seemed that if mutations occur at a rate that is common in healthy cells, the average human lifespan might not be long enough to accumulate the mutations required to develop, for example, a colorectal tumor. Therefore, the so-called mutator model was proposed, which explained the development of tumors by an increased mutation rate

(Loeb, 1991). Later, another model was proposed based on the assumption that some driving mutations, in addition to their contribution to cell transformation, confer such growth advantages to cells that they induce their clonal expansion (see chapter 2A, Fig. 7). This ensures that a sufficiently large cell clone is generated for another necessary mutation, even if the probability of its occurrence is physiologically low (Tomlinson et al., 1996).

Genetic information is constantly damaged

The fact that the mutation rate in healthy cells is low may be surprising, considering that DNA in human cells is constantly damaged by various factors. Either the bases in the DNA (adenine, guanine, cytidine, thymine) are damaged or the DNA strands are interrupted or broken. These DNA breaks can occur in either strand. DNA double-strand breaks are particularly dangerous because they cause extensive rearrangements of the genome. Sources of DNA damage (physical, chemical, biological) include external sources such as various types of radiation (ionizing radiation, UV light, X-rays), chemicals (dioxins, tar, heavy metals), or certain viruses, but also endogenous ones resulting from normal cellular metabolism, such as free oxygen radicals that are inevitably generated in mitochondria as a byproduct of respiration. Exposure to external factors can be controlled, i.e. eliminated or reduced to some extent, while endogenous factors cannot be completely avoided because they are related to the normal physiological function of cells.

Genetic information of multicellular organisms must be sufficiently stable

The genome (the sum of genetic information) of multicellular organisms is something like a blueprint for their structure and functioning. All 3.72×10^{13} cells of the human body are formed by the repeated division of an initial cell and carry the same genetic information, the same genome – a common, uniform, shared blueprint. If too many errors – mutations – were to occur between or during each cell division, the genome would change too much, jeopardizing development and the functioning and survival of the organism. Random mutations can occur anywhere in the genome. In general, they can have both positive and negative effects on the functioning of the cell, although they are much more likely to be harmful. It is easier to damage the harmonious functioning of the cell than to improve it (Chandhok, Pellman, 2009; Breivik, 2005). Therefore, the genetic information of multicellular organisms must be sufficiently stable and resistant to mutations. But even in healthy, so-called genetically stable cells, the mutation rate is not zero. Complete, absolute genetic stability would make adaptation and evolution impossible, or unduly restrict them; moreover, maintaining absolute accuracy and stability of DNA would probably be too great a burden on the cells. The physiological degree of genetic stability therefore has the character of a carefully balanced equilibrium (Fig. 21) (Friedberg, 2001).

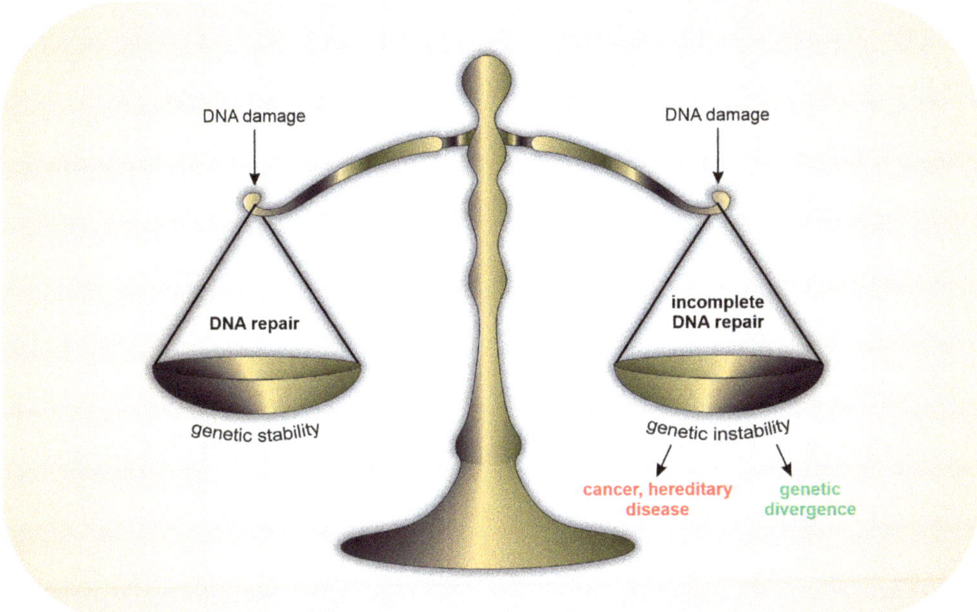

Fig. 21 Balance of genetic instability
Life requires fine-tuning the level of genetic instability. A certain level of genetic instability is the source of the genetic diversity that underlies adaptation and evolution. The rate of genetic instability, that is, the rate at which mutations occur in the genome, must not be too high. Germline mutations are associated with the risk of congenital diseases. Mutations in somatic cells can manifest in many ways, the most serious being the development of tumors (adapted from Friedberg, 2001, modified).

DNA is constantly being repaired

The mutation rate per nucleotide per cell division has been determined to be 10^{-9} in healthy human cells (Albertini et al., 1990). The rate at which mutations occur in a gene during cell division is roughly estimated to be 10^{-6}. That is, a new mutation in a given human gene is found in about one cell in a million (Fig. 7).

How do cells manage to maintain such a low level of genetic instability? They are very well adapted to cope with constant damage to their DNA. They are equipped with a number of effective detection and repair mechanisms. The detection mechanisms constantly monitor DNA integrity, and when deviations from the standard structure are detected, they send out a signal that forces the cell to respond accordingly. A cellular "DNA damage repair program" is triggered. This program involves a complex and coordinated network of cellular signaling pathways and mechanisms. It includes cell cycle checkpoints at which the cycle can be halted to allow the damaged DNA to be repaired. Once activated, the repair mechanisms are directed to the site of the damage, where they provide correction. The optimal solution, of course, is a complete return to the original, standard state. When this is not possible, the cell strives for maximum approximation. None of the repair mechanisms is completely error-free, and even after their intervention, a small amount of DNA damage may remain. The physiological rate of genetic instability then depends on how much DNA damage remains undetected, unrepaired, or imperfectly repaired. If the damage to the genome is too extensive and the cell is unable to repair it efficiently enough, it may respond by inducing apoptosis or cellular senescence, as described in chapters 4A and 5A.

An example of a protein involved in the immediate recognition of DNA damage is ATM. Among the proteins that subsequently mediate an appropriate cellular response to DNA damage, the tumor suppressor p53 occupies a prominent position (Kastan, Bartek, 2004). In the previous chapters, we have seen that p53 arrests the cell cycle at the restriction point and can also effectively induce apoptosis and senescence. In addition, it can activate a number of repair mechanisms and help arrest the cell cycle at other checkpoints as well. Because of its complex role in the DNA damage repair program and maintaining genetic stability, it is sometimes referred to as the "guardian of the genome" (Lane, 1992).

Genetic instability and its types

Genetic instability is manifested by a high frequency of mutations in the genome of cells. The occurrence of a mutation in the genome in itself does not indicate genetic instability. Increased genetic instability is indicated by the increased rate at which mutations occur in the genome. Mutations cause changes in nucleotide sequence, by the exchange of one nucleotide for another, a short deletion (loss of one or more nucleotides) or insertion (of one or more additional nucleotides), but also chromosome rearrangements (e.g. chromosome translocations, in which sections are exchanged between two chromosomes), aneuploidy, i.e. changes in the number of whole chromosomes, and gene amplification (multiplication of genes). The mere occurrence of any of the above types of mutations does not indicate the degree of genetic instability. All of these types of errors occur randomly in the genome. Some of them may be beneficial to the affected cell and are therefore selected and retained. Genetic instability is simply indicated by an increased probability, i.e. an increased frequency, an increased rate at which these mutations occur in the genome. In the case of genetic instability, mutations are not a completely random phenomenon, but the result of a systemic error that causes mutations to occur much more frequently compared to the physiological state (Lengauer et al., 1998).

Repair mechanisms

Because there are different types of DNA damage, cells are equipped with several different repair mechanisms (Lengauer et al., 1998; Beckman, Loeb, 2006). Some errors in DNA occur during its replication in each cell cycle. The enzyme complexes that perform replication, called DNA polymerases, are not perfectly accurate and occasionally insert the wrong nucleotide or the wrong number of nucleotides into the daughter strand when copying DNA. DNA polymerases do have proofreading mechanisms to check the accuracy of nucleotide insertion, but even when they are used, some errors remain in the newly synthesized DNA strand. After replication is complete and inspection of the daughter strand by comparison with the parent strand reveals the error, it is corrected by a mechanism known as mismatch repair. During DNA replication, errors occur most frequently in those regions of the genome (coding and noncoding) where multiple identical nucleotides follow each other. This can result in the insertion of an incorrect stretch of nucleotides of different lengths. These errors are typical of microsatellites, noncoding regions of the genome in which short sequences of two to five nucleotides are repeated. When mismatch repair fails, most mutations are found in microsatellites. The increased genetic instability caused by the lack of proper mismatch repair function is therefore referred to as microsatellite instability (Lengauer et al., 1998).

Nucleotide excision repair removes and replaces nucleotides damaged by exposure to external mutagens such as UV radiation or chemical agents. In nucleotide excision repair, a portion of the DNA strand containing a damaged nucleotide is excised and removed. Then, the missing section is synthesized according to the undamaged sequence of the complementary DNA strand and joined to the original strand to restore the correct nucleotide sequence (Friedberg, 2001). Imperfect nucleotide excision repair leads to genetic instability (Lengauer et al., 1998).

Double-strand breaks are a particularly serious kind of damage to cellular DNA. To repair them, the cell can use several types of repair mechanisms. The most accurate of these is homologous recombination. In this process, the part of the damaged strand containing the break, including the adjacent regions, is removed and the DNA is repaired according to the complementary, undamaged double-stranded sister DNA. Homologous recombination results in perfect repair of the damage.

However, the presence and availability of an undamaged template is a prerequisite for the sister chromatid repair. Therefore, this correction can only be made in the late S phase and G2 phase of the cell cycle. In addition, cells are equipped with another mechanism called non-homologous end joining. This mechanism simply joins and heals breaks in DNA, but does not ensure the correct restoration of the original nucleotide sequence. Compared to homologous recombination, this repair is faster, it repairs DNA breaks, but it can leave errors. However, unlike homologous recombination, the mechanism of non-homologous end joining works at any phase of the cell cycle, i.e. even when cells are in a long quiescent state (van Gent et al., 2001; Jackson, Bartek, 2009; Bunting, Nussenzweig, 2013).

Mitotic spindle checkpoint

Cell cycle checkpoints are the mechanisms that the cell uses to actively suspend entry into the next phase of the cell cycle until it is certain that all processes associated with the previous phase have been completed and other requirements have been met (see chapter 3A). The same mechanisms can be used by the cell when confronted with DNA damage. For example, a cell with damaged DNA must interrupt its replication to avoid copying DNA with errors and passing it on to daughter cells (Bartek, Lukas, 2007; Shaltiel et al., 2015). A prominent role in maintaining genetic stability is played by the mitotic cell cycle checkpoint, which acts before the onset of anaphase. It is a mitotic spindle checkpoint. Failure of this checkpoint is associated with chromosomal instability. Mitosis is the most dramatic and a potentially very risky phase of the cell cycle. During mitosis, sister chromatids are irreversibly separated and pulled apart into daughter cells. Even a small mistake during this division can have far-reaching consequences: the daughter cells may gain an extra chromosome or, on the contrary, lose a chromosome. Preventing such a serious error is the task of the mitotic spindle checkpoint.

During metaphase, all chromosomes consisting of two sister chromatids arrange themselves in the equatorial plane of the nucleus and attach to the microtubules of the mitotic spindle with their kinetochores, i.e. multiprotein complexes bound to the centromeres of the chromosomes. When all sister chromatids are properly attached to the mitotic spindle, anaphase follows, which begins with the separation of the sister chromatids of the chromosomes and continues with their equal segregation to the poles of the cell. The even distribution of chromatids to the opposite poles of the cell is ensured by the mitotic spindle checkpoint before the onset of anaphase. This fascinating, complex mechanism controls the proper attachment of all chromatids to the mitotic spindle. Unattached chromatids trigger a signal that delays the onset of anaphase, so that attachment can be completed. When a component of the mitotic spindle checkpoint is damaged and fully or partially nonfunctional, the accuracy of chromatid distribution to daughter cells is compromised, significantly increasing the risk of aneuploidy formation and chromosome instability (Cortez, Elledge, 2000; Jallepalli, Lengauer, 2001; Fukasawa, 2002; Chandhok, Pellman, 2009).

Tumor cells exhibit increased genetic instability

It takes a relatively long time to accumulate a sufficient number of mutations for the development of a fully malignant tumor because the probability of new mutations in genetically stable cells is low. The occurrence of random mutations poses some risk, but it is not high and is not threatening to multicellular organisms. Mistakes happen, and even the formation of a cell with some potential for tumor transformation does not mean that it will actually undergo and complete such transformation. On the way to complete transformation into a tumor, the cell encounters a whole series of obstacles

– protective barriers that the multicellular organism uses to prevent tumor formation. However, a completely different situation arises when we are dealing not with random and unlikely errors, but with systematic errors, which are random in the sense of their specific position in the genome, but not random in the sense of inevitability, or the probability with which they occur. Thus, mutations that occur because of increased genetic instability are so common that the probability that the tumor transformation process will be completed is dramatically increased. The induction of genetic instability fundamentally alters the dynamics of the carcinogenesis process – it greatly accelerates it (Lengauer et al., 1998; Beckman, Loeb, 2006; Negrini et al., 2010).

Sources of genetic instability in tumors: Tumor syndromes and mutator mutations

One of the strongest evidences that increased genetic instability extremely increases the probability of tumor development is the existence of a whole series of congenital tumor syndromes related to congenital disorders of repair mechanisms.

Historically, the first recognized example was a rare disease called xeroderma pigmentosum. It is characterized by extreme sensitivity to UV radiation and a high probability of developing skin tumors at a young age on exposed areas of the body. This disease is caused by congenital mutations in genes encoding proteins involved in nucleotide excision repair (Cleaver, 1968).

Lynch syndrome is associated with a high probability of developing colorectal cancer and a lower probability of developing other tumor types. It is caused by mutations in genes whose products are involved in mismatch repair (Karran, Bignami, 1994). Indeed, mutations in microsatellites are frequently found in these tumors.

Congenital tumor syndromes caused by mutations in genes encoding proteins involved in double-strand break repair include some hereditary forms of breast and ovarian cancer, Fanconi anemia, and others (Wang, 2000; Narod, Foulkes, 2004; Curtin, 2012). Other tumor syndromes are associated with congenital disorders caused by failure to recognize DNA damage, e.g. ataxia telangiectasia caused by mutations in the ATM gene or other components of the DNA damage response (Bloom syndrome, Werner syndrome, and many others). The cause of the rare Li-Fraumeni syndrome, in which there is an increased risk of a wide range of tumors at a young age, is congenital mutations in the tumor suppressor gene p53 (Wang, 2000; Cobb et al., 2002; Hickson, 2003; Shiloh, Ziv, 2013).

The most common form of genetic instability in tumors is chromosomal instability. Approximately 85% of all solid tumors have a high degree of aneuploidy, meaning that the cells of these tumors contain an abnormal number of chromosomes, which may also have major structural rearrangements. This is due to mutations in genes encoding some components of the mitotic spindle checkpoint. Although chromosomal instability is very common in tumor cells, congenital predispositions to its development are rare. One of these is a very rare but very serious mosaic variegated aneuploidy syndrome, which is associated with a high risk of developing some type of tumor and death in early childhood (Matsuura et al., 2000).

DNA replication stress

Congenital mutator mutations explain the cause of increased genetic instability in hereditary forms of tumors and contribute to the development of a general concept of the role of genetic instability in the process of carcinogenesis. The mutator hypothesis explains the mechanism by which the accumulation of mutations necessary for tumor development, and thus the process of carcinogenesis, is accelerated by mutations occurring in genes encoding participants in DNA damage repair. These mutations lead to genetic instability by impairing the ability of cells to repair DNA damage. At the same time, this hypothesis assumes that mutator mutations are more likely to occur in the early stages of tumor development (Beckman, Loeb, 2006).

Increased genetic instability is not only seen in hereditary tumor types, but does occur in most sporadic tumors that develop independently of a congenital predisposition. Genetic instability in sporadic tumors was thought to be caused by mutations similar to those seen in hereditary tumors, except that they occur randomly in somatic cells. Surprisingly, however, it turned out that the frequency of mutator mutations in sporadic tumors is much lower than expected. Moreover, in sporadic tumors, these mutations occur only at advanced stages of carcinogenesis. At the same time, double-strand breaks in DNA are often detectable at the early stages of tumor development. Thus, it seemed that there must be other sources of genetic instability in sporadic tumors. Gradually, it was hypothesized that a common cause of increased genetic instability in sporadic tumors is DNA replication stress triggered by oncogenes (Halazonetis et al., 2008). According to this hypothesis, deregulated rapid cell division caused by unphysiologically high activation of proto-oncogenes or inactivation of some tumor suppressors leads to increased genetic instability (mainly of chromosomes). Rapid DNA replication induced by deregulated stimulation of cell division leads to frequent occurrence of DNA double-strand breaks, resulting in increased chromosome instability (Beckman, Loeb, 2005; Halazonetis et al., 2008; Negrini et al., 2010; Orr and Compton, 2013). According to some authors, in hereditary tumors (and generally in cases where the mutator pathway is used to induce genetic instability), a high degree of genetic instability precedes the activation of proto-oncogenes and silencing of tumor suppressors and the resulting acceleration of cell division, whereas in sporadic tumors the induction of genetic instability is the consequence of these events (Halazonetis et al., 2008; Negrini et al., 2010).

Critically short telomeres: A source of genetic instability in tumors

Another "non-mutator" source of increased genetic instability, which, like DNA replication stress, is directly related to the process of carcinogenesis and in a sense is its by-product, could be critically short telomeres. Telomeres and telosomes function to protect the ends of linear chromosomes from being perceived by cellular mechanisms as DNA double-strand breaks. Limiting replication potential by progressively shortening telomeres with each cell division acts as a protective barrier against tumor formation by limiting the lifespan of cell lines and thus their potential for success in completing the overall process of tumor transformation (see chapter 5A). In a healthy cell, a critically shortened telomere triggers signals that lead to cell cycle arrest, initiation of senescence, or induction of apoptosis. However, if some damage interferes with the cell signaling pathways (particularly those associated with the tumor suppressors p53 and p16/Rb) that mediate the appropriate cellular response to a critical telomere shortening described above, the cell can divide again. This also perpetuates telomere shortening. Chromosomes that lack protection of their end structures activate the DNA repair program for double-strand breaks and, as a result, may fuse to form dicentric chromosomes and other irregular structures. This increases the likelihood of translocations, recombinations, and the risk of loss of chromosome parts (Fig. 18). Thus, these processes contribute significantly to the increase in genetic, especially chromosomal, instability. This phase is, of course, temporary for the cells and ends either with their complete collapse and death or with the activation of one of the mechanisms for telomere maintenance or extension. In this way, a cell with a severely damaged genome acquires immortality (see chapter 5A). Thus, while relatively undamaged cells with inactive telomerase and short telomeres provide an effective protective barrier against tumorigenesis, elimination of the relevant signaling pathways and critically short telomeres contribute, at least transiently, to tumorigenesis by inducing chromosomal instability (Fig. 22; Hackett, Greider, 2002; Deng, Chang, 2007).

Heterogeneity of cell clones and tumor evolution

An increase in genetic instability alters the dynamics of tumor development, significantly accelerating it, and supporting its further evolution and adaptation. Indeed, with increasing genetic instability, heterogeneity and diversity within the tumor increases, as more and more mutations may occur randomly in some (but not all) cells of a growing tumor. Thus, a tumor is composed of several subpopulations of tumor cells that, in addition to the basic genetic alterations common to all tumor cells (because they arise early in tumor development), carry their own mutations, i.e. mutations that are restricted to subclones or even single cells. These subpopulations differ not only in their genotype, i.e. the set of mutations they carry, but also in their phenotype, i.e. their properties (Fig. 23A). Cell clones differentiated in this way can then be subjected to selection. During tumor development, the subclones that are best equipped and adapted for the given circumstances can further expand.

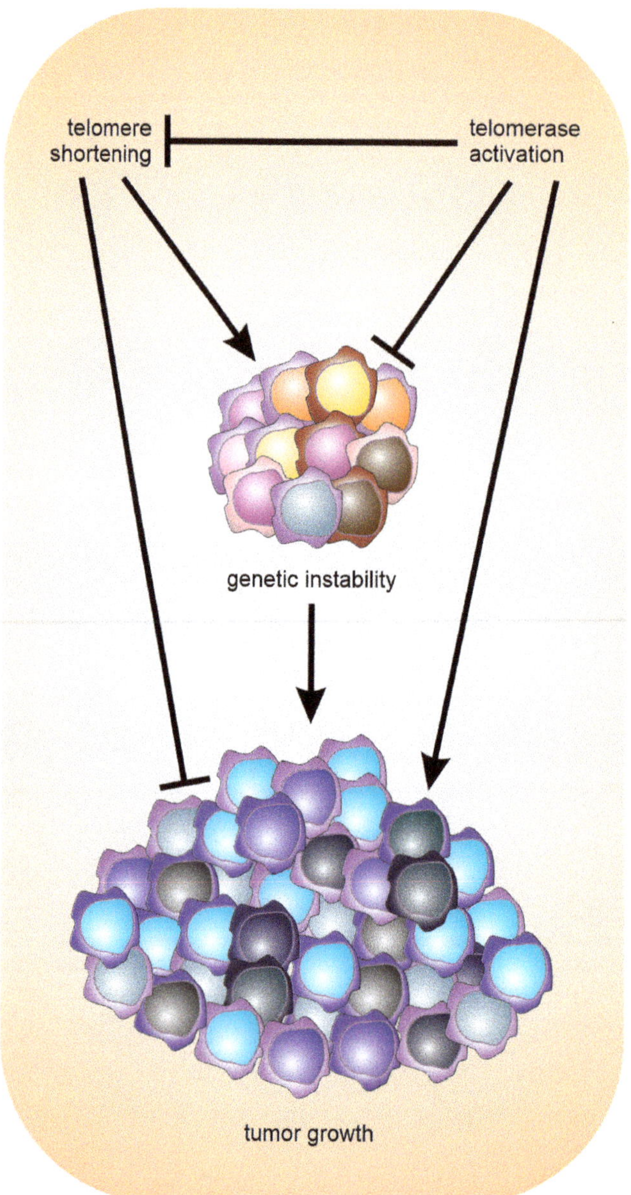

Fig. 22 Dual role of telomerase and dysfunctional telomeres in increased genetic instability and tumor growth
Shortening of telomeres in the early stages of tumor development may contribute to genetic instability that drives carcinogenesis, but may also limit tumor growth by activating checkpoints and initiating cellular senescence. Activation of telomerase at later stages of tumor development enables tumor growth while limiting genetic instability (adapted from Hackett, Greider, 2002, modified).

One of the serious consequences of tumor heterogeneity is complications in tumor treatment. A heterogeneous tumor cell population may not respond uniformly to therapy. Therapy may be effective on some or even most cell types in the tumor, but if it leaves even a small clone of tumor cells unaffected because of a particular mutation that makes them resistant, the therapy will select those cells. They may begin to form a new tumor mass that is more resistant and aggressive (Fig. 23B) (Gerlinger, Swanton, 2010; Turner, Reis-Filho, 2012; Marusyk et al., 2012).

Similarly, tumor heterogeneity is also important in metastatic processes. In this case, selection pressure comes not from a therapeutic agent but from a changing environment through which metastatic tumor cells move and in which they form secondary tumors (see chapter 9A). Again, selection comes into play, "choosing" from a heterogeneous cell population the one that best adapts to and is best suited for the new environment. This, in turn, can lead to complications in choosing an effective therapy, as secondary tumors (metastases) can differ significantly from primary tumors. Furthermore, tumors with high genetic instability and heterogeneity continue to develop, and even with massive therapy, there is always a significant chance that they will "succeed" and escape the effects of therapy (Marusyk, Polyak, 2010; Marusyk et al., 2012).

The "just right" principle

Increasing tumor heterogeneity is associated with higher tumor aggressiveness and thus greater "success." One might think that high genetic instability (the rapid accumulation of more and more mutations) is favorable for tumors. However, it turns out that successful tumors probably balance on an optimal level of genetic instability, an instability that is "just right." Too low a level of genetic instability would result in slow tumor development, inadequate ability to overcome the protective barriers of the organism, and also in the inability of the tumor to adapt to certain selection constraints. However, too high a level of instability can severely impair cell viability. It can increase cell death, either through a form of programmed cell death or through a form of mitotic catastrophe caused by loss of cell integrity and an excessively chaotic genome (Fodde et al., 2001; Komarova, Wodarz, 2004; Beckman, Loeb, 2006; Breivik, 2005; Weaver et al., 2007; Gaspar et al., 2009; Albuquerque et al., 2010).

The "just right" principle applies not only to the degree of genetic instability but also to other factors, such as the activity of certain oncogenes. Even in the case of oncogenes, one might think that the higher the activity of their products, the more pronounced the stimulation of cell division and thus a greater contribution to tumor development. However, excessive activity of oncogenes can also be counterproductive. It can directly induce apoptosis, as mentioned in chapter 4, or trigger a high rate of cell proliferation that can lead to severe DNA replication stress. As a result, an excessive level of genetic instability occurs, which in turn also triggers some form of cell death.

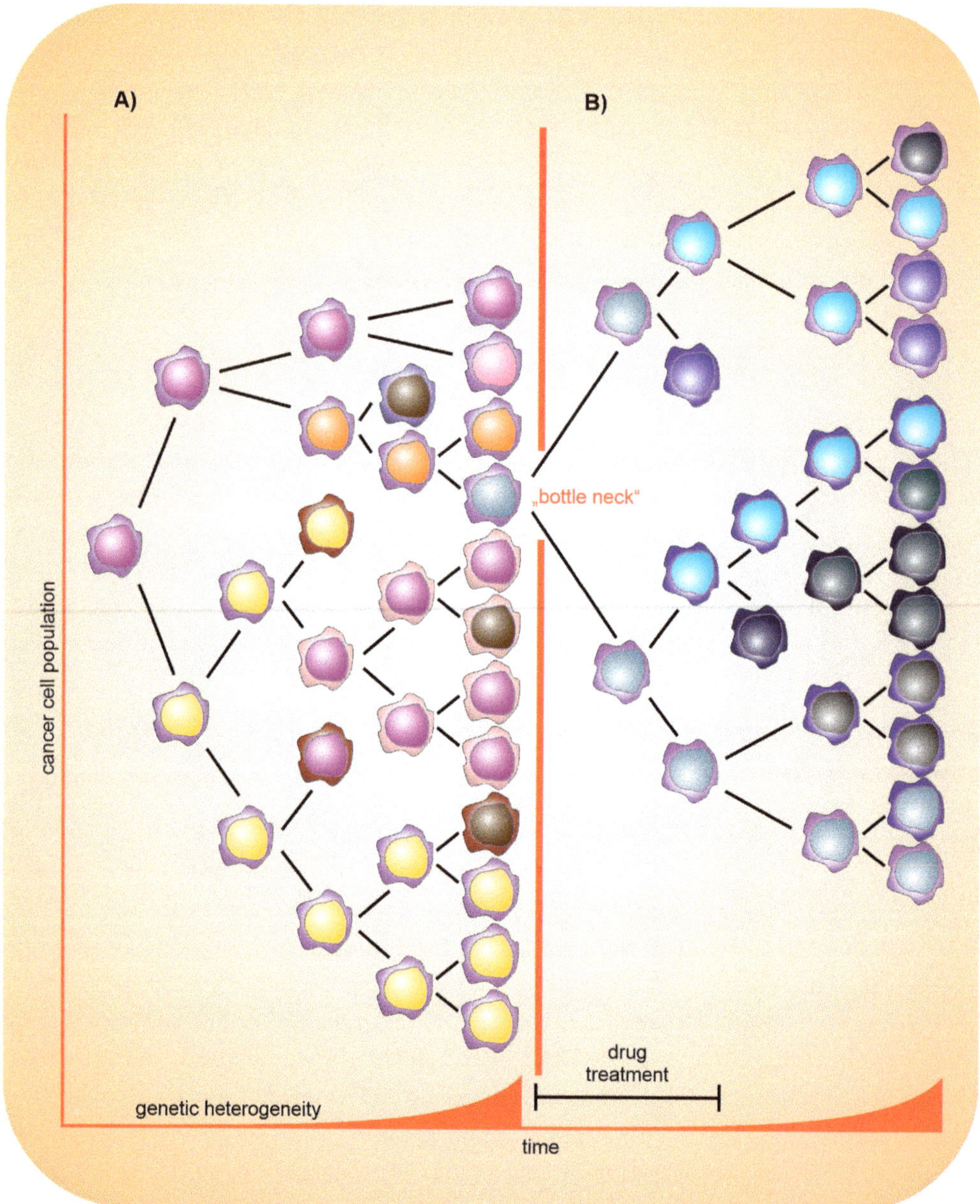

Fig. 23 Tumor heterogeneity during tumor progression and its treatment
(A) A tumor develops clonally from a single renegade cell, but genetically distinct subclones arise during its development. They coexist in the tumor, some expanding and transforming rapidly, while others change and expand slowly or disappear. (B) Selection imposed by a therapeutic agent allows only the resistant clone to survive. Thus, the therapeutic is an evolutionary bottleneck that causes a temporary reduction in genetic heterogeneity. In the course of further tumor development, which is accompanied by the acquisition of additional mutations, the clone rapidly evolves (adapted from Gerlinger, Swanton, 2010, modified).

Clearly, the "just right" principle has broader application. And apparently another circumstance involving this principle has a broader validity. How much is "just right"? It turns out that there is no simple answer to this question. The "just right" value is dynamic. Over time, as the tumor develops, it changes. As the tumor progresses and the tumor cells transform, their tolerance to the activity of oncogenes rises and genetic instability increases. The tumor cells become more resistant and can withstand more and more damage (Fig. 24) (Fodde et al., 2001).

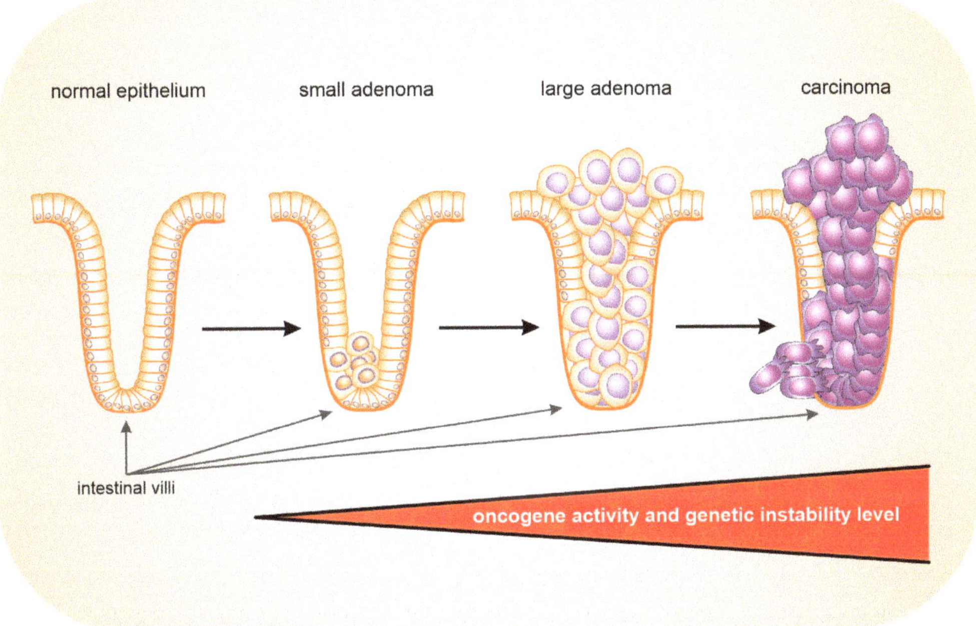

Fig. 24 Increase of tolerance to damage
As tumor development progresses and oncogenic tumor transformation deepens, the tolerance of tumor cells to the degree of oncogenic activity and the degree of genetic instability increases. The tumor cells become more and more resistant, they can withstand more and more damage. The development of colorectal carcinoma is given as an example: the intestinal mucosa consists of a single layer of epithelial cells that form folds called villi. The initial stages of colorectal carcinoma development represent small adenomas that gradually develop into carcinomas via medium and large adenomas (adapted from Fodde et al., 2001, modified).

Complete repair versus damage adaptation

There are two opposing forces at work in cells. One constantly drives them to proliferate, the other is strictly concerned with maintaining the integrity of the genome, i.e. genetic stability. The moment these forces collide, i.e. the moment DNA damage occurs in a cell, repair takes priority and the cell cycle is halted at one of the checkpoints. Only when the corrections are complete is the feedback mechanism, checkpoint signaling, turned off, and the cell cycle is reactivated and can continue. The process of cell cycle reactivation is referred to as renewal after DNA damage.

However, under certain circumstances, despite DNA damage, proliferation may be favored and the cell cycle may resume at the checkpoint instead of activating DNA repair. This phenomenon is referred to as adaptation to DNA damage. The adaptation program appears to be activated in response to different types of DNA damage, but always depends on the extent of that damage. It occurs only in the presence of mild or moderate DNA damage (Syljuåsen, 2007; Bartek, Lukas, 2007; Serrano, D'Amours, 2014; Shaltiel et al., 2015).

There is no doubt that the DNA damage adaptation program, which tolerates the proliferation of cells with damaged DNA, may increase genetic instability and thus contribute to the development of tumors (Syljuåsen, 2007). Therefore, the question arises as to the purpose of this mechanism and why it evolved in cells. Several possible explanations have been proposed. Reactivation of the cell cycle could allow damaged cells to reach a stage from which they can enter the cell death program (Syljuåsen, 2007) or die by mitotic catastrophe (Bartek, Lukas, 2007). Progression to further phases of the cell cycle may also allow cells to repair DNA, as some repair mechanisms are tied only to specific phases of the cell cycle. Another explanation for the importance of adaptation is related to the observed ability of cells to segregate the damaged DNA into one of the daughter cells, possibly allowing the survival of a second, relatively healthy daughter cell. In addition, it is possible that adaptation creates cells that are damaged enough to trigger an immune response – a reaction that eliminates them. Finally, the possibility that the adaptation process reduces the inflammatory response, which also contributes to tumor development, is also being considered (Serrano, D'Amours, 2014). In any case, it appears that under certain circumstances, the survival benefit of mildly damaged cells may outweigh the risk associated with their damage.

Genetic instability and the environment: Don't stop for repairs in a war zone!

A different perspective on the possible meaning or purpose of survival of damaged cells is offered by a model presented by an author under the metaphorical title "Don't stop for repairs in a war zone" (Breivik, 2001, 2005). This model is based on the idea that tumors are considered genetic diseases because they are based on the accumulation of mutations, but they are also related to the environment because some environmental factors (smoking, diet, radiation) clearly trigger or stimulate the development of tumors. Environmental factors, on the one hand, increase the frequency of mutations, but also subsequently influence their fate and help determine their significance and impact through the mechanism of natural selection. One might think that the more aggressive and mutagenic the environment, the more important the role of DNA repair and the more stringent the requirement for its full functionality and high efficiency.

However, neither observation nor experimental data confirm this assumption. Just the opposite seems to be the case. An environment containing multiple mutagens favors cells lacking DNA repair mechanisms. The explanation is provided by the "Don't stop for repairs in a war zone" model, which describes cells and their viability using the metaphor of two racing car teams (Fig. 25).

Whenever the car is damaged, the blue team stops and repairs the damage. The red team ignores the flashing red lights on the dashboard and continues driving. Under normal circumstances, the cautious blue team would win because it always has a flawless, optimally running car, while the damage to the red car adds up. But under the harsh, mutagenic "war" conditions, the situation reverses. The defects in the blue car accumulate faster than they can be fixed, and the blue team is stuck in the "depot" of the checkpoint. It is caught in the pitfall of its diligence. The red car squeals and bounces around the track as it gradually becomes more and more damaged, but still has a reasonable, non-zero chance of crossing the finish line (Fig. 25).

This model offers an explanation for why a mutagenic environment favors genetically unstable cells and suggests that in an environment where cells are exposed to high levels of damage, cell survival may come at a higher price than perfect repair of their DNA (Breivik, 2001, 2005).

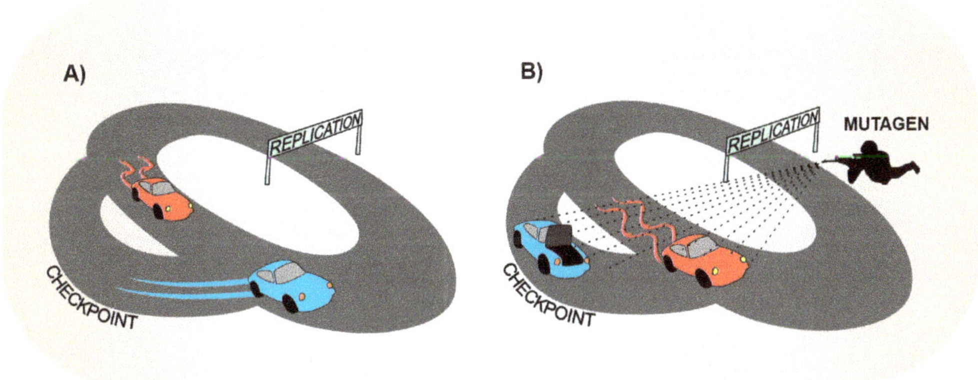

Fig. 25 Don't stop for repairs in a war zone! Pros and cons of DNA repair
The problem of maintaining an optimal level of genetic instability/stability can be seen as analogous to auto racing strategies. The blue car stops and repairs each time it is damaged, while the red car ignores the warning lights and continues. (A) Consistent repair is a successful tactic under calm conditions. The careful blue driver wins because he always has a flawless, optimally running car, while the red car accumulates more and more damage. (B) Under "wartime" conditions, however, the situation is reversed. The blue car's damage accumulates faster than it can be fixed, and so the blue team is stuck in the depot. The red car is damaged and slow, but eventually makes it to the finish line. The "do not repair" tactic is the more successful alternative under these conditions (adapted from Breivik, 2001, 2005, modified).

6B. Non-compliance with the rules

Genetic stability, which prevents sufficiently rapid accumulation of genetic changes required for complete tumor transformation of cells, is an effective barrier protecting the organism from tumor development. When cells "play by the rules," it means that they are genetically stable and use all the repair mechanisms and other tools to correct errors in their genome. As a result, mutations in such cells occur only occasionally. It is therefore almost impossible for a completely malignant tumor to develop during the lifetime of the organism. In humans, however, tumors occur quite frequently, and most of them actually exhibit increased genetic instability. Tumor cells simply do not follow the rules. Disregarding the rules thus represents another, the fifth, of the non-deadly sins.

Rules and their meaning

According to the Cambridge Dictionary, a rule (norm, standard) is an accepted principle or instruction that states the way things are or should be done, and tells you what you are allowed or are not allowed to do. It either prescribes what must be done or not done, or takes the form of what is generally considered normal, acceptable, or customary. Thus, there are both written and unwritten rules, characterized by different degrees of obligation, different scopes of validity, and different degrees of enforceability. What are rules good for? Do they constrain us, or do they help and support us? Are the rules of the road a problem for drivers? Do they limit us from taking full advantage of our abilities and the performance of our cars? Or do they help us to survive in heavy traffic amid the confusion of many other drivers with different abilities operating vehicles with different capabilities?

Is musical notation and the prescribed ways of performing a musical work an irritating limitation on the individual abilities of great virtuosos, or a help and sure guide to achieving a beautiful, harmonious orchestral sound? Are we limited by God's Ten Commandments in accepting and fulfilling our desires, and maximizing our personal development, or do they protect us and give us a safe guide in life and in finding our rightful place in the community?

Chaos and order

Genetic diversity and variability are among the universal characteristics of a natural population. The presence of genetic variability is a necessary condition for adaptability and evolutionary change. There is a direct relationship between the degree of genetic variability and the rate of evolution and degree of adaptation. On the other hand, genetic stability is also of great importance. Organisms must maintain and reproduce their genetic information in a nearly error-free manner to ensure the identity of their species and their own survival. New mutations are a main source of genetic variability in organisms, while elimination and arrangement of variants is preferably done by natural selection (Relichová, 1997, 2009). The need to keep genetic information sufficiently stable and, at the same time, the need for variability is one of the fundamental biological paradoxes. It is comparable to many other systems where chaos and order, two opposite and at the same time complementary states, coexist.

Chaos is characterized by randomness, high degrees of freedom and indeterminacy. A state without inner structure. Order is regularity, harmony, rhythm and periodic arrangement. Chaos lacks rhythm, chaos is the absence of order. It is incomprehensible and disturbing. Order is the arrangement of a system that allows for prediction of its future dynamics with sufficient probability. Chaos makes such predictions impossible. Order is something we understand. Chaos is represented by the reservoir of possibilities. In a functioning system, chaos and order balance each other; there is a dynamic equilibrium between the production of chaos and its mitigation, usually maintained by a network of negative and positive feedback loops. If the production of chaos in a system is predominant, disorganization increases and leads to decline. On the other hand, if order is too high, there is a risk that the system will become isolated and lose its dynamism. It is the transitional phase between chaos and order, their equilibrium, which is optimal for the existence of systems (Veverka, 2013, pp. 340–352).

In the cells of multicellular organisms, mutations are a dominant source of genetic variability. Mutations contribute to chaos. Conversely, all components of the cell response system to DNA damage (recognition systems, repair mechanisms, cell cycle checkpoints, senescence, cell death) are involved in reducing chaos. In human cells, equilibrium is established at a rate of 10^{-6} mutations per gene and cell division. This rate of incidence of new mutations is apparently optimal for the life of cells in the human body, in the system to which they belong. And it is a truly dynamic equilibrium because it is the result of the rate of DNA damage, i.e. the generation of variability and chaos, and the rate and efficiency with which that damage is repaired or otherwise eliminated, i.e. with the reduction of variability and chaos.

The inevitability of DNA damage

Constant damage to DNA is an integral, inescapable, and unavoidable part of life. The causes of DNA damage are both external and internal. External factors can often be controlled to some degree (e.g. we can choose not to smoke, to change our diet and lifestyle, etc.), but they cannot be completely avoided. DNA damage caused by internal factors cannot be avoided at all because it is related to the normal, physiological function of cells.

Free oxygen radicals are an example of an endogenous factor that has considerable potential to damage DNA (and other structures as well) while being an inevitable part of cellular life. They are formed in mitochondria during respiration. However, they are not only a passive by-product of this vital process, but also play a very important role in its regulation. Mitochondria operate according to the principle of sensitive feedback, in which free radicals are the crucial signals. They are involved in regulating the expression of some components of the respiratory chain, but they can also trigger mitochondrial degradation or cell death, apoptosis, by releasing cytochrome c into the cytoplasm (see chapter 4A). Free oxygen radicals thus contribute to the long-term maintenance of cellular respiration, which is tailored to the needs of the cell (Lane, 2012, pp. 198–202). On the one hand, they are a source of damage to DNA and other cellular structures; on the other hand, they perform an irreplaceable, vital regulatory function in the cell.

The processes that occur in lymphocytes during their maturation and differentiation are another example. B lymphocytes are cells of the immune system responsible for the production of specific antibodies. It is estimated that the human organism produces about 10^8 to 10^{11} antibodies with different specificities (Krejsek, Kopecký, 2004; Jílek, 2014). This large number makes impossible the option that each antibody is encoded by a specific gene. In fact, there are only three genes in the human genome that code for antibodies. The discrepancy between the wide range of antibodies produced and the limited coding capacity of the genome is resolved by the unique arrangement of genetic information for antibodies and the mechanisms that generate different variants of antibodies.

Genes encoding antibodies are arranged in the form of gene segments that can rearrange and assemble into new combinations. In maturing B lymphocytes, mechanisms are activated that both effectively combine these segments and further differentiate antibodies by inducing mutations in their antigen-binding site. These mechanisms operate with high precision and specifically target antibody-encoding genes. In rare cases, they can also induce mutations in other genes. For example, if proto-oncogenes or tumor suppressor genes that regulate proliferation or cell death are mutated in this manner, they can trigger the tumor transformation process of the affected lymphocyte and initiate the development of lymphoma (Wang, 2013; Robbiani, Nussenzweig, 2013; Keim et al., 2013). Thus, on the one hand, these unique mechanisms are life-sustaining and life-saving, as they allow the formation of a broad spectrum of antibodies; on the other hand, they pose a life-threatening risk, as they carry the possibility of developing a severe form of cancer, lymphoma.

A double-edged sword

The two examples described above show that there are processes in cells that are necessary for their survival and proper functioning, but at the same time have significant potential to damage their structures, including DNA, increase genetic instability, and thus indirectly jeopardize their survival and good functioning. They are examples of dichotomy, a principle that applies generally and not just to cells. Double-edged swords have two edges, symbolizing advantages and disadvantages, benefits and risks, positive and negative sides. In both cases described above, the advantages outweigh the disadvantages. Effective regulation of respiration and efficient energy production in mitochondria make the slow accumulation of DNA damage seem less significant in the long run. Similarly, immune system activity, including effective protection against pathogenic infections, is far more relevant to human survival than perfect protection against a slow increase in genomic instability in lymphocytes, which can lead to the development of lymphoma in some people.

Exposome

The extent and significance of persistent and lifelong DNA damage is reflected, among other things, in the exposome concept, first formulated in 2005. The exposome represents the sum of all the influences that the organism has been exposed to throughout its life, including the prenatal period. Three overlapping components of the exposome have been described: (1) the general environment, which includes, for example, climate, urbanism, social status, and stress; (2) the specific environment, i.e. specific pollutants, diet, physical activity, smoking, drug use, infections, and the like; and (3) the internal environment, which includes endogenous biological factors such as metabolism, microflora, inflammation suffered, oxidative stress, and the like. This third domain also includes the body's responses to external factors and internal metabolic processes that can alter or process chemicals and factors to which the body is exposed (Wild, 2005; Miller, 2014).

The exposome concept was first proposed to emphasize the need for better and more complete data on the importance of the environment in the development of diseases, including tumors, and thus added genetic data, the importance of which is generally not doubted. If we understand the genome (the total genetic information stored in the DNA of a given organism, including all genes and noncoding sequences, i.e. the complete DNA sequence) as a set of instructions and rules that govern the life of an organism, then the exposome concept was formulated largely as a complement to the existence of the genome, which addresses how instructions and rules are corrupted and how this affects the life of the organism.

The genome is a list of rules, and genetic stability is a guarantee that they will be maintained in unchanged form. Together, they help create and maintain order. The exposome summarizes the disruptive factors and damage to the established rules. Increasing genetic instability reflects a decreasing ability to enforce adherence to these rules, which means an erosion of order and increasing chaos. In his book *Evoluce svým vlastním tvůrcem*, Miroslav Veverka describes and generalizes the circumstances of the emergence of chaos and associates it with tumors and other unfavorable directions of development: "Chaos arises where free energy has undermined the balance of feedback loops and the stability of homeostatic mechanisms. The whole system enters a state far from equilibrium. Tumor growth, epidemics and pandemics of infectious diseases, fires, blood feuds, national or international crises, religious, political, social unrest, pogroms, looting and pillaging break out. Chaos erupts, however, only if the growing variance is not reduced in time by effective selection or choice" (Veverka, 2013, pp. 350–351).

Carcinogenesis: An evolutionary process in the body

Biological evolution is a long-term, spontaneous process in the course of which life forms develop and differentiate. The driving forces are variability and diversity on the one hand, and natural selection on the other. Increasing diversity means an increase in chaos; natural selection is a force that maintains order. Natural selection is a process that selects individuals (genes, alleles) from diverse pools according to various criteria and either suppresses or favors them. It helps to maintain the fittest individuals, best adapted to the present living conditions, in the population. Sufficient diversity is a necessary condition for ecosystem stability and its potential for further adaptation and evolution. So why should genetic instability be a problem? If it is a source of diversity, it should or could be a positive element in living systems. So what is the problem? Is there anything to learn from tumors? The development of tumors can be seen as an evolutionary process within a multicellular system (Breivik, 2005, 2006).

The value of an individual from an evolutionary point of view is expressed by biological fitness. This is a set of traits that affect an individual's chances of passing on its genes to the next generation. In single-celled organisms, selected fitness is primarily based on the ability to grow and divide as fast as environmental conditions allow. French geneticist and Nobel laureate François Jacob remarks that "the dream of every bacterium is to become two bacteria" (Lane, 2005, p. 114). The bacterium *Escherichia coli*, for example, divides every 20 minutes under optimal conditions.

Multicellular organisms, however, are more complicated. Genes that are inherited in the germline are selected for their ability to maintain control and stability (harmony) among somatic cells throughout the organism. In other words, the genes that are most reliable in ensuring that a well-functioning and harmonious system develops are also the best guarantors that the genes will be passed on to the next generation as successfully as possible. However, the situation is different for somatic cells, where the ability to grow and divide rapidly is still favored, as it was for all cells when they lived independently as single-celled organisms (Breivik, 2005, 2006). This largely contradictory situation of somatic cells within a multicellular organism is expressed by Nick Lane in his book *Power, Sexuality, Suicide*: "The individual is an organism composed of genetically identical cells, which are specialized to perform diverse tasks for the good of the organism as a whole. From an evolutionary point of view, the question is, why did these cells subordinate their selfish interests to collaborate so altruistically in the body? Inevitably there were conflicts between the various levels of organization in the body, between genes, organelles, and cells, but paradoxically without these internecine battles the strong bonds that forge the individual might never have evolved. Such conflicts spurred the evolution of a molecular 'police force' that curbs selfish interests much as the legal system enforces acceptable behaviour in society. In the body, programmed cell death, apoptosis, is central to the policing of conflicts" (2005, p. 191).

If a multicellular organism is to succeed as a whole, the "selfish interests" of somatic cells must be effectively held in check. Lane compares apoptosis, whose role in maintaining the integrity of the multicellular organism and in tumor development was discussed in chapter 4B, to a police force. The degree of genetic stability indicates the extent to which rules are followed. From its role in tumor development, we know that a high degree of genetic stability (i.e. a high degree of compliance with the rules) is also necessary for the integrity of the system. And it also provides some answers to the question of what it means to follow the rules. Yes, the traffic rules certainly constrain me as a driver. Because of the rules, I may never be able to use all my driving skills and abilities, my full driving potential. If I want to drive really fast, perhaps I can try to drive alone on a closed stretch of road (and follow the given rules) so as not to endanger the community of other drivers and road users. The soloist can certainly play his instrument according to his own feelings, with his own phrasing, his own rhythm, with a personal interpretation, perhaps even improvise freely and look for new ways of performance. But as a musician in an orchestra, as a member of a community, he must follow the conductor and seek harmony and unison with the other players. Rejecting God's Ten Commandments may be possible on a desert island (although who knows?), but living well and harmoniously in a community of other people without generally following moral rules is not.

It can be argued that when we accept and follow the rules, we give up our own freedom, our possibilities to find new ways, new solutions, new manners, and thus we give up very general and natural possibilities of diversification. And this is generally true. But we should not forget this: "There is no sense in which a malignant tumour is making a bid for freedom. It is simply a ghost in the machine, a pointless reversion to an earlier cell type, which ruled before the evolution of the 'individual'—that of cells doing their own thing. In this sense, cancer gives a dull and empty sense of the sheer meaninglessness of evolution. It's hard to think of any deeper meaning for cancer"

(Lane, 2005, p. 202). Miroslav Veverka puts it even more generally: "Spontaneous organization in itself is far from any ethical value. Nature remains indifferent to the human understanding of progress. It looks only for opportunities, for free niches. Even undesirable phenomena can organize spontaneously. Cancer outbreaks, parasite development, epidemics, marital disruption, drug addiction, psychopathic disposition, criminal careers, political tensions, economic crises and warlike conflicts. They disrupt the identity of the metasystem and threaten its survival" (2013, p. 453).

DNA replication stress, or the praise of slowness again

Chapter 3B was about senseless speed and man's race with himself as the first non-deadly sin. In this chapter, we see that rushed, uncontrolled cell division, accompanied by too rapid DNA replication, *per se* promotes genetic instability – without further mutational intervention. In other words, too much speed hinders compliance with the rules and increases chaos! Carl Honoré, the "expert on slowness," illustrates this statement: "Modern capitalism speeds up more than it thrives, because the pressure to win leaves no time for checking quality. In recent years, software vendors have become accustomed to rushing their products to market before they could be fully tested. The result is an epidemic of computer failures, glitches and malfunctions that cost companies billions of dollars every year. And then there are the human costs of turbo-capitalism. Medical offices are flooded with people suffering from stress-related diseases: insomnia, migraines, hypertension, asthma, gastrointestinal problems, burnout syndrome" (2012, p. 12).

A similar statement, but from a completely different field, comes from Dan Ariely. Ariely is a psychologist, a scientist who studies irrational decisions in business and personal life (he and his findings are discussed in more detail in chapter 12B). He has developed procedures to experimentally determine how our decision making is influenced. Among other things, he has found that a person who is rushed, tired, and overwhelmed by having to concentrate on several things at once is much more likely to cheat and break rules. This is not the result of bad character, reluctance or lack of motivation. It is the consequence of being in a hurry and acting too quickly (Ariely, 2012).

Some authors studying DNA replication stress in tumor cells consider the link between the rate of cell proliferation and increased genetic instability to be so fundamental that they propose that genetic instability, one of the features typical of tumor cells and almost universally present in them, should be considered a secondary feature arising from the primary feature – the deregulated cell cycle (Negrini et al., 2010). At first glance, this looks like a funny play on words. However, it may reflect a deep insight and understanding of context, as well as a practically important conclusion. Our fight against breaking the rules, cheating and fraud will not be successful as long as we try to fight only with a system of penalties and fines or with other prohibitions and commandments. Perhaps we need to examine whether our desire for possessions and advancement, constantly moving faster and farther, and our demand for unending growth is not the origin of the difficulties.

Just right

Although one might think that maximizing genetic instability would be beneficial from the tumor's point of view, this is not the case. The optimal "just right" level of genetic instability generates such a degree of chaos that the process of carcinogenesis can proceed effectively enough, but without compromising cell viability. It is therefore a state of equilibrium between chaos and order.

The "just right" principal is quite general, as an example from a completely different field shows. The child psychologist Zdeněk Matějček writes: "One of the Ten Commandments for parents of children with severe mental disabilities is: 'Take care in the right way and at the right time!' We often see excessive (completely understandable, but wasteful, purposeless, often even harmful) efforts by good parents to promote the development of a disabled child as much as possible and as quickly as possible. Instead of developmental progress, defense mechanisms are then awakened, which become a serious obstacle to the child's further 'learning.' The child is then not even able to learn what should be 'learnable' from the point of view of his developmental maturity" (2004, p. 148). When something exceeds the optimal level, it triggers defense mechanisms, cell death, etc. This principle conveys a clear message: everything in moderation! Proceed step by step! Hurry up slowly!

A slippery slope

The "just right" principle could also be called the principle of adequacy. And if something is adequate, we might ask the question, what is meant by adequacy? Just as different stimuli are appropriate for different children, different degrees of genetic instability are adequate for tumor cells at different stages of the carcinogenesis process. Tumors show us that adequacy (like the right speed – see chapter 3B) is dynamic and susceptible to change. The greater the degree of transformation of tumor cells, the greater their tolerance to genetic instability (Fig. 24). It appears that tumor cells learn to cope with increasing chaos during their development. This also appears to be a general principle, sometimes referred to as a "slippery slope." A slippery slope is a process that, once started, is difficult to stop or control and usually leads to a progressively worse state. It is a sequence of steps that inevitably lead from one event to the next with unexpected, unintended consequences.

In medical ethics, for example, the term "slippery slope" is used to explain attitudes toward euthanasia: "Common experience teaches us that if we allow ourselves to do something we did not want to do or should not do, the question immediately arises whether we could afford to do something else that is even more problematic. Once a person or an institution moves into a certain slippery slope, gravity pulls them further and further down. In the case of euthanasia, the step that allows doctors to end the lives of dying people at their own request will not be final. It will drag us further and further down, and gradually this possibility will be expanded. One day we will reach the point where we can end the lives of people who are seriously ill but not yet dying, the lives of people with incurable diseases but who could live for a long time, the lives of severely handicapped newborns, of senile people or of people with various mental disabilities, etc. Numerous documents from the Netherlands prove that the 'slippery slope' has become a reality there and that intentional and unintentional euthanasia (or even euthanasia against the patient's will) are only a small step apart" (Munzarová, 1996).

Further evidence of how "small" changes to the rules, i.e. "just right" changes, and their gradual escalation due to increasing resistance can lead to a horrible end is provided by Stefan Zweig in his book *The World of Yesterday: Memoirs of a European*. He describes Austria and the events leading up to World War II: "National Socialism, in its unscrupulous technique of deceit was wary about disclosing the full extent of its aims before the world had become inured. Thus they practised their method carefully: only a small dose to begin with, then a brief pause. Only a single pill at a time, and then moment of waiting to observe the effect of its strength, to see whether the world conscience would still digest this dose. And since the European conscience – to the hurt and shame of our civilization – eagerly accented its unconcern because, after all, these atrocities occurred 'beyond the border,' the doses became progressively stronger until all Europe finally perished from them" (Zweig, 1943, p. 251).

We have already mentioned Dan Ariely and his experiments in which he studied the influences on our decision making. He also found that when a person violates his own standards (whether in dieting or for financial reasons), he is much more inclined to give up further attempts to control his actions and much more likely to make further missteps. He slides off the slope (Ariely, 2012).

Don't stop for repairs in a war zone! Even the desire for perfection must be reasonable

The cell responds to various forms of DNA damage with this sequence of events: cell cycle arrest, repair, and resumption of proliferation. For a long time, it was assumed that cell cycle renewal occurs only after DNA repair is completed. However, it was later shown that adaptation to damage is an alternative to complete DNA repair. Continuing cell division before DNA repair is completed, of course, increases the risk of the gradual accumulation of defects in cells. Nevertheless, cells have this mechanism and use it under certain circumstances. There are situations in the life of cells where survival and continuation are more important than perfection. It is as if the "don't stop for repairs in a war zone" model teaches us: in an imperfect world, you cannot always insist on perfection. Those who insist on perfection may never get off the ground and never move. And so, despite everything written in this chapter about the importance of following rules and maintaining order, and despite the clear lessons from tumors that breaking rules is a non-fatal sin – without diluting the appeal for compliance – we must agree with Albert Einstein, who said, "Not perfection, but imperfection characterizes the universe." We should strive for accuracy and perfection, but we must not cling to them and insist on them.

7A. Induction of angiogenesis

The story of Judah Folkman: The beginnings

Judah Folkman was given an interesting task in 1961 as a young doctor, a surgeon in the U.S. Navy. He was asked to find a way to provide a long-term supply of fresh blood to the aircraft carrier. He developed an experimental test to check the viability and stimulate the proliferation of blood cells. He used rapidly growing mouse tumor cells. He found that he could produce only tiny, pinhead-sized tumors in his experimental setup. Paradoxically, after injection into the mouse body, the same cells formed macroscopic, or much larger, tumors. And why? The only visible difference Folkman noted was that the tumors formed in the mice's bodies were densely packed with blood vessels (Schreiber, 2010).

After leaving the Navy, Folkman returned to his original specialty and worked as a pediatric surgeon. During operations, he noticed that the tumors he removed from children's bodies were always densely packed with blood capillaries as well. In his mind, he kept returning to his experiments in the Navy and spent his nights and weekends in the laboratory conducting further experiments. On this basis, he made the hypothesis, bold for the time, that tumors could not grow larger than the size of a pinhead unless they were supported by a vascular system. He called his hypothesis "angiogenesis," an older term for the growth of new blood vessels. He assumed that tumors secrete a mysterious factor that stimulates angiogenesis and triggers the growth of new blood vessels to supply and nourish them. He concluded that tumor growth could be inhibited by preventing the growth of blood capillaries. He published his work in 1971, but it did not receive much support; indeed, it was strongly opposed (Folkman, 1971; Schreiber, 2010).

Vascular system

The vascular system (vasculature) in the human body is a network of flexible tubes that serve to transport and exchange nutrients, oxygen and other gases, hormones, other signaling molecules, and waste products of cellular metabolism. There are many different cells flowing in the vascular system – various blood cells and also cells of the immune system, for example. The vascular system and the heart are formed during embryogenesis as one of the first organ systems and are a necessary in adulthood for the maintenance of metabolism and homeostasis of all tissues, as well as their healing and repair. The development of the vascular system must be adequate in each tissue. Both insufficient and excessive vascular growth is pathological and poses a significant danger to the organism. Blood vessels are mainly composed of two types of cells. The endothelial cells, which form the lining of blood vessels and are in close contact with the blood. The basement membrane separates the endothelium from the pericytes, smooth muscle cells that surround the endothelial cells and provide strength, elasticity and tension to the blood vessels (Fig. 26). In blood capillaries, the pericytes surround the endothelium in a single layer; in larger vessels, the smooth muscle cells form multiple layers (Shima, Ruhrberg, 2006).

Development and growth of the vascular system

During embryogenesis, the development and growth of the vascular system, called the process of neovascularization, occurs rapidly and in all types of tissues. Adequate access of all cells to the blood system is ensured by the development of blood vessels, which is closely coordinated and associated with the growth and development of all other tissue components. During embryogenesis, two mechanisms are involved in the development of the vascular system. Vasculogenesis is the formation of blood vessels *de novo* from endothelial cell precursors, angioblasts. Angiogenesis is then the process by which new blood vessels are formed from existing vessels by budding and remodeling. The maturation of the vasculature also involves the establishment of interactions of the endothelium with pericytes and other smooth muscle cells, increasing stability and strength, and the acquisition of properties that distinguish arteries from veins. All of this ensures that the vasculature is impermeable, thus blood does not penetrate surrounding structures and flows smoothly. In contrast to the embryonic vascular system, the mature vascular system is quiet, stable, and resistant to change. Once the tissue completes its development and growth, the process of neovascularization is halted, and if reactivated under certain circumstances, it is always temporary and tightly regulated.

In an adult, under normal circumstances, the vasculature represents an extremely stable, quiescent cell population. Endothelial cells are among the longest living cells in the body, apart from the cells of the central nervous system. The lifespan of endothelial cells forming an adult capillary is approximately 1,000 days. In comparison, the lifespan of the epithelial cells that make up the intestinal mucosa is only two to three days. While only one in 10,000 endothelial cells (i.e. 0.01%) in the adult vasculature is in the phase of active cell division, a full 14% of intestinal epithelial cells are actively proliferating. The physiological growth of new blood vessels occurs in the adult body only under very specific circumstances, such as hormonal stimulation during the female

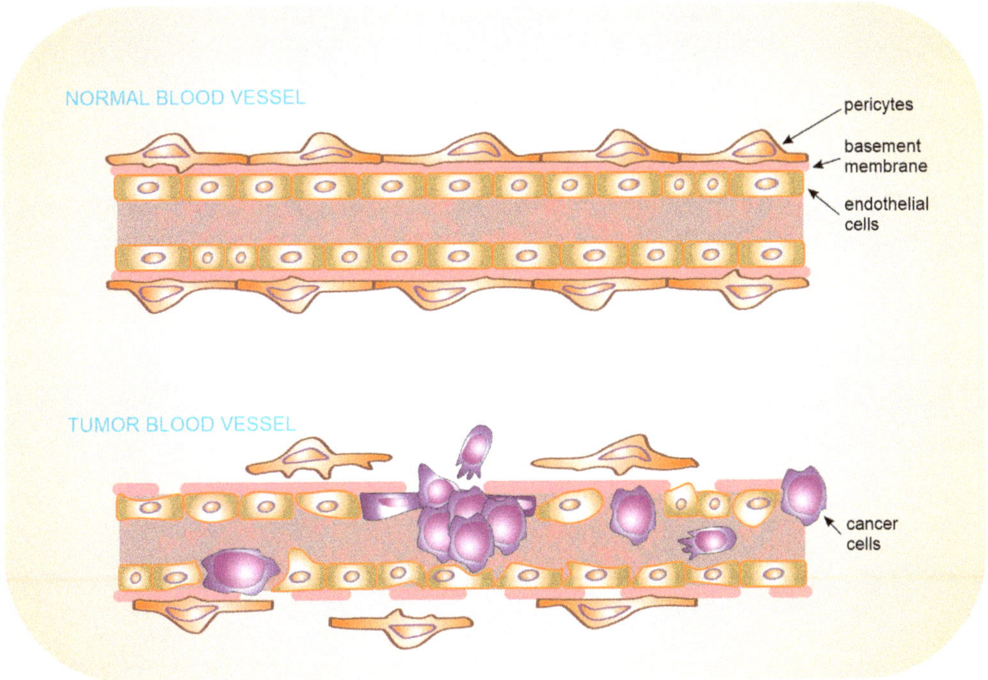

Fig. 26 Structure of the tumor vasculature
Tumor blood vessels differ structurally and functionally from healthy, physiological vessels. Physiological vessels grow in an orderly fashion, in harmony and in close cooperation with other tissue components to which they belong. The tumor vasculature is quite chaotic and dysfunctional. Tumor vessels lack a continuous sheath of supporting smooth muscle cells and pericytes, and the basement membrane is discontinuous. Tumor vessels are often referred to as mosaic vessels. Tumor cells occur in both the wall and lumen of blood capillaries (adapted from Carmeliet, Jain, 2000, 2011).

reproductive cycle or during wound healing. It is mediated mainly by the mechanism of angiogenesis. At the site where the vasculature is stimulated to form new capillaries, the basement membrane surrounding the endothelial cells is temporarily degraded, the endothelial cells change shape, begin to proliferate, and invade the surrounding stroma. Firm adhesions form between the nascent endothelial cells, and the lumen of the nascent capillaries is formed. The germinating capillaries fuse, close into loops, and blood begins to flow through them (Hanahan, Folkman, 1996; Carmeliet, Jain, 2000; Shima, Ruhrberg, 2006).

Tumor angiogenesis

The ability of tumors to exceed the size of a proverbial pinhead (i.e. a cell mass up to 2 mm^3 in diameter) depends on their ability to access the blood system. Only in this way can tumor cells be adequately supplied with vital nutrients and oxygen and the waste products of their metabolism be removed. The distance that can be bridged for this purpose by simple diffusion of substances and oxygen is no more than 100 to 200 μm. At greater distances, the exchange of molecules is no longer efficient enough (Carmeliet, Jain, 2000; Pecorino 2012; Shima, Ruhrberg, 2006).

Tumor cells growing in tissues seem to have an inherent ability to induce growth of blood vessels. However, this is not the case. Proliferating premalignant lesions are initially avascular, which limits their ability to expand. The acquisition of an angiogenic character is a peculiar and specific step in their development (Fig. 27). Gradually, it became clear to researchers that the transition from a nonangiogenic to an angiogenic state typically occurs in the early stages of carcinogenesis and not by a gradual evolution. It is not a slow transition but an abrupt change (Hanahan, Folkman, 1996).

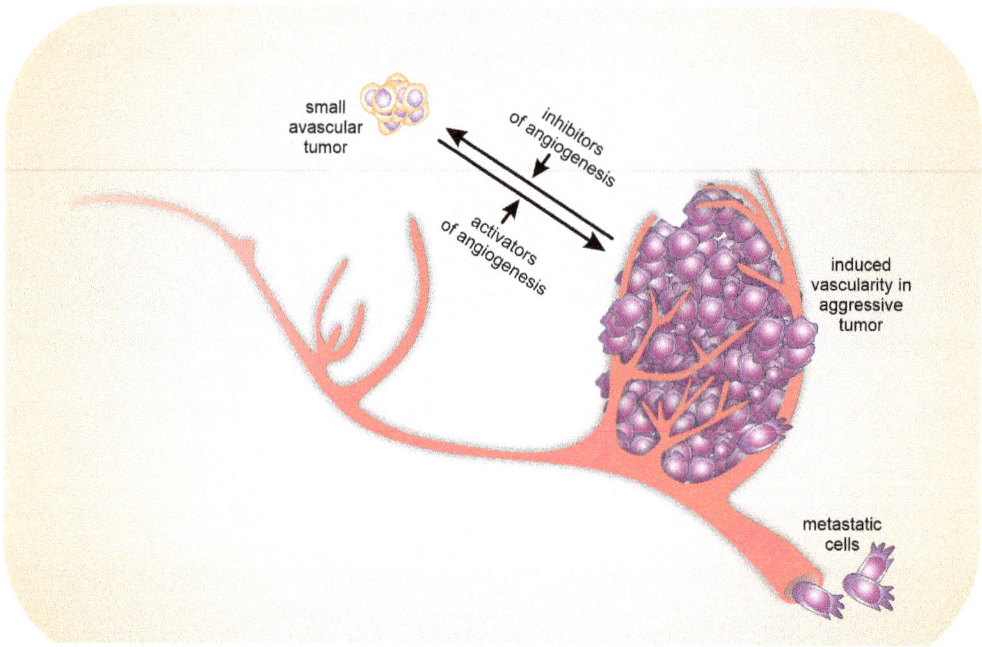

Fig. 27 Angiogenic and non-angiogenic tumors
Premalignant lesions are initially avascular, which limits their ability to expand. They cannot reach a size larger than a pinhead. Induction of angiogenesis and the growth of new blood vessels provide the tumor with nutrients, oxygen, and growth factors, enabling its expansive growth and further development.

Dormant tumors

The induction of angiogenesis is not an entirely obvious, "natural" step in the development of tumors, as indirectly evidenced by the existence of dormant tumors. These are microscopic foci of tumor cells that neither grow nor spread. They were first discovered at the beginning of the 20th century, mainly in connection with the growing number of traffic accident victims. Evidence was mounting that many people who had never been diagnosed with signs of cancer had significant numbers of microscopic tumors in their bodies. Microscopic tumors, which consist of about 10^5 cells, are found in various organs and are apparently dormant. This means that they are quiescent, literally "asleep," not growing, not developing in any way, and not causing any problems for their carriers. They remain undetected, hidden. For this reason, they are sometimes called occult tumors. Many never develop, in others it takes years or even decades. The reason for dormancy is usually the inability to induce angiogenesis (Bissell, Hines, 2011; Hesketh, 2012, p. 131).

Activators and inhibitors of angiogenesis

Tumors implanted in avascular areas usually result in rapid growth of the vasculature into the tumor tissue. The tumors appeared to release some potent activators of angiogenesis that stimulate dormant capillaries to sprout. In fact, an intensive search soon found these. The most important is the vascular endothelial growth factor VEGF, which stimulates endothelial cell proliferation, migration, and survival. Other proangiogenic factors, angiopoietins, influence the interaction of endothelial cells with pericytes and are involved in vascular maturation. The distribution of ephrins and their receptors makes a difference between arteries and veins (Yancopoulos et al., 2000). On the other hand, effective inhibitors of angiogenesis were gradually described over time. The first were angiostatin and endostatin, isolated in the laboratory of Judah Folkman (O'Reilly et al., 1994, 1997). Among the important inhibitors of angiogenesis is thrombospondin-1, which blocks endothelial cell adhesion, proliferation, and survival.

A number of pro-angiogenic factors are stored in the extracellular matrix from which they are released upon its proteolytic degradation. This ensures their rapid release. Some angiogenesis inhibitors are also constantly available, but use a different strategy. Many inhibitors result from cleavage of precursor molecules that do not themselves regulate angiogenesis. For example, the potent angiogenesis inhibitor angiostatin is formed by proteolytic cleavage of the plasmin gene, and endostatin is a cleavage fragment of collagen XVIII. The precursors of some inhibitors, e.g. endostatin, thrombospondin-1, and others, are part of the extracellular matrix, whereas other precursors are not. For example, plasminogen, the precursor of angiostatin, is a blood protein (Hanahan, Folkman, 1996; Folkman, 2003).

Angiogenic switch

Regulation of angiogenesis must meet two seemingly contradictory requirements. On the one hand, it must reliably and constantly stabilize the vascular system, i.e. keep it in a quiescent, unchanging state and allow only a slow exchange of endothelial cells. On the other hand, it must be able to activate rapidly when the tissue requires renewal or restructuring of the vascular system, whether physiologically, e.g. during the menstrual cycle, pregnancy and the like, or in the event of injury. The response must be immediate. A slow shift to active mode could endanger life. A regulatory mechanism that meets both requirements must be very robust and durable on the one hand, but also sensitive and flexible on the other. How can these seemingly incompatible requirements be met?

The equilibrium angiogenic switching hypothesis provides the answer. It proposes that the regulation of angiogenesis depends on a dynamic balance between its activators and inhibitors. An increase in the activity of activators or a decrease in the activity of inhibitors tilts the equilibrium toward angiogenesis, whereas a decrease in the activity of activators or an increase in the activity of inhibitors switches the equilibrium to the "off" position (Fig. 28). Thus, it appears that under normal circumstances, inhibitors predominate in tissues that reliably maintain the vasculature at rest. A significant increase in activator concentration or decrease in inhibitor activity (or a combination of both) is required to initiate angiogenesis. Both activators and inhibitors of angiogenesis are always "on hand." They are maintained in an inactive form (either by retention in the extracellular matrix or as part of their own precursors – zymogens), but are ready for very rapid activation (by release from the extracellular matrix or cleavage of the precursor) in response to the appropriate stimuli (Hanahan, Folkman, 1996).

Turning on tumor angiogenesis I: Oxygen deprivation

Several different mechanisms seem to be involved in the induction of tumor angiogenesis. Therefore, this process is quite complex. However, two of the mechanisms are predominant. The first of the most important triggers of tumor angiogenesis is a low oxygen concentration known as hypoxia. Maintaining an optimal oxygen concentration is essential for all cells in the body. They use a very effective and relatively simple mechanism based on hypoxia-inducible factor (HIF). This factor can effectively stimulate angiogenesis, in part by directly inducing VEGF in response to oxygen concentration. Although this factor is referred to as "hypoxia induction," which might suggest that it forms only when a cell is deprived of oxygen, it actually functions in the opposite manner. HIF, a complex of the two proteins HIF-1α and HIF-1β, is constantly and constitutively produced in cells. However, at normal oxygen concentrations (normoxia), the HIF-1α subunit is modified by oxygen immediately after its synthesis and then efficiently degraded. When the oxygen concentration in cells decreases, no modification or degradation of HIF-1α occurs, a functional HIF dimer is formed, which contributes significantly to tipping the balance of the angiogenic switch toward "on." This ensures that the vasculature remains intact and quiescent at a normal oxygen concentration, but that the cell can immediately respond to a decreased oxygen concentration by increasing HIF levels, thereby activating angiogenesis (Carmeliet, Jain, 2000; Pecorino, 2012).

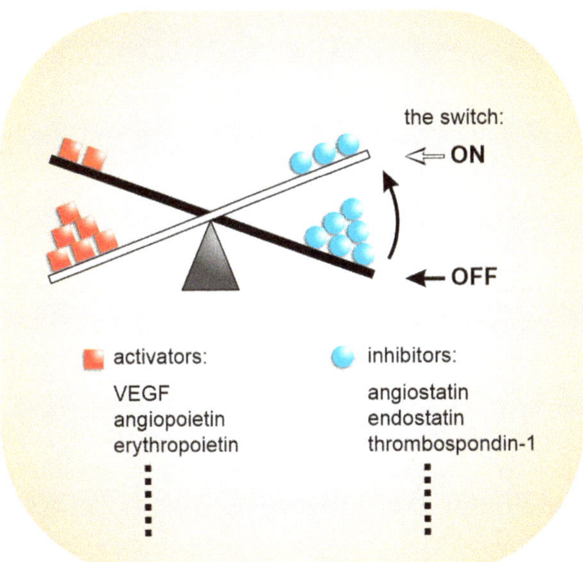

Fig. 28 The balance of the angiogenic switch
The regulation of angiogenesis is based on a dynamic balance between activators and inhibitors of angiogenesis. An increase in the activity of activators or a decrease in the activity of inhibitors shifts the equilibrium toward turning on angiogenesis, whereas a decrease in the activity of activators or an increase in the activity of inhibitors shifts the equilibrium to the "off" position (adapted from Hanahan, Folkman, 1996, modified).

Turning on tumor angiogenesis II: The role of oncogenes and tumor suppressors

The second predominant mechanism that triggers tumor angiogenesis relies on the activation of proto-oncogenes and the inactivation of tumor suppressors. Unphysiologically high activity of proto-oncogenes and low activity of tumor suppressors leads to deregulated, hasty cell division and/or failure of programmed cell death and is thus one of the major factors in tumor development (see chapters 3A and 4A). Rapidly growing and dividing cells form a mass that can easily reach a limiting size of 100 to 200 µm, which cannot grow further due to the oxygen starvation. However, tumor cells with damaged oncogenes and tumor suppressors cannot respond appropriately to the lack of oxygen by ceasing cell division, inducing senescence, or initiating programmed cell death. If cell division continues, areas of low oxygen concentration, hypoxia, develop in the tumor mass, which in turn – through HIF activity – causes a shift in the balance towards activation of angiogenesis.

This is not the only role of oncogenes and tumor suppressors in the induction of angiogenesis. Many oncogenes can directly stimulate angiogenesis by activating the expression of certain pro-angiogenic factors, most commonly VEGF. At the same time, some oncogenes reduce the expression of angiogenesis inhibitors, most commonly thrombospondin-1. One of the most important tumor suppressors, the p53 protein, activates the expression of thrombospondin-1 and is involved in the maintenance of the quiescent vasculature under physiological conditions and thus in tumor suppression. When p53 is inactivated in tumor cells, it loses this ability, and thus the balance of the angiogenic switch shifts significantly toward "on." The effect is the same as when protooncogenes are activated (Kerbel, Folkman, 2002; Folkman, 2003; Pecorino, 2012).

Turning on tumor angiogenesis III: A complex process

Activation of tumor angiogenesis is a truly complex process. Lack of oxygen alone increases the likelihood that angiogenesis will be initiated. Activated proto-oncogenes and inactivated tumor suppressors in tumor cells support the initiation of angiogenesis through at least two independent pathways. Indirectly by increasing hypoxia in growing tumor tissue, and directly by activating activators and suppressing inhibitors of angiogenesis. In addition, other mechanisms are also involved in tumor vasculature development. Non-tumor cells can also participate in stimulation of angiogenesis, particularly fibroblasts, which form part of tumor tissue and produce a number of pro-angiogenic factors (Kerber, Folkman, 2002). Pericytes are also actively involved in the process of tumor angiogenesis by secreting growth factors that stimulate endothelial cell proliferation, affecting the surrounding extracellular matrix and thus promoting endothelial cell migration (Ribeiro, Okamoto, 2015).

Although the development of tumor vasculature is termed tumor "angiogenesis," and it was long believed that tumor blood vessels were formed only by this angiogenic process, it has been found that tumors can stimulate and force endothelial cell progenitors derived from bone marrow or that are circulating in the bloodstream to participate in the development of tumor vessels. Thus, part of "tumor angiogenesis" is the process of vasculogenesis. In some tumors, up to 40% of the tumor endothelium arises from endothelial cell precursors (Carmelit, Jain, 2000; Kerbel, Folkman, 2002; Pecorino, 2012).

Structure of tumor vessels, mosaic vessels, vasculogenic mimicry

In contrast to physiological neovascularization, in which vessels form in perfect harmony, organization, and close cooperation with the other components of the tissue, tumor vasculature is quite chaotic and dysfunctional. Tumor vessels are tortuous, crossed, and irregular in shape and size. They lack continuous coverage by supporting smooth muscle cells and pericytes, and the basement membrane is also discontinuous. The individual components of the vessel walls are incorrectly arranged. The tumor vessels often lack a clear distinction between the arterial and venous systems, tend to be permeable, and blood flow through the tumor is generally chaotic and insufficient in some places (Fig. 26). The cause of this irregular, chaotic structure of the vasculature is constitutively activated pro-angiogenic signaling and thus the inappropriately high activity of some pro-angiogenic factors. A reduction of their activity may result in the formation of a less chaotic, less damaged vasculature (Carmeliet, Jain, 2000, 2011; Hanahan, Coussens, 2012).

Tumor vessels are often termed mosaic vessels. This means that their walls are partially formed by tumor cells. Sometimes tumor blood capillaries are even formed exclusively by tumor cells, even without the participation of endothelial cells. This phenomenon is called vasculogenic mimicry. It was first observed in highly aggressive malignant melanomas, but later described in other tumor types as well. On the one hand, vasculogenic mimicry proves the high plasticity and adaptability of tumor cells. On the other hand, it documents the strong pressure on the development of the vascular system within the tumor (Carmeliet, Jain, 2000, 2011; Shima, Ruhrberg, 2006; Klener, 2010).

Concomitant tumor resistance: Continuation of Judah Folkman's story

Sometimes dormant metastases are activated after surgical or radiological removal of the primary tumor: they trigger angiogenesis and begin to grow. This phenomenon, called concomitant resistance, occurs in only a small fraction of tumors, but it worried doctors and researchers for a long time. Surgeons were also concerned because they were sometimes accused of activating the dormant metastases. Judah Folkman discovered the basic laws of tumor angiogenesis. In his laboratory, the first regulators of angiogenesis were found and isolated. He was not only a great scientist, but also an experienced physician and surgeon. In a radio interview years later, he recalled the moment when all the information and knowledge came together in his mind and gave him a deep insight and understanding of the laws of angiogenesis induction. He realized that some primary tumors very likely inhibit the growth of distant dormant micrometastases by releasing angiogenesis inhibitors that reach the micrometastases through the bloodstream. After removal of the primary tumor, these inhibitors also disappear as well, and in the micrometastases the relative activity of proangiogenic factors increases and angiogenesis is stimulated. The surgical procedure itself, or the healing processes that the surgical procedure triggers in the body, also appear to induce pro-angiogenic factors and thus may contribute to the malignant process. Folkman also recognized that the relatively high concentration and stability of angiogenesis inhibitors is likely primarily responsible for maintaining the resting state of the physiological, healthy vasculature (Schreiber, 2010; Pecorino, 2012; Hesketh, 2012).

8A. Reprogramming of the energy metabolism

In normal cells, metabolic processes are tightly regulated. The cell grows and survives in harmony with its environment. The availability of nutrients and their abundance promotes the synthesis of proteins, lipids and nucleic acids, while starvation triggers processes that maximize the production and economical use of energy in the cell.

ATP: Adenosine Triphosphate

Energy transfer in the cell occurs through adenosine triphosphate – ATP (Fig. 29). It is the most abundant energy currency in living systems. A large amount of energy is stored in the structure of ATP, which can be released by splitting the molecule (mostly into ADP – adenosine diphosphate and phosphate, rarely into AMP – adenosine monophosphate and two phosphate residues) and used to drive various cellular processes. These include, for example, various biosynthetic processes such as the synthesis of nucleic acids and proteins, intracellular and membrane transport, cell migration, and many others (Fig. 30). In addition, ATP also serves as a source of phosphate groups for phosphorylation, as one of the nucleotides for nucleic acid synthesis, and as the initial substrate for the synthesis of an important signaling molecule, cyclic AMP (cAMP).

ATP formation

Unlike glycogen or lipids, ATP does not serve as a long-term cellular energy store, but as an energy source and carrier for immediate use. If ATP constantly provides energy for a whole range of biochemical reactions and cellular processes, then it must also be constantly regenerated by the breakdown of energy-rich organic substances. If ATP supplies were not constantly replenished, the average mammalian cell would run out of ATP in one to two minutes. The most common energy source for ATP regeneration is sugar – glucose.

The breakdown of glucose occurs in two steps: glycolysis and oxidative phosphorylation. Glycolysis takes place in the cell cytoplasm and converts glucose

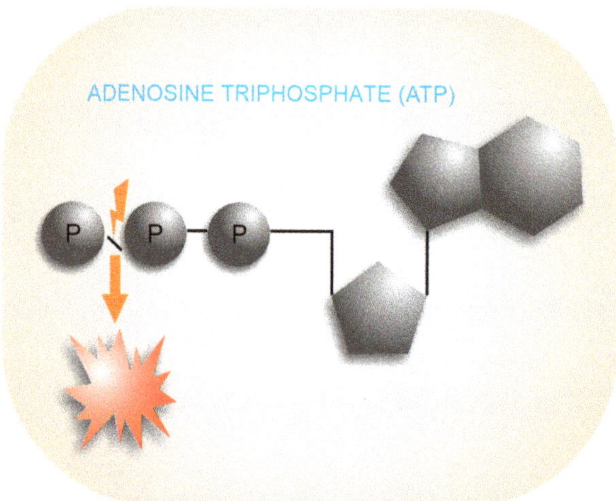

Fig. 29 Structure of ATP

ATP is composed of adenosine and a triplet of phosphates (P). This structure stores a large amount of energy, which can be released by dissociation (most commonly into ADP – adenosine diphosphate and phosphate).

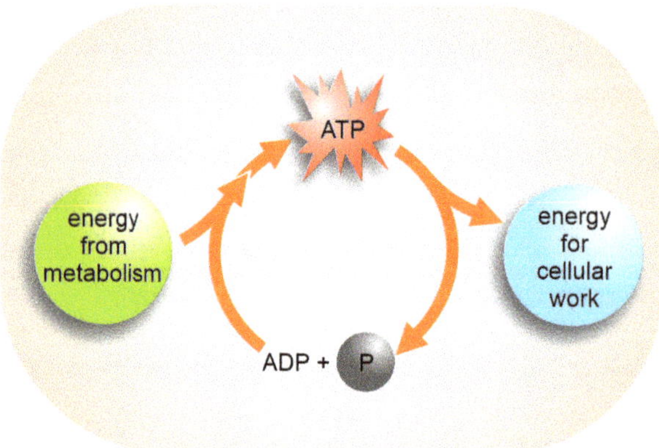

Fig. 30 The role of ATP in the cell

The energy released by dissociation of ATP can be used for a variety of cellular processes, including biosynthetic processes, transport, and cell movement. The cell regenerates ATP by degradation of energy-rich organic substances.

into two molecules of pyruvate. In oxidative phosphorylation, pyruvate is transported into the mitochondria and oxidized to carbon dioxide (CO_2) leading to the formation of electron donors via the Krebs cycle. With the help of respiratory chain enzymes located at the inner mitochondrial membrane, these electron donors are transferred to oxygen, resulting in the formation of water molecules and the release of energy used to synthesize ATP. Glycolysis results in formation of two ATP molecules from the decay of one glucose molecule. Glycolysis can occur both in the presence of oxygen (under aerobic conditions) and in the absence of oxygen (under anaerobic conditions). In addition to its independence from oxygen, another advantage of glycolysis is its speed. ATP formation by glycolysis is about 100 times faster than that by oxidative phosphorylation. Moreover, oxidative phosphorylation occurs only in the presence of oxygen, i.e. under aerobic conditions. On the other hand, oxidative phosphorylation is much more efficient in ATP regeneration. In oxidative phosphorylation, the decay of one molecule of glucose produces an additional 36 ATP molecules (Fig. 31).

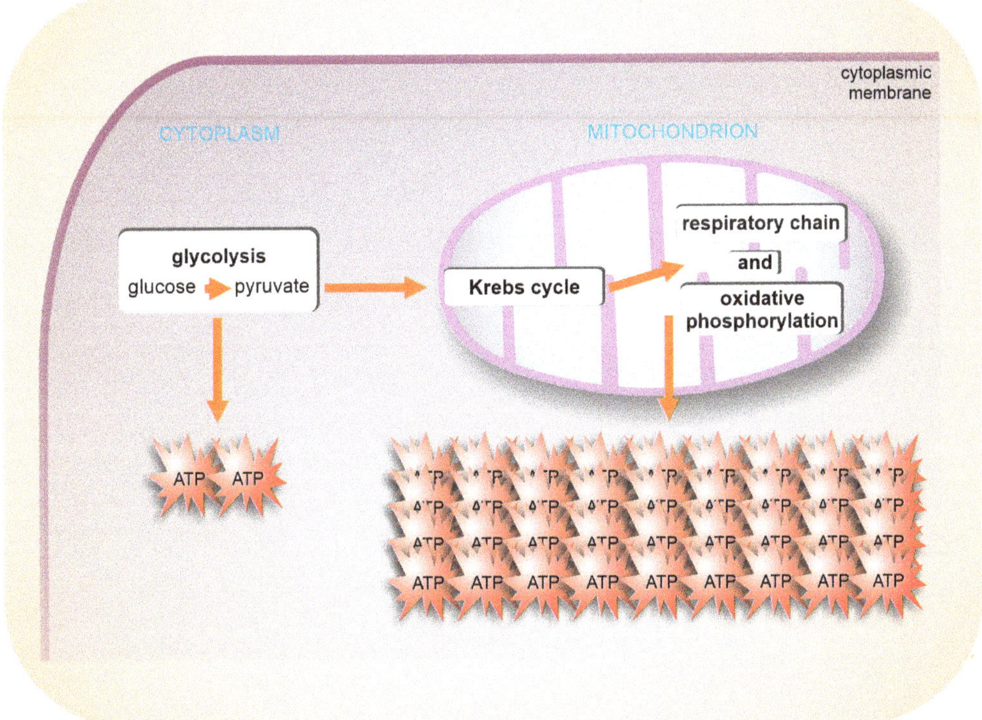

Fig. 31 Glycolysis and oxidative phosphorylation

Degradation of glucose occurs in two steps. In glycolysis, glucose is converted into two molecules of pyruvate and two molecules of ATP. In oxidative phosphorylation, the pyruvate is transported to the mitochondria, where its breakdown in the Krebs cycle produces electron donors that are transferred to oxygen with the participation of respiratory chain enzymes located at the inner mitochondrial membrane. The energy released during this process is used to synthesize ATP. During oxidative phosphorylation, the breakdown of a glucose molecule generates an additional 36 molecules of ATP.

Tumor cells: The Warburg effect

In the presence of oxygen (i.e. under aerobic conditions), healthy cells normally convert one molecule of glucose into water and carbon dioxide, producing 38 molecules of ATP. However, under anaerobic conditions, glycolysis is favored, and only a minimal amount of pyruvate is transported to the mitochondria for further decay. In the absence of oxygen, pyruvate is reduced to lactic acid, lactate, which is exported from the cell. Tumor cells are characterized by a high rate of growth and division and therefore rapidly consume energy – ATP. One might expect them to use the most efficient way of regenerating ATP through oxidative phosphorylation.

As early as the first half of the 20th century, German biochemist and Nobel laureate Otto Warburg discovered that tumor cells promote glycolysis even in the presence of oxygen (Warburg, 1927). This phenomenon is known as aerobic glycolysis or, after its discoverer, the Warburg effect. Tumor cells apparently reprogram their metabolism and, instead of efficiently breaking down glucose or pyruvate, into water and carbon dioxide in the mitochondria, perform only glycolysis, i.e. the breakdown of glucose to pyruvate (and lactate). Thus they are able to obtain only two molecules of ATP from each molecule of glucose, instead of the possible 38 molecules. Otto Warburg even believed that cells become tumors precisely because they have damaged mitochondria and cannot perform oxidative phosphorylation (Warburg, 1956). Warburg's observations and descriptions of the biochemical behavior of tumor cells were accurate, but the correct interpretation of the possible causes and consequences of this phenomenon was many years in coming.

Import of glucose into cells

By preferring glycolysis, tumor cells obtain about 19 times less energy from glucose than if they used oxidative phosphorylation. At the same time, their energy consumption is high. How do they resolve this apparent contradiction? They compensate for the low efficiency of energy production by increasing the import of glucose from the external environment into the cytoplasm. First, tumor cells increase the expression of glucose transporters (GLUT), which ensure efficient import of glucose into cells, and second, they increase the expression of lactate transporters, which export products of glycolysis from the cells. Cells that have high expression of these transporters can therefore easily consume large amounts of glucose and obtain enough energy from it, albeit inefficiently.

Why do tumor cells prefer glycolysis?

One possible reason why tumor cells prefer glycolysis could be hypoxia, which occurs in tumors whether they are angiogenic or not (see chapter 7A). Because of the chaotic, poorly functioning, and irregular tumor vasculature, and the uneven and precipitous growth of tumors, the more or less hypoxic sites occur repeatedly even in angiogenic tumors. Therefore, the reprogramming of energy metabolism and the switch to aerobic glycolysis could be considered as "logical" adaptations to these circumstances. Moreover, hypoxia also activates the glucose transporter expression by activating the transcription factor HIF, thereby increasing the import of glucose into cells under oxygen deprivation. HIF also increases the expression of transporters that ensure the efficient export of lactate from cells. There may even be a "symbiosis" between hypoxic tumor cells, which mainly perform aerobic glycolysis and release lactate into the microenvironment, and cells with sufficient oxygen, which are able to import lactate and obtain energy from it by oxidative phosphorylation. A similar "symbiosis" has been described, for example, between neurons and astrocytes in the brain (Hesketh, 2012; Potter et al., 2016).

The import of glucose into the cells of a multicellular organism is strictly controlled

Unlike bacteria and unicellular eukaryotic organisms, animal cells that are part of a multicellular organism are autonomous in the uptake of nutrients. They cannot absorb nutrients from their environment simply because they are available. Rather, the uptake of nutrients into cells is controlled by growth factors. The amount of nutrients is limited, and the cells of a multicellular organism compete for them. Thus, direct control by growth factors is a mechanism by which energy consumption in the body is regulated and is also linked and coordinated with the control of cell division.

In order to survive even with limited energy sources, cells switch to a catabolic (degrading) metabolism. This aims to produce ATP from a limited amount of intracellular nutrients with maximum efficiency. This is the case in most of differentiated, non-proliferating cells that are not stimulated by growth factors. These cells metabolize glucose to pyruvate by glycolysis and oxidize most of the pyruvate to CO_2 and water by oxidative phosphorylation.

On the contrary, when growth factors are available or even in excess, cells increase the uptake (import) of nutrients, especially glucose and glutamine, switch to anabolic (synthetic) metabolism, and gain energy mainly by intense glycolysis. Originally, it was thought that growth factors, by stimulating cell proliferation that requires energy, lead to a rapid depletion of the cell's energy reserves and a decrease in the ATP : ADP ratio. The increased ATP demand subsequently, i.e. secondarily, stimulates its rapid regeneration by aerobic glycolysis. In fact, this is not the case. The switch in energy metabolism is primary, i.e. a direct response to high growth factor activity and is not caused by reduced ATP levels. It is low glucose in the cell (not ATP) that forces the cell to use energy resources efficiently and stimulates oxidative phosphorylation in the mitochondria. Once glucose uptake increases, the cell switches to a less efficient way of obtaining energy through aerobic glycolysis, and metabolic activity in the mitochondria is reduced. The key to switching energy metabolism in the cell is therefore an increase in glucose uptake.

Growth factors that stimulate cell proliferation act by directly controlling the cell cycle and cell division (see chapter 3A). In addition, they also act directly on cell metabolism to adapt it efficiently to the needs of dividing cells. They increase the import of glucose into cells and stimulate aerobic glycolysis. The switch to aerobic glycolysis leads not only to a rapid, albeit inefficient and seemingly uneconomical regeneration of ATP, i.e. for energy production, but also to an intensive production of precursors, building blocks for the synthesis of macromolecules: nucleic acids, proteins, lipids, which are required for the formation of daughter cells during rapid proliferation (Ward, Thompson, 2012). In the context of tumor development, it has also been shown that reprogramming of energy metabolism to a longer-term anaerobic variant may play an important role in the generation and development of the "first renegade cell," which becomes the immediate onset of tumor development (Zhang et al., 2013).

Why tumor cells prefer glycolysis: The role of oncogenes and tumor suppressors

In addition to the hypoxia present in tumors, which undoubtedly stimulates glycolysis, it is the action of growth factors that causes tumor cells to reprogram their energy metabolism and switch to less efficient aerobic glycolysis. Similarly, as the unphysiologically high activity of oncogenes and insufficient activity of some tumor suppressors cause the unphysiological, rapid stimulation of tumor cell proliferation (see chapter 3A), this is also the primary cause of the switch in energy metabolism. Indeed, many oncogenes and tumor suppressors play critical roles in controlling energy metabolism in both healthy and tumor cells by, for example, stimulating the expression of GLUT glucose transporters (Bayley, Devilee, 2012; Ward, Thompson, 2012; Zhang et al., 2013). One example is the tumor suppressor p53, the most important guard in cells against tumor transformation. As the "responsible" tumor suppressor, it presents a significant obstacle to the conversion of the cell's energy metabolism to aerobic glycolysis. In fact, it uses several coordinated mechanisms to block glycolysis and, on the contrary, stimulate oxidative phosphorylation. The specific mechanism by which it prevents the Warburg effect is inhibition of glucose transporter expression and activation of the expression of genes involved in oxidative phosphorylation. The tumor suppressor p53 also significantly regulates autophagy, which is closely related to cellular energy metabolism (see below) and plays an important role in tumor development. It is symptomatic that tumor suppressor p53 can both stimulate and inhibit this process, depending on the circumstances (Vousden, Ryan, 2009; Wang et al., 2012; Berkers et al., 2013).

Autophagy: Death or survival?

Autophagy can be considered an extreme form of exploiting one's own internal resources and maximizing the efficiency of ATP production in the presence of limited nutrients. Autophagy was already briefly mentioned in chapter 4A, which deals with programmed cell death. Autophagy (from the Greek "self-consumption") is the process by which a cell breaks down its own structures (large macromolecular complexes and organelles) into simple building blocks that can then be recycled. This degradation takes place in lysosomes, cell organelles that contain a variety of different enzymes that can break down all types of macromolecules – proteins, nucleic acids, carbohydrates, and lipids into amino acids, nucleotides, simple sugars and free fatty acids. The cell may initiate autophagy in response to nutrient, oxygen, and energy deficiencies, or to high levels of damage to its own structures due to aging or oxidative stress.

Autophagy is a form of programmed cell death. It is defined as cell death that occurs without chromatin condensation and is accompanied by the formation of massive autophagic vacuoles. Thus, it is quite different from apoptosis. And there are other differences between autophagy and apoptosis. Apoptosis is accompanied by relatively early cytoskeletal collapse, cleavage of DNA, and early onset of caspase activity. Organelles are not degraded in apoptosis until later stages (see chapter 4A). In autophagy, the reverse is true. Organelles are degraded in the early stages, whereas the cytoskeletal system remains intact for a long time and caspases are also activated later in the autophagy process.

This difference in the dynamics of degradation of cellular structures may be related to the fact that autophagy has a special position among the other forms of programmed cell death. When only some initial stages of the whole process occur, autophagy does not necessarily lead to cell death; on the contrary, it can significantly strengthen the long-term survival of the cell. By enabling the degradation of severely damaged cellular structures and contributing to the restoration of macromolecule synthesis and cellular energy metabolism, it is involved in the maintenance of cellular homeostasis and represents an adaptive mechanism necessary for survival under adverse conditions. Long-term deprivation allows cells to use the autophagy mechanism to exploit their own resources, causing them to gradually enter a dormant state: they shrink to one-third of their original size but retain the ability to regenerate, regain their original size, and return to active cell division (White, DiPaola, 2009; Kroemer et al., 2009; Zhou et al., 2012).

Autophagy in the process of carcinogenesis: A double-edged sword

Under physiological conditions, the induction of autophagy can be, on the one hand, a mechanism that saves the life of the cell, but on the other hand, a mechanism that leads to cell death. And everything indicates that the impact of autophagy on the process of tumorigenesis is dual. The increased stress to which tumor cells are exposed and which damages cellular structures triggers autophagy. The degradation of destroyed structures by autophagy then reduces the extent of cellular damage or leads to their death. In this way, the process of carcinogenesis is inhibited or slowed. Autophagy also allows tumor cells that are under metabolic stress (especially tumor cells with impaired mechanisms of cell death by apoptosis) to survive and enter a resting state and regenerate until the stress is reduced or eliminated. In addition, tumor cells in a dormant state are more resistant to the effects of antitumor therapies. All of this promotes tumor development.

Autophagy is disturbed in many tumors. The effects of this disturbance on tumor development are dual, as is the effect of autophagy itself. The development of some tumor cells is slowed by the failure of autophagy, which reduces their chances of survival under conditions of metabolic stress. In contrast, failure of autophagy stimulates the development of other tumor types. Weakened autophagy reduces cell repair mechanisms, and surviving cells, especially those with damaged genomes, may be stimulated to further transform (see chapter 6A). Autophagy is thus another example of a double-edged sword, a cellular "weapon" that can sometimes act like Dr. Jekyll and sometimes like Mr. Hyde (White, DiPaola, 2009; Zhou et al., 2012).

Mitochondria

Mitochondria have already been mentioned in chapter 4A in the context of apoptosis as organelles involved in the processes of programmed cell death and tumor development. However, the role of mitochondria in the life of the cell is much broader. Mitochondria are instrumental in the energy metabolism of the cell and play a key role in biosynthetic processes, as the mitochondrial Krebs cycle generates substrates for the biosynthesis of amino acids, nucleotides, and lipids. Mitochondria are also involved in maintaining the redox balance of the cell, cell signaling, innate immunity and other processes.

Mitochondria are cytoplasmic organelles that evolved through the process of endosymbiosis. They are semi-autonomous: most mitochondrial proteins are encoded by nuclear genes, but mitochondria also have their own small genome of 37 genes, 13 of which encode 13 proteins that are part of the respiratory chain. Each mitochondrion contains 2 to 10 copies of its genome, and each cell contains 200 to 2,000 mitochondria. The number of mitochondria in a cell is determined by two processes. One is biogenesis, which is complexly regulated and reflects the metabolic state and energy requirements of the cell; the other is mitophagy, a specialized form of autophagy that selectively degrades excess and damaged mitochondria. Mitophagy thus regulates the number of mitochondria in the cell and keeps the mitochondrial population healthy. The mutation rate of mitochondrial DNA is about ten times higher than the mutation rate of nuclear DNA. This is due to the fact that mitochondrial DNA is physically much closer to the source of harmful oxygen radicals generated at the inner mitochondrial membrane during respiration. In addition, it is less protected by proteins and less efficient repair mechanisms than those in the nucleus (Czarnecka et al., 2007; Barbosa et al., 2012; Lane, 2012; Stefano, Kream, 2015; Vyas et al., 2016; Zong et al., 2016).

Reprogramming of cellular metabolism – reprogramming of mitochondria

In tumor cells, mitochondria are involved in the same cellular processes as mitochondria in healthy cells. During the tumorigenic transformation of cells, mitochondria undergo profound changes. In terms of cellular metabolism, these changes lead to a suppression of oxidative phosphorylation and an increase in aerobic glycolysis, i.e. the Warburg effect, but also to other significant biochemical changes. These include, for example, an increase in the consumption of the amino acid glutamine. Increased glutamine consumption was the second significant metabolic change found in tumors after the Warburg effect, leading to the realization that the propensity for aerobic glycolysis is only one part of a complex metabolic reprogramming of tumor cells and metabolic processes that take place in the mitochondria.

In addition, mitochondrial remodeling triggers other metabolic changes. Mitochondrial metabolism is closely linked to the production of oxygen radicals. These, in turn, are involved in the regulation of mitochondrial metabolism, regulation of the number of mitochondria and regulation of cell death by apoptosis. Metabolic reprogramming in tumor cells and their mitochondria generally leads to increased production of DNA-damaging oxygen radicals, which increases genetic instability (see chapter 6A). Mitochondrial DNA itself, which contains only 37 genes, is particularly vulnerable, but mutations occurring in it can affect the process of mitochondrial remodeling and thus the process of cell transformation. In general, mitochondrial remodeling is also associated with increased resistance to induction of cell death (see chapter 4A; Barbosa et al., 2012; Vyas et al., 2016; Zong et al., 2016).

Tumor suppressor p53 and mitochondria

We have already mentioned the tumor suppressor p53 several times, and there is much more to say about it (see chapter 13A). In this chapter alone, we have seen that it strongly interferes with the switch of the cellular energy metabolism to aerobic glycolysis and is involved in the regulation of autophagy. The p53 protein generally achieves these effects by regulating the expression of its target genes in the nucleus, the products of which then interfere with specific regulatory processes. Moreover, the p53 protein acts directly in the mitochondria, where, for example, it can intervene in the regulation of apoptosis and where it is also involved in various mechanisms to maintain the stability of the mitochondrial genome. In chapter 6A we noted that it is referred to as the "guardian of the genome" because of its extensive role in the cellular response to DNA damage and in maintaining genetic stability. For some of its functions in mitochondria, it is also referred to as the guardian of the mitochondrial genome (Berkers et al., 2013; Vyas et al., 2016; Park et al., 2016).

7B. Selfish misuse of resources
8B. Wasting

Tumor cells rely on angiogenesis and misuse the body's resources

Non-angiogenic tumors grow slowly and linearly, reaching a size of only a few millimeters. They can survive and metabolize for a long time, but they cannot grow and enlarge. Proliferation of tumor cells is always balanced by cell death. The spread of non-angiogenic tumors is limited by the availability of oxygen, energy sources, nutrients and growth factors. In contrast, angiogenic tumors grow rapidly and exponentially and can reach considerable size. Thus, activation of angiogenesis represents a fundamental step in tumor development and alters the dynamics of further development. Activated angiogenesis provides the tumor with access to the blood system, i.e. to all the resources of the body. The tumor consumes the resources of the system, but it does not contribute anything back, does not participate in the body's healthy functioning. The tumor is not in harmony with the body and egoistically misuses resources only for its own development. The egoistic misuse of the resources thus represents another, the seventh, non-deadly sin.

The Warburg effect: Wasting tumor cells

Tumors don't just selfishly misuse resources. They actually waste them! They prefer aerobic glycolysis to oxidative phosphorylation, gaining only two molecules of ATP from each molecule of glucose instead of the possible 38 molecules. The less efficiently they use the energy stored in glucose, the more glucose they take from the external environment and wastefully consume. They're wasting! Wasting is the eighth non-deadly sin.

Control of bloodstream: Graded inequality in resource allocation

The function of the blood system is to supply the entire body with an adequate amount of blood. This sounds like a relatively simple and straightforward task. In fact, the blood flow to individual parts of the body changes constantly and dynamically. It

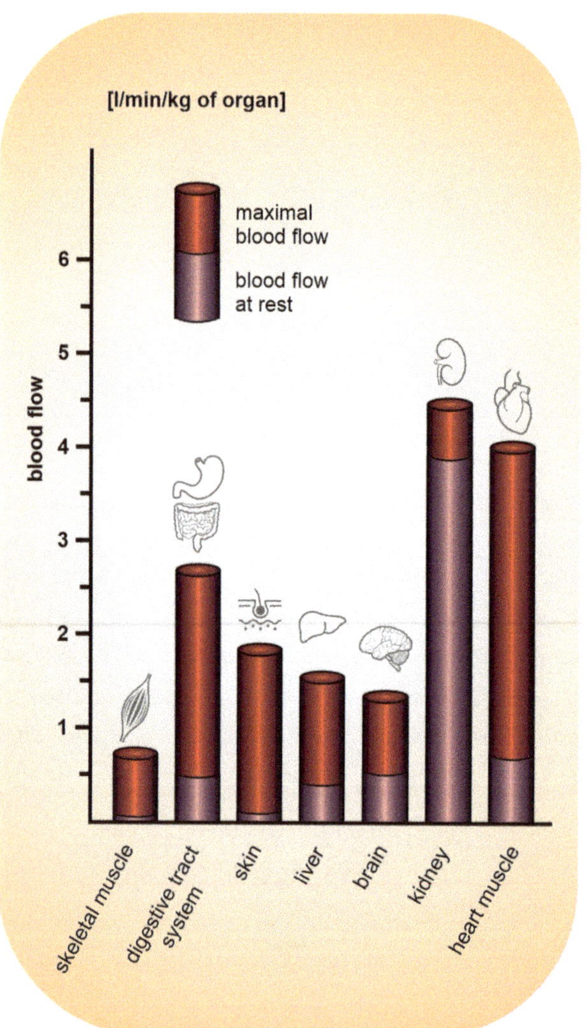

Fig. 32 Blood flow through the human body

The distribution of resources is based on vital importance and immediate demand: uneven, but with meaningful distinctions. The brain is supplied preferentially; it is a vital organ that is particularly sensitive to oxygen deficiency. Blood flow to the heart muscle must not be allowed to decrease, since heart failure would cause the entire circulation to collapse. The kidneys mainly have control and excretory functions; only a fraction of the blood flow in them is used for the kidney tissue's own needs. During heavy physical work, up to 75% of the circulatory system's output flows to the skeletal muscles. During digestion, on the other hand, the digestive system is supplied preferentially. Blood flow through the skin is mainly for heat dissipation (adapted from Silbergnal, Despopoulos, 1993, modified).

must respond to both the fluctuating external environment and the variable activity of individual body organs. All of this occurs within two everlasting, inviolable rules.

The first rule is to maintain minimal blood flow to all organs and tissues in the body. No part of the system should be left without blood supply. Interruption or reduction of the supply of oxygenated blood to the tissues would lead to their death, which is called an infarction. It is obvious that in case of a heart attack, pulmonary embolism or stroke, the life of the entire organism is threatened. In fact, the organism is fatally threatened by an infarction in any tissue. Anemic, necrotic tissue can develop into gangrene (mildew), which is susceptible to bacterial infection. This is life-threatening because the bacteria can spread to other tissues, produce toxins, and cause complete sepsis and death of the entire organism. Therefore, at least a minimal supply of oxygenated blood to all organs and tissues of the body is essential for the life of the entire system.

The second limiting circumstance in the control of blood circulation is the maximum possible cardiac output. It must not be exceeded. Maximum blood flow to all organs at the same moment would almost certainly exceed the capacity of the heart. Therefore, blood flow is preferentially distributed to active organs and less to resting organs. At the same time, the requirements of the organs for blood supply are not completely equal. For example, the brain receives a preferential supply. Not only because it is a vital organ, but also because it is particularly sensitive to oxygen deficiency and nerve cells once destroyed cannot usually be replaced. The blood supply to the heart muscle must also be stable, because in the case of heart failure all blood circulation would cease. The blood flow through the kidneys is enormous – in proportion to their weight. However, this blood flow serves mainly for control and excretory functions, and only a fraction of it is needed for the kidney tissue's own needs. In emergencies, blood flow to the kidneys may be temporarily restricted in favor of the heart and brain. During hard physical work, up to 75% of the circulatory output is needed to supply the skeletal muscles. Similarly, during food digestion, the digestive system is preferentially supplied.

The main purpose of blood flow through the skin is heat dissipation. Therefore, blood flow through the skin is particularly high when heat production is increased or when ambient temperatures are high. The blood flow per unit mass of several organs of the human body is shown in Fig. 32 (Silbergnal, Despopoulos, 1993), demonstrating the basic principles of resource allocation within the system described above. The limited resources are distributed among the organs according to their importance and immediate needs. The resource consumption per unit weight of each organ changes dynamically and varies according to the actual work intensity. However, even at rest, consumption is not the same for all organs. Moreover, the consumption of one organ is precluded from spoiling at least the minimum supply of another organ. Otherwise, the survival of the entire system would be jeopardized, including, of course, the best-supplied organs (Silbergnal, Despopoulos, 1993).

Gradual inequality in the distribution of resources: The Old Testament

In his book *Economics of Good and Evil*, Tomáš Sedláček attempts to answer the question of whether it is worth doing good and whether there is a benefit to doing good. Among other things, he points out that the Old Testament describes a remarkable set of socioeconomic provisions with anti-monopolistic and social implications. These measures aim to prevent the concentration of property and power and, on the contrary, to ensure a minimum level of provision for the poorest. "Every 49 years there was a 'summer of pardon' in which real estate was returned to the original owners according to the original distribution. Debts were to be forgiven, and Israelites who had fallen into slavery because of debt were to be set free. The generation that succeeded a debt-ridden or poor father received their land back and had a chance to start farming again" (Sedláček, 2009, pp. 63–65). In addition, almsgiving and other acts of mercy are commanded in order to support poor people. Every Israelite is obligated to tithe from his entire harvest, a portion of which is to be distributed to the poor and needy. Every seven years the land is left to rest, and what grows in the uncultivated fields is distributed to the poor. The poorest have the right to glean. Field owners are duty bound not to reap the harvest to the last grain, but to leave the leftovers in the field for the poor (Sedláček, 2009, pp. 64–66). In this respect, the Old Testament seems to copy the rules that apply to the distribution of resources in the human body: it does not establish equality, but tries to ensure that inequality is proportionate and reaches – at least to a minimum – everyone.

Positron Emission Tomography (PET): Imaging inequality in resource use

PET is a modern imaging technology widely used in medicine and oncology. It displays the body's tissues and organs based on their varying ability to absorb a radioactive substance with a very short half-life. In oncology, this radioactive substance is usually a modified glucose molecule, 18-fluorodeoxyglucose (FDG), which differs from normal glucose by an incorporated radioactive fluorine isotope, ^{18}F. FDG is related to glucose in such a way that cells take it up from the blood and degrade it in the same way as glucose. However, this works only in the first step of the metabolic chain. The next step cannot be completed. The molecule formed from FDG is different from that formed when glucose is broken down, and does not undergo further degradation. It is temporarily retained in the cell, where it emits radioactivity that is detected by PET camera detectors. FDG-PET thus distinguishes tissues based on how intensively they take up and consume glucose. Those that consume a lot "glow" brighter than the surrounding tissue (the brighter, the more glucose consumed), while quiescent tissues that are not working and do not consume glucose are not displayed (Fig. 33). Tumor cells divide rapidly, and the Warburg effect forces them to take up and consume a large amount of glucose from the environment. Therefore, they also take up FDG intensively and "glow" clearly on PET images compared to the surrounding tissue. PET has been

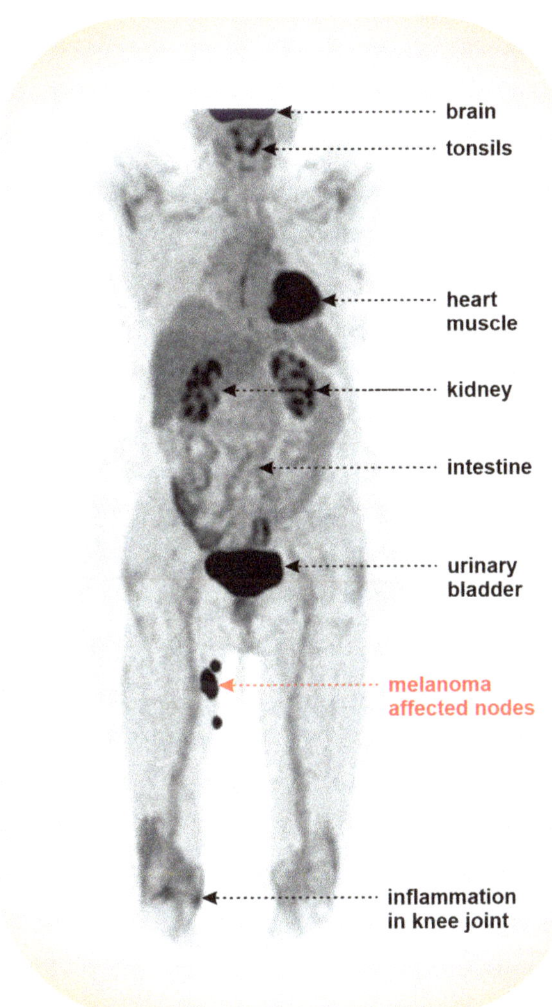

Fig. 33 Positron emission tomography (PET) reveals inequalities in the use of resources in the human body
PET provides an image of the body's tissues and organs based on their differential uptake of an administered radioactive substance, in this case modified radioactive glucose FDG. In this image of the body of a 47-year-old woman, a high physiological accumulation of glucose can be seen in the brain, tonsils, heart muscle and intestines. Radioactive glucose is excreted in the urine, and therefore the kidneys and bladder are clearly visible. A significant tumor was noted in the right thigh, namely a nodule affected by melanoma. An accumulation of radioactive glucose was also noted in the right knee joint, indicating ongoing inflammation.

used very successfully to detect tumors, both primary and metastatic (Fig. 33). It is hard to imagine a more vivid demonstration of how intensely tumors consume the body's resources than the images from FDG-PET.

The post-industrial society: From graded inequality to absolute incommensurability

Much has changed since the days of the Old Testament and its anti-monopolistic and social strictures. In his book *Three Social Worlds*, sociologist Jan Keller looks at the changing social structures of contemporary society. He notes that since the last quarter of the 20th century, economically developed countries have entered what he calls a post-industrial phase in which the differences between the various social classes are critically intensifying: "We are leaving a society of continuously graded inequality and entering a society in which property, power, and social status are absolutely incommensurable" (Keller, 2011, p. 9). Jan Keller sees no problem in the fact that wealth and poverty coexist. They have always existed. "The problem is that the hitherto continuous pyramid of wealth, power, and prestige is disintegrating and special social worlds are separating at its two ends. And they are moving further apart." He demonstrates this with the example of the richest Frenchman, whose annual income is equal to that of 15,700 of his fellow citizens who receive the minimum wage (Keller, 2011, pp. 11–12).

In its report on the state of global wealth in 2020, the world-renowned Credit Suisse Research Institute notes that the wealth gap between the richest and the poorest is considerable (Credit Suisse Global Wealth Report 2020). The top one percent own 30.5% of the world's wealth; the richest ten percent own 69% of the world's wealth. According to Jan Keller, this is a consequence of the fact that since the 1980s, in virtually all economically developed countries, the proportions of social wealth distribution have changed significantly in favor of the upper class. As a result, the gap is not even, but is widening. Plato, who lived from 427 to 347 B.C., believed that the common good requires that the wealth gap between the richest and the poorest not exceed the ratio of 5:1. The American banker John P. Morgan, who lived and worked at the turn of the 19th and 20th centuries (1837–1913), still believed that the head of a company should not receive more than twenty times the income of his employees (20:1). In 2009, this ratio was already 464:1 (Keller, 2011, p. 21).

Absolute incommensurability: The success of *Capital in the 21st Century*

It seems that the topic of "disproportionate inequality" or "absolute incommensurability" as Jan Keller (2011) called it, its sources, causes and effects, is extremely timely. It interests and worries many. This is evidenced, for example, by the success of Thomas Piketty's book *Capital in the 21st Century*. This comprehensive work by the French economist is about the deepening of global wealth inequality, which he considers one of the greatest dangers to the current system. Inappropriate inequality, however, is not an achievement limited to the historical present. In his book, Piketty opposes, among other things, the so-called Kuznets curve, conceived by Simon Kuznets in the 1950s and 1960s. The Kuznets curve illustrates the relationship between economic growth and social inequality. According to Kuznets, capital behaves rationally. When it is concentrated in the hands of the rich minority, it promotes economic growth.

When the concentration of capital in the hands of a small group of people reaches an unhealthy level, it spreads by itself to broader layers of society on its own. Social inequality automatically decreases through self-regulating forces, and the resulting economic growth benefits everyone, including the poor. Thus, one only needs to be patient with capitalism, because it has a natural ability to keep inequality in check. But Piketty disagrees with Kuznets' conclusions. He shows that the value of capital increases much faster than economic growth. The greater the wealth, the faster it valorizes. The net result is that capital is and will remain concentrated in fewer and fewer hands. Inequality will continue to rise. And Piketty points out that too much inequality is dangerous for society and can threaten the entire system (Piketty, 2014; Šimečka, 2014; Bajgar, 2014).

Thomas Piketty's book was a great success and became a world bestseller (Šimečka, 2014), triggering criticism and numerous discussions (Schneider, 2014; Škabraha, 2015). Whatever his admirers and supporters or his critics have true, one thing is absolutely clear: the theme of deepening social inequalities poses a significant threat, and is hot and very urgent topic today.

Socioeconomic inequality and the Gini coefficient

And this is no wonder! The unequal distribution of resources, or the extent of this inequality, has an enormous impact on the life of society. This can be seen, for example, in the fact that in order to assess poverty, which can be measured with several scales, the so-called Gini coefficient and the human poverty index (HPI) are the most common. While the HPI is calculated based on factors like the number of people in a given community who die at a young age, lack education, lack access to drinking water, lack health care, etc., the Gini coefficient determines the degree to which the distribution of resources in a given community (e.g. a territorial unit) deviates from a perfect distribution. If the Gini coefficient is zero, resources are evenly distributed among all segments of the population. At a value of one, one person receives all the resources, while everyone else has nothing (Koukolík, 2010, p. 121). Therefore, it is clear that poverty can mean both real and absolute deprivation and misery (as reflected in the HPI), and "disproportionate inequality," i.e. relative poverty, compared to an ideal distribution. Economist Branko Milanović, an expert on inequality, confirms that interest in the issue is indeed increasing nowadays. And he acknowledges that if the sharp increase in the salaries of a narrow and already rich elite causes discontent, this need not be due to envy, but may be an expression of the "normal human sense of justice" (Lindner, 2015). František Koukolík, a neuropathologist who has long studied the intersection of his field with social issues, believes that in a society where inequality in the distribution of power and property is too great, i.e. where the socioeconomic gap is too wide, those who are worse off experience increasing stress: "The world as a whole is getting richer, but what matters is not how rich or poor someone is, but how steep the social ladder is. The steeper it is, the more signals are triggered in the brain that something is wrong" (Koukolík, Riebauerová, 2013).

Tumors: Graded inequality or complete incommensurability?

In chapter 7A on the control of blood circulation, it is explained that the consumption of individual tissues and organs in the human body changes over time and even at rest consumption is not uniform. Distribution is uneven, but graded according to the importance of the organs and their actual work. What about tumors? What do they mean in terms of consumption in the system in which they occur? We already know that tumors corrupt the system of resource allocation to serve themselves selfishly and without benefit to the system. What about the extent of their consumption and waste? And what are the consequences for the body? From the FDG-PET images, it is clear that tumors place a greater or lesser burden on the body depending on their size and rate of growth, but their level of consumption is not fundamentally life-threatening (Fig. 33). After all, the control of blood circulation constantly responds to changing internal and external conditions and the overall condition of the organism and can provide adequate redistribution of blood flow to the individual organs. With the presence of a tumor, the circulatory system thus has "only" one more "item" in the body to control, albeit a very greedy one that can hardly be satisfied from limited resources. All the organs of the body have to "just" shrink a little for its sake. But is it really that simple?

Tumor cachexia

Cachexia or breakdown of the organism is dangerous and rapid weight loss accompanied by considerable weakness. Cachexia is defined as the loss of 10% of body weight within six months (Penet, Bhujwalla, 2015). It occurs in association with many diseases, but affects more than 80% of cancer patients. It decreases quality of life and the chances of recovery. Cachexia is directly responsible for 20% of cancer patient deaths. Severe cachexia can occur in patients with a tumor not exceeding 1% of their total body weight. The development of cachexia is not dependent on tumor size (Loberg et al., 2007; Fearon et al., 2012; Petruzzelli, Wagner, 2016).

Thus, cachexia is certainly not the result of simple consumption by the tumor itself. It does not occur because of the energy diverted by the tumor for its immediate survival. In general, one of the signs, and at the same time one of the possible causes, of cachexia is anorexia, a problem of food intake. However, the mechanism of tissue loss in cancer patients is completely different from that caused by simple starvation. While starvation mainly affects adipose tissue, cancer cachexia mainly affects skeletal muscle. Moreover, treatment for anorexia, such as intravenous nutrition, usually cannot reverse cachexia in cancer patients. This is further confirmation that reduced caloric intake is not the main cause of cachexia. In tumor cachexia, skeletal muscle is primarily affected, but other organs are also significantly affected: the liver, pancreas, heart, adipose tissue and brain. Tumor cachexia is therefore defined as a systemic, complex, multifactorial metabolic syndrome associated with profound alterations in energy balance and sugar, fat and protein metabolism. A catabolic metabolism develops that is focused on the breakdown, decomposition of more complex substances into simpler ones, and release of energy. Moreover, this state is very resistant to anabolic signals that could reverse it. Why does it all happen? How can a tumor that represents less than one hundredth of the total body weight fundamentally change the metabolism of the entire body, forcing it to function in its favor and subordinate the system to its own interests?

It has been found that these effects can be triggered by molecules that tumors release into their environment. The first molecule capable of inducing cachexia was isolated from a medium in which tumor cells were grown and was named cachectin (Beutler et al., 1985). Cachectin belongs to the cytokines, the signaling proteins that mediate communication between cells of the immune system and that are involved in the immune response (chapters 9 and 10 discuss the role of the immune system in tumorigenesis in more detail). It gradually became apparent that the key factors leading to the development of cachexia are, first, cytokines produced and released by the tumor and, second, cytokines released by the body's immune system in response to the tumor. Cachexia thus illustrates that tumors which initially arise as small, localized foci of uncontrollably dividing cells, gradually develop into systemic disease. That is, a disease that affects the entire organism. (These stages of tumor development will be discussed in more detail in later chapters.) Tumor cachexia and the associated initiation of a catabolic state result in a negative energy balance and the release of nutrients into the bloodstream. As the system is gradually degraded and destroyed, the highly proliferating and energy consuming tumor cells consume the nutrients released into the bloodstream and use them for their own support (Loberg et al., 2007; Penet, Bhujwalla, 2015; Petruzzelli, Wagner, 2016; Porporato, 2016).

In Crow Canyon: The Demise of the Anasazi

The Anasazi lived in the southwestern United States and northeastern Mexico. They were small farmers who grew corn and occasionally hunted. They wove cloth and made elaborate baskets. They were also capable of building large pueblos and complex ceremonial spaces. Their culture lasted for about 700 years. Throughout their existence, they had to deal with severe droughts. Therefore, granaries built into high rock overhangs were among the most common structures of the Anasazis. They were used to store grain supplies to be distributed in times of shortage. In this way, Native Americans created a society based on sharing and mutual aid; the society was poor and relatively homogeneous. At the end of the 13th century, the dense Anasazi settlement disappeared rather suddenly and the cornfields turned into a semi-desert. The traditional explanation for the demise of this culture was drought, which was frequent at the time. However, an analysis of tree growth rings revealed that the area had been hit by droughts more than a hundred times in its seven hundred-year history. Thus, Native Americans were able to effectively cope with recurrent drought. According to a recent explanation, at the end of their cultural period, it was not only the climate that contributed to the Anasazis' demise, but also their changing social model. "The Anasazi began to build large buildings and form elites. The climate became more chaotic, but they themselves also lost the ability to share equitably. At the end of the 13th century, a great drought came, some people moved away, others joined with surrounding tribes, but most of the population died out. The poor died first, and a few years later the elites died out too" (Cílek, 2015).

Angiogenesis: Cooperation between tumor and non-tumor cells

Activation of angiogenesis represents a crucial turning point in tumor development. Non-angiogenic tumors grow slowly and reach the size of a pinhead, while angiogenic tumors grow rapidly and can reach a considerable size. Judah Folkman found that tumors are dependent on angiogenesis. Tumor properties such as deregulated cell division, resistance to programmed cell death, immortality, and high genetic instability can be explained at the tumor cell level. These are properties of the tumor cells themselves. On the other hand, tumor angiogenesis cannot be explained without also considering the direct involvement of non-tumor cells, e.g. endothelial cells, pericytes, and many others. This implies that the involvement and cooperation of non-tumor cells is an essential prerequisite for further tumor development. Tumors depend on the cooperation of non-tumor cells. If the non-tumor cells do not cooperate with the tumor, it cannot develop into a dangerous form. The organism is not threatened by the presence of clones of tumor cells. However, cooperation with tumor cells can be lethal for the organism. Why does this happen? What makes non-tumor cells cooperate with the tumor? What role do the tumor cells themselves play in this? Do they learn to communicate with non-tumor cells and signal them to tolerate cooperation with their tumor sisters and brothers? Or is there a weakness in the vascular system itself?

What makes the vascular system cooperate with the tumor? Is there a weakness in the regulation of angiogenesis? Sensitivity of bottom-up management?

The main function of the vascular system is to distribute resources. In a well-functioning system of a multicellular organism, resources are not distributed evenly, but are graded according to immediate need and importance. The whole system with all its cells is adequately supplied. It is reasonable to expect that a prerequisite for successful performance of such a task is constant monitoring of the supply status of all the "customers" to find those that are insufficiently supplied and to quickly restore or increase the supply to them. And this is exactly what happens. When oxygen consumption in tissues increases, its concentration decreases, resulting in local hypoxia.

Hypoxic cells send the appropriate signals about this state of emergency; for example, they secrete the vascular endothelial growth factor VEGF, which stimulates the ingrowth of capillaries into the tissue – angiogenesis. Thus, oxygen deprivation itself directly increases the likelihood that angiogenesis will be initiated. In this way, the physiological balance in healthy tissue is maintained or restored.

But the same mechanism is used or rather abused by the tumor. The rapid expansion of tumor cells with uncontrolled cell division and low propensity for programmed cell death leads to an accumulation of cell mass, with many cells too far from the blood capillaries. This creates many "dissatisfied" hypoxic cells that send out distress signals. It is almost only a matter of time before the blood capillaries begin to grow through a tumor, before the vascular system completes the "recipient" network even at this newly formed, inadequately supplied site. This is an abuse of the feedback-based system that otherwise seems to work well under physiological conditions: sufficient oxygen inhibits further development of the vascular system, while lack of oxygen promotes it. This ensures constant optimization of resource supply. The supply of resources is the response to demand.

It seems that this system is vulnerable. Would it not be better if the distribution of resources were fixed, as well as the structure of the vascular system and the density of the blood capillaries? This would effectively prevent the development of abundant, non-physiological structures. Resource consumption and structure development would be limited by supply. According to evolutionary biochemist Nick Lane, "the network is always controlled by demand, and not the other way around" (Lane, 2005, p. 173), and one possible reason is oxygen toxicity. Too much oxygen in tissues forms reactive free radicals that cause significant damage to cellular structures (see chapter 6A). The situation where supply exceeds demand is risky. "The best way to prevent such free radicals from forming is to keep tissue oxygen levels as low as possible." Therefore, the density of the capillary network is constantly adapting flexibly and responding to the demands of the tissue (Lane, 2005, p. 172). Thus, the good functioning of this system is based on the "solidarity" and the adequacy of the demands of the parallel network members. The mechanism is flexible and quite "democratic," it is an example of management from below. But at the same time, it is vulnerable and exploitable, since it depends largely on balancing the demands of all system members. And the tumor, the selfish, disordered tissue, consumes the resources ruthlessly.

What makes the vascular system to cooperate with the tumor: The contribution of the tumor cells themselves

The contribution of tumor cells to the induction of angiogenesis thus seems clear. By their uncontrolled proliferation, tumor cells create a demand that is sooner or later satisfied. This is an indirect but effective pathway based on how supply and resource allocation work in a multicellular organism. However, tumor cells have been found to stimulate angiogenesis in a different and more direct way. There are a number of oncogenes and tumor suppressors that are able to control both the cell cycle (see chapter 3A) and angiogenesis by regulating the expression of the corresponding control genes. The regulation of cell division is linked to the regulation of angiogenesis. The same factors that stimulate cells to proliferate also increase the likelihood that the cell mass thus formed will become cross-linked with blood capillaries. Under physiological conditions, this is a "logical" relationship. For example, if cells proliferate rapidly during tissue development or healing, it is quite "reasonable" that the vasculature would develop simultaneously to immediately supply newly formed tissue with nutrients. Unfortunately, this "preset" coupling persists under non-physiological conditions when the continuous stimulation of cell proliferation is not due to a genuine need reflecting the interest of the whole system, but due to damaged oncogenes and tumor suppressors, and the likelihood of initiation of angiogenesis increases.

Complicating matters further, stimulation of proliferation is similarly linked to control of energy metabolism. While resting cells have only a small "store" of glucose and are therefore forced to use it efficiently by oxidative phosphorylation or to use their internal resources efficiently (e.g. by inducing autophagy), cells that are induced to proliferate rapidly are coupled and supported by stimulated glucose import. This allows them to switch to aerobic glycolysis, which is faster and is also a source of building blocks for the construction of new cells. This is not a waste, but a wise investment in developing or restoring the structure and function of tissues and organs, which is in the interest of the whole body, the whole system. However, when this switching is due to damaged oncogenes and tumor suppressors, it becomes a meaningless support for selfish, exploitative, and wasteful tumor cells. It does not serve the system in any way but, on the contrary, threatens it to a considerable extent.

Excessive speed – an excuse for abuse and waste?

In chapter 6 we saw how rapid, uncontrolled cell division increases genetic instability. Too much speed (the first non-deadly sin) leads to rule violations through what is called replication stress and increases chaos (the fifth non-deadly sin). We now see that rapid cell division is also associated with the stimulation of angiogenesis and the switching of energy metabolism. Thus, disproportionate, excessive speed carries the danger of misuse of resources (sixth non-deadly sin) and the danger of waste (seventh non-deadly sin), in addition to the danger of disobeying rules. Similar connections are reflected in the already mentioned quote by Carl Honoré: "Modern capitalism accelerates more than is appropriate, because the pressure to be the first at the finish line does not leave enough time for quality control. In recent years, software vendors have become accustomed to rushing their products to market before they could be fully tested. The result is an epidemic of computer failures, glitches and malfunctions that cost companies billions of dollars every year. And then there are the human costs of turbo-capitalism. Doctors' offices are flooded with people suffering from stress-related illnesses: insomnia, migraines, hypertension, asthma, gastrointestinal problems, burnout syndrome" (Honoré, 2012, p. 12).

In short, excessive, inappropriate speed is insidious and risky. If we do not recognize in time the self-serving mentality and egocentricity of an inappropriate expansion of one part of the system, there is a risk that this "renegade" component will begin to misuse resources, waste them, grow at the expense of other components, and jeopardize the survival of the entire system. We can – like the Anasazi – lose everything! And the elites' share in this loss is apparently quite logical, perhaps even legitimate. Elites who control the distribution of resources and can claim a large portion of them for themselves are at the same time – protected by their advantages and luxuries – much less exposed and sensitive to the growing problems in society, and can be almost blind to the impending collapse (Parsley, 2014).

Excessive speed – the source of all problems?

Carcinogenesis is a multistep process. Damage to genes and development of the typical features of tumors occur gradually. In the Introduction, we noted that none of the eleven typical features of tumors causes a tumor alone and is therefore not lethal by itself to the affected organism (chapter 2B). However, it appears that damage to cell division regulation is more important than other tumor features and has more far-reaching consequences. Loss of cell division regulation is not only itself associated with two typical features of tumors (acquisition of self-sufficiency in growth signals and insensitivity to cell cycle arrest signals), but also contributes significantly to the development of three other features. Thus, even if we can answer the question posed in the heading of this paragraph in the negative, i.e. that excessive speed is not the cause of all tumor-like problems, it does carry the risk of developing several of them. It is therefore not surprising that senseless speed, man's race with himself, is "only" one of the non-deadly sins in the context of our overlaps, but for Konrad Lorenz it is a deadly sin (Lorenz, 1974, pp. 26–31).

Haste makes waste, or the value of harmony

Reasonable, appropriate, "just right" speed has another extremely valuable effect besides those already mentioned: it is a prerequisite for harmony with the natural environment, with one's surroundings, with other parts of the system. In the Introduction to this book, we quoted Konrad Lorenz, who said that we often recognize the importance of an organic system only after a pathological disturbance has caused its disease (Lorenz, 1974, p. 5). This is perfectly illustrated by the vascular system of tumors. The tumor vascular system is chaotic, highly disorganized, irregularly structured, and has unreliable function (Fig. 26). It is when we look at the blood capillaries of a tumor that we often become aware of the perfection and harmony with which the normal, physiological, healthy vascular system functions. It develops in a coordinated manner in tandem with the needs of other tissue components. The coupling of cell division regulation, angiogenesis induction, and control of energy metabolism, as well as the feedback system that reinforces them (e.g. through the induction of hypoxia), illustrate how closely the development and growth of all tissue components, often very different from each other, are interconnected. Contrarily, the irregular, chaotic vascular system in tumors illustrates and reminds us what the "fast and furious" proliferation of "elite" tumors leads to. It effectively forces the vascular system to "cooperate" and access resources, but produces only chaos, confusion, inefficiency and waste. It is disharmony!

9A. Formation of metastases

Metastases are the main cause of death in cancer patients

Tumors initially arise and develop as small, localized foci of redundant cells that accumulate due to excessive cell division and/or insufficient death (Fig. 16). They gradually acquire features typical for tumors and can develop from a localized focus into a systemic disease affecting the entire organism. Benign tumors remain at their site of origin; they cannot grow through the surrounding tissue. They are usually slow growing, encapsulated, and morphologically resemble the original tissue, and their cells are usually well differentiated. Malignant tumors invade surrounding tissue from the site of origin and may also form distant secondary deposits – metastases – in other parts of the body. Their boundaries are difficult to define, and their cells often grow rapidly, spread and gradually lose resemblance to the original tissue.

Metastases, then, result from the spread of tumor cells from the site where they first appeared and begin to develop in other parts of the body. During this process, tumor cells released from the primary tumor travel through the vascular or lymphatic systems or through body cavities and form new tumor foci in other parts of the body. These new tumors are formed by tumor cells of the same type as the primary tumor. For example, if a breast tumor spreads and metastases form in the lungs, the tumor cells in the lung metastases still have the characteristics of breast cells. They may be significantly changed compared to the cells in the primary tumor, but they are not lung cells.

About seventy percent of cancer patients have detectable metastases at the time of diagnosis, and about 90% of all cancer-related deaths are due to metastases. However, even tumors that do not spread to surrounding tissue and do not metastasize can kill the patient. Death can result directly from failure of the organ in which the primary tumor develops. For example, if the organ is an endocrine gland, the developing tumor may significantly affect hormone production leading to failure of the entire organism. In general, however, tumors are less dangerous and easier to treat as long as they are localized and encapsulated (Loberg et al., 2007).

The metastatic cascade

The formation of metastases is a multistage, complex and dynamic process. Its course may vary depending on the type of tumor, the patient, and the interactions between the tumor and the organism, but it can generally be divided into several typical steps. The first step is (1) the release of tumor cells from the primary tumor and their migration into the surrounding area (local invasion). In the next step, the cell invades the surrounding tissue and passes through the vessel wall into the blood (or lymph) stream, which is called (2) intravasation. Then, the tumor cell must be able to (3) circulate and survive in this stream and later (4) leave it, penetrate the vessel wall by extravasation and invade the surrounding tissue. At its destination, it must be able to (5) survive, form a micrometastasis, and then divide to form a clinically detectable, vascularized, growing secondary tumor focus, called a macrometastasis, by a process we refer to as (6) colonization of a new environment (Fig. 34). These events need not be in smooth succession; the dynamics and even the course of the metastatic cascade can vary widely among different tumor types. In many tumors, the ability to form metastases is the result of a long development and progressive transformation of tumor cells (gradual progression). In others, however, cells migrate very early, and metastases develop independently from various cells, forming cells that are distinctly different from the primary tumor (parallel progression). While most tumors spread through the blood or lymphatic systems, some malignancies metastasize in other ways. For example, malignant gliomas arising from glial cells of the central nervous system tend to spread within normal brain tissue rather than metastasize to distant parts of the body. While many tumors metastasize rapidly after the metastatic cells reach a new environment, other tumor cells may remain in a dormant state at secondary sites for a long time (even many years). Thus, metastasis is not only a very complex process, but also a multiform one that can proceed with different dynamics and in different ways (Cotran et al., 1999; Weinberg, 2008; Hesketh, 2013).

Cancer: Invasive carcinoma

The process of invasion and metastasis also has its peculiarities in carcinomas. Carcinomas are tumors that develop from epithelial cells. They account for more than 80% of all tumors in humans and occupy a certain special position among them. Thus, the term cancer originally did not refer to a tumor disease. It referred only to carcinomas, i.e. epithelial tumors, and only to those of an invasive nature. Today, however, this delineation is blurred, and in common usage the word cancer is increasingly interchanged with the terms tumor or neoplasm (Žaloudík, 2008, pp. 12–13), although the original, narrow definition of cancer as invasive carcinoma is still retained by some pathologists and surgeons today. Other authors argue that the word cancer should refer to any malignant (i.e. metastatic) tumor (Lazebnik, 2010).

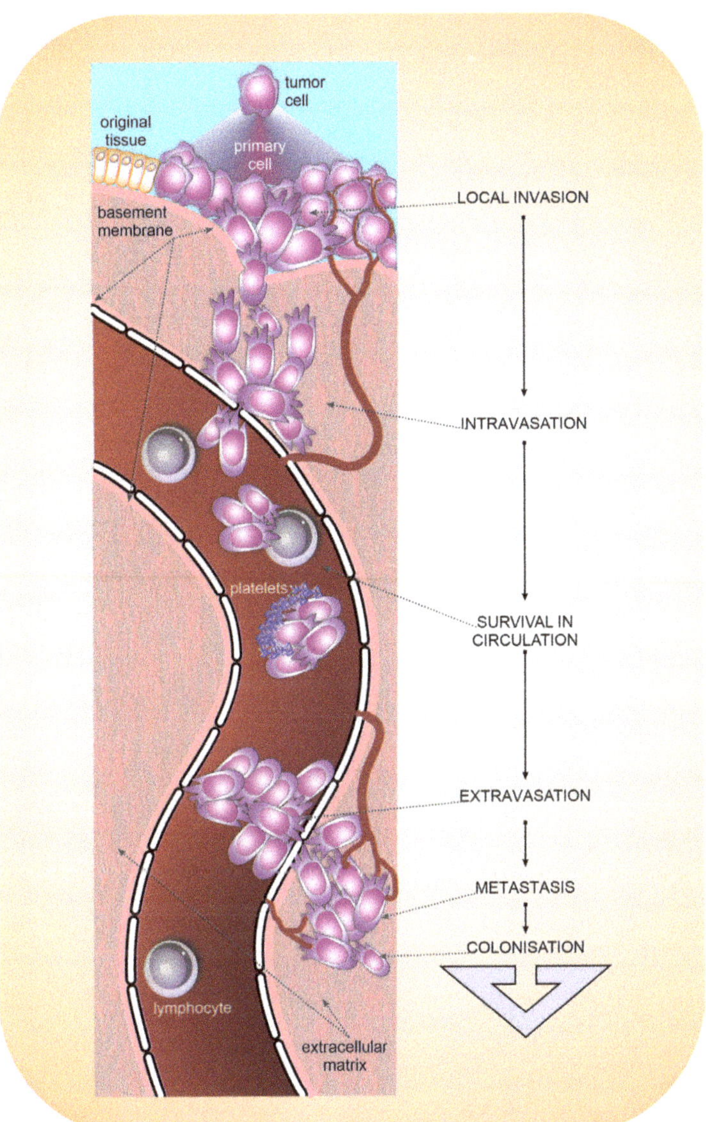

Fig. 34 Metastatic cascade

The formation of metastases is a multistep process. The first step is local invasion, in which the primary tumor cells are released and migrate into the immediate environment. In the next step, the metastatic cell invades the surrounding tissue and penetrates through the vessel wall into the blood capillary (intravasation). It then migrates through the vasculature, exits by extravasation, and invades the surrounding tissue. At the target site, it must be able to survive and form a micrometastasis. Colonization is the stage at which tumor cells can divide and form a clinically detectable, vascularized secondary tumor focus, a macrometastasis (adapted from Buck, 1995; Cotran et al., 1999, modified).

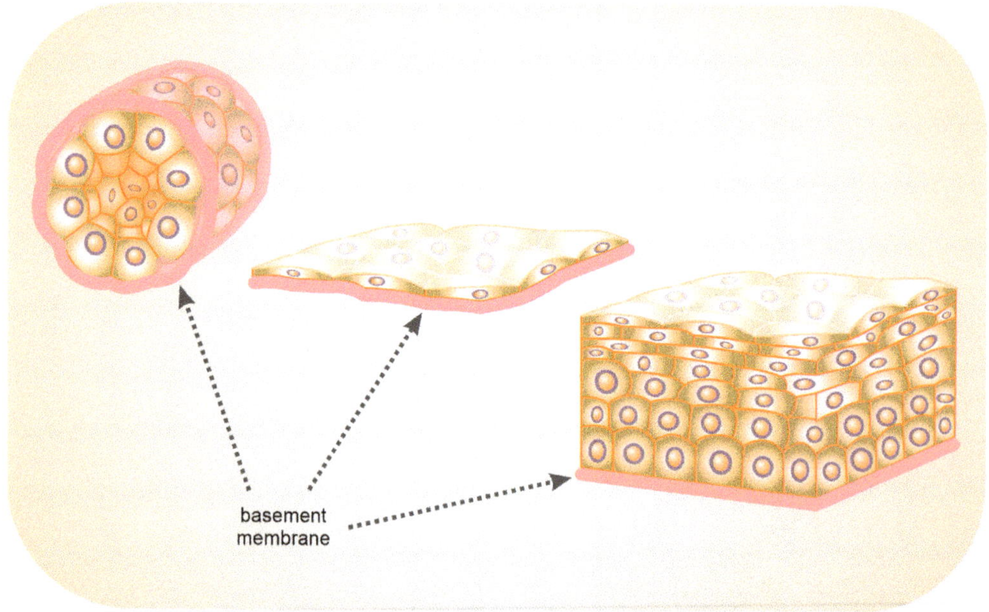

Fig. 35 Structure of the epithelium

The epithelium consists of a continuous layer or layers of epithelial cells and only a minimum of extracellular matrix. Epithelial cells are firmly connected attached to each other. They sit on the basement membrane, which separates them from a layer of supporting connective tissue. The basement membrane is the scaffold on which the polarized epithelial cells are arranged.

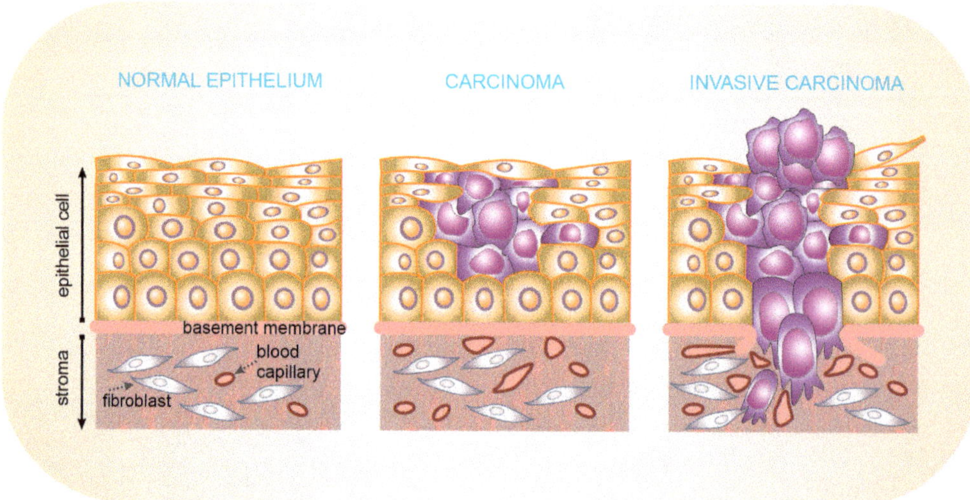

Fig. 36 Carcinoma

The development of a carcinoma begins on the epithelial side of the basement membrane. As long as it stays on this side, it is considered benign. Malignant tumors break through the basement membrane and translocate it, and the released cells migrate and invade the surrounding stroma (adapted from Hanahan, Folkman, 1996, modified).

Carcinomas: Tumors arising from epithelial cells

Epithelia are one of the four basic types of animal tissue (along with connective tissue, muscle tissue, and nervous tissue) and cover most of the external and internal surfaces of the body. The epithelium has a typical structure (Fig. 35). It consists of a continuous layer (or layers) of epithelial cells and only a minimum of extracellular matrix. The epithelial cells are tightly attached to each other, which gives mechanical strength to this tissue. They adhere to the basement membrane, which separates them from the supporting connective tissue layer, the stroma. The basement membrane is a special type of extracellular matrix that serves as a kind of scaffold on which polarized epithelial cells are arranged. It is composed of proteins secreted mainly by epithelial cells, but also by stromal cells. Epithelial tissue is completely avascular, i.e. it lacks any vascular supply, and its cells are nourished by diffusion from the stromal side through the basement membrane.

The development of carcinomas always begins on the epithelial side of the basement membrane. As long as the tumors remain on this side, they are considered benign. However, many carcinomas sooner or later acquire the ability to break through the basement membrane and invade, and individual released cells or small clusters of cells begin to migrate and invade the surrounding stroma. Thus, the tumors become malignant (Fig. 36). Disruption of the basement membrane removes a physical barrier that impedes tumor invasion and thus greatly increases the chance of tumor cells successfully passing though the next stages of the metastatic cascade.

By settling in the stroma, tumor cells also gain direct access to blood and lymphatic vessels. Transforming epithelial cells can stimulate angiogenesis on the stromal side even before they cross the basement membrane. They can affect endothelial cells by releasing angiogenic factors that can cross the basement membrane. By breaking through the basement membrane and gaining direct access to blood vessels, tumor cells not only gain an easier supply of oxygen and nutrients, but also direct contact with endothelial cells and the possibility of intravasation, i.e. entry into the vascular system. During intravasation, at least two cell types cooperate significantly with tumor cells, namely endothelial cells, which form the inner lining of blood vessels (see chapter 8A), and macrophages, which belong to the immune system.

Once the tumor cell enters the vascular system – blood or lymph – it becomes a circulating tumor cell and can be transported by blood or lymph to other parts of the body. However, this long-distance transport is associated with great risks to the tumor cell. Attachment to the extracellular matrix and/or neighboring cells may still be required by tumor cells, and the loss of connection may trigger anoikis, programmed cell death induced by loss of cell adherence (see chapter 4A). Primary tumor cells also often depend on various types of stromal support, such as growth factors and mitogenic or survival signals provided by their specific microenvironment. When they leave the primary tumor, the migrating cells lose this support. In addition, the blood itself represents an actively hostile environment for tumor cells, in which they must resist the physical forces associated with blood flow and pressure as well as the action of cells of the immune system. The likelihood of circulating tumor cell survival may increase if they manage to obtain support from platelets that surround them, forming a protective shield.

Circulating tumor cells often get stuck in tiny blood capillaries smaller than their diameter. This brings them into close contact with endothelial cells, which can initiate the process of extravasation by which they leave the vasculature. This process also requires complex interactions of tumor cells trapped in capillaries with other cells and structures. Platelets and endothelial cells are involved, the basement membrane of the capillary must be breached, and only then can the tumor cells begin to invade the surrounding structures.

Only a minimum of invaded tumor cells can survive in the newly penetrated tissue because the foreign tissue microenvironment is not particularly hospitable for newcomers, lacking the right composition of growth and survival factors. Stromal cells and ECM components are also hostile to the invaders, and the architecture of the tissue itself may present challenges. Therefore, metastatic cells rarely survive without the ability to grow and proliferate. They initially form only small clusters of cells, micrometastases. Only some micrometastases develop into metastases, macroscopic, detectable deposits of tumor tissue. The development of micrometastases into macroscopic, detectable metastases is called colonization. It is a specific, unique stage of the metastatic cascade that is rather inefficient.

The efficacy of the metastasis cascade

The entire metastasis process is very inefficient at the cellular level. Some phases of the metastatic cascade are more efficient (e.g. survival in the bloodstream, extravasation), but once the tumor cell leaves the vasculature and invades the foreign tissue, the efficiency of the entire process is greatly reduced. The least effective phase is probably colonization, which is thus the limiting step of the entire cascade (Valastyn, Weinberg, 2011). For example, mouse tumor experiments have shown that approximately one million cells are released into the vasculature daily from a tumor weighing approximately 1 gram and containing of 10^9 cells. At the same time, the visible metastases can be counted on the fingers of one hand. Therefore, the probability of a single tumor cell successfully undergoing cascading metastasis is very low. This is because most cells in the adult body are part of a specific tissue throughout their lives, they do not move, and this location in the body is carefully controlled. The tumor cell that successfully goes through the entire metastatic cascade must undergo significant changes during this process and acquire a number of new properties. This includes damage to the mechanisms that control the correct position of the cells in the organism. Loss of position control is a necessary prerequisite for the initiation of a malignant state. It is necessary for tumor cell migration from the primary tumor, intravasation, transport through the vascular system, extravasation, and colonization. Position control is closely related to the function of cell adhesion complexes.

Cell adhesion

Cell adhesions are responsible for the binding of cells to the extracellular matrix and/or to neighboring cells. This process is mediated by molecules called cell adhesion molecules, adhesion receptors, or cell adhesion complexes that penetrate the cytoplasmic membrane. These include, for example, cadherins, integrins, selectins, and many others. Cell adhesions are absolutely irreplaceable for the formation and maintenance of multicellular structures; they are essential for the organization of tissues into functional systems. They serve to physically connect cells and transmit signals, to communicate. The cell can only be part of the tissue and survive in this place if it is sufficiently connected to the structures surrounding it – the neighboring cells and the extracellular matrix. Without this connection, it cannot survive and dies by programmed cell death, anoikis.

A typical adhesion receptor (Fig. 37) consists of an extracellular, a transmembrane, and a cytoplasmic part. The extracellular part binds the ligand. This can be a structural motif of a component of the extracellular matrix, an extracellular part of the adhesion receptor of the neighboring cell, or a soluble signal molecule. The transmembrane part of the receptor anchors it to the cytoplasmic membrane. The cytoplasmic intracellular part of the receptor interacts with the cytoskeleton. The cytoskeleton is a network of interconnected fibers and tubes that penetrates the cytoplasm from the nucleus to the plasma membrane. Among other functions, the cytoskeleton provides firm support for the cell, determines its shape, and protects it from mechanical damage.

Adhesion receptors connect the cell cytoskeleton to extracellular structures – the ECM and surrounding cells – and are critical for the formation of tissue strength and resistance. In addition, the cytoplasmic part of the adhesion complex also contains the molecules involved in the transduction of signals from the receptor to the interior of the cell. Signal transduction between a cell and its immediate environment provides information about its position and location in the organism. Only a correctly anchored cell that receives the "right" survival signals can survive. In the wrong location and without a supportive environment (which can be stimulating to other cells), the cell dies by anoikis. Only the cell that can attach and receive supportive signals is in the right place in the organism. It is "at home." This fact is well illustrated by the word "anoikis," which means "homelessness" in Greek. A homeless cell that wanders around in the wrong place in the organism and is lost "in the world" dies (see chapter 4A; Buck, 1995).

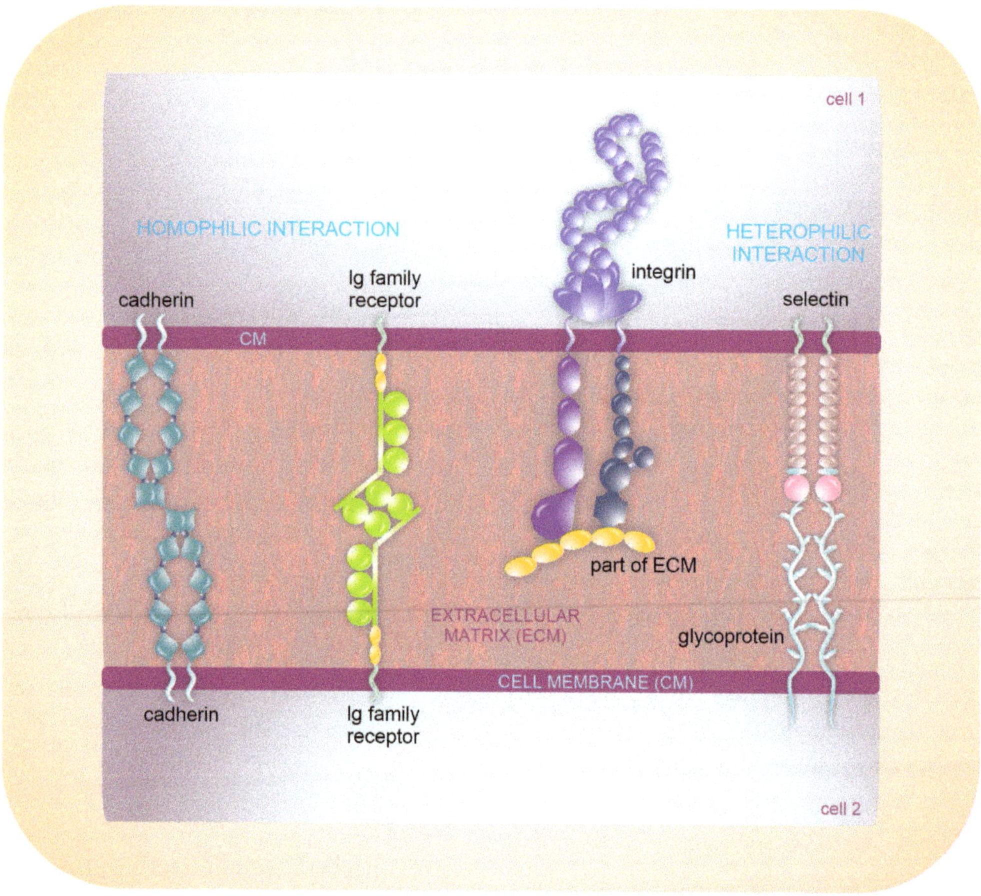

Fig. 37 Cell adhesion molecules

Cell adhesion molecules (cadherins, receptors of immunoglobulin family, integrins, selectins) are involved in the formation of cell adhesions. Cell adhesion molecules enable binding of cells to the extracellular matrix or to neighboring cells. The extracellular part of the adhesion receptor binds a ligand, which may be a structural motif of some parts of the extracellular matrix, the extracellular part of the adhesion receptor of a neighboring cell, or a signaling molecule. The transmembrane part of the receptor provides its anchorage in the cytoplasmic membrane.

The metastatic cascade: The continuous change in cell adhesions

The metastatic cascade begins with the release of cells from the primary tumor and ends with the formation of a secondary focus at another site, often far from the primary tumor. For the cell, the metastatic cascade is a long and difficult journey through many different and unfamiliar "landscapes," and at each stage of this journey, the cell must receive signals from the external environment that will enable it to survive. It must develop a "sense of home." Each step of the metastatic cascade therefore involves, on the one hand, the removal of some existing adhesion interactions that allow the cell to leave its current position and, on the other hand, the creation of new adhesions that allow it to survive in the environment into which it is moving.

The very first stage of the metastatic cascade of carcinomas – the migration of epithelial tumor cells from the primary tumor – is due to the loss of their adhesion to other cells of the epithelium, by loss of E-cadherin expression. Cadherins are adhesion molecules that form homophilic bonds, i.e. bonds between two identical molecules that connect homotypic, i.e. identical, cells. The extracellular part of a cadherin anchored in the cytoplasmic membrane of a cell interacts with the extracellular part of a cadherin of the same type on the surface of a neighboring cell (Fig. 37). By interacting with each other, cadherin molecules hold neighboring cells tightly together, linking them and ensuring that the same cells "recognize" each other. Cadherins are involved in tissue morphogenesis and in maintaining tissue integrity. The names of the individual cadherin types reflect the type of cell producer in which they were first discovered. The aforementioned E-cadherins were discovered on the surface of epithelial cells, N-cadherins are produced by neurons, R-cadherins by retinal cells, and so on. Thus, E-cadherin is responsible for the tight association of epithelial cells in epithelial tissues mentioned above (Fig. 35). In most epithelial tumors, carcinomas, the function of E-cadherins is impaired. Loss of E-cadherin function allows carcinoma cells to break free from their attachment to other epithelial cells, thus initiating the metastatic cascade. In addition to the absence of E-cadherins, invasive carcinomas often begin to express N-cadherin, which is normally produced in migrating neurons and mesenchymal cells during organogenesis. In general, cadherin expression is tightly regulated during the morphogenesis of the developing embryo. Changes in cadherin expression are always associated with morphogenetic movements. When cells move from one cell layer to another, this is always associated with the loss of expression of the cadherin that bound the cells in the original cell layer and the onset of expression of another cadherin that ensures cell interactions in the new tissue. Cadherins thus serve as anchors that hold cells together and also as morphoregulatory molecules that determine the positions of cells in the body.

Another change typical of invading and metastatic cells is alteration in the expression of integrins. Integrins are adhesive receptors that anchor cells to the extracellular matrix (Fig. 37). Invading and metastatic cells are exposed to a changing tissue microenvironment during the metastatic cascade, including the changing extracellular matrix (ECM). Migration through the variable ECM requires adaptation, which is achieved in part by the exchange of integrins that the migrating cell exposes on its surface. Integrins are the molecules that mediate anoikis. The cell can avoid cell death due to loss of binding to the ECM only if it produces matching integrins. As the cell moves through the shifting ECM, it must change the spectrum of integrins exposed on its surface. By gradually exchanging surface integrins, the cell moves step by step through the environment, slowly progressing in the metastatic cascade journey.

Cadherins and integrins are not the only two categories of adhesion molecules involved in the metastatic cascade. Thus, the interactions of tumor cells with endothelial cells during intravasation, transport, and extravasation are also mediated by selectins and a number of adhesion molecules from the immunoglobulin family (Fig. 37). The dynamics of expression of each type of adhesion receptor corresponds to the individual phases of the metastatic cascade (Buck, 1995; Hanahan, Weinberg, 2000).

Basement membrane degradation and extracellular matrix remodeling: Metalloproteinases

In addition to changes in adhesion receptor expression, the metastatic cascade is accompanied by other processes based on extracellular proteases, the enzymes that cleave proteins. Thus, the ability for local invasion as well as intra- and extravasation is always dependent on their production and release. Proteases are essential for the degradation of basement membranes and ECM components because space must be made in the ECM for invading, moving cells. There are a number of proteases involved in this process, but matrix metalloproteinases (MMPs) are the most important of these. MMPs are produced by cells in the form of zymogens, inactive proenzymes that are activated in the extracellular environment surrounding them. This ensures that the remodeling of the ECM does not originate from the ECM itself, but is closely linked to the cell, its properties and behavior. Activation of metalloproteinases, i.e. conversion of zymogen to active enzyme, occurs by proteolytic cleavage of the zymogen, and most metalloproteinases are activated by another active metalloproteinase or another protease. In some carcinomas, the protease producers are mainly the epithelial cells themselves; in others, this function is performed by stromal cells, which are, however, stimulated by tumor cells to produce metalloproteinases.

Activated metalloproteinases are responsible not only for the degradation of structural components of the ECM but also for a significant change in cell-to-cell signaling. In addition to structural proteins, the extracellular matrix also contains various signaling molecules, such as growth factors, survival factors, mitogenic signals, and more. Signaling molecules can be activated by metalloproteinases, either indirectly by being released from the ECM as a result of their degradation, or directly by proteolytic cleavage of their precursors. Some cell surface adhesion molecules, such as E-cadherin, can also be targeted by MMPs. MMP activation can therefore significantly affect metastasis formation, as well as angiogenesis, cell division, differentiation, apoptosis, and immune cell function. It is also significantly involved in epithelial-mesenchymal transition and in the formation of pre-metastatic niches (see below). MMP production and activity is increased in almost all tumor types and correlates with their stage, invasiveness, and aggressiveness (Egeblad, Werb, 2002; Shuman Moss et al., 2012; Weinberg, 2014).

Epithelial-mesenchymal transition (EMT)

During malignant transformation, cancer cells undergo a very dynamic process that alters their phenotype. The morphology, polarity, mobility, metabolism, and gene expression profile of these cells change. Carcinoma cells at different stages of development can range from fully differentiated epithelial cells (in the primary tumor and also in metastases) to more or less typical mesenchymal cells (in invasion, intravasation, transport and extravasation processes). The difference between epithelial and mesenchymal cells is considerable. Epithelial cells form compact cell surfaces. They are characterized by tight binding to each other, firm attachment to the basement membrane, polarity, and immobility. At the molecular level, they are characterized by the expression E-cadherin. In contrast, mesenchymal cells are less interconnected, they are motile, have migratory and invasive potential, are more resistant to apoptosis, produce significantly more components of the ECM, and express N-cadherin (Fig. 38).

The transition of the epithelial phenotype to the mesenchymal phenotype is not only a necessary part of the initial stages of the metastatic cascade, but is also associated with physiological processes. These include morphogenesis during embryonic development as well as healing processes. As discussed in chapter 1A, during gastrulation there is the formation of the endoderm, ectoderm, and mesoderm germ layers and subsequently, during organogenesis, there is a very extensive rearrangement of cell groups, the formation of organs, and the basic body plan. During these processes, the transitions of epithelial into mesenchymal cells occur repeatedly. During healing processes, when injuries are repaired and damaged tissue is regenerated and reconstructed, fibroblasts and related mesenchymal cells are produced in abundance.

Thus, the transition from epithelial cells to mesenchymal cells occurs repeatedly and for different purposes during the life of an organism, and this transition is associated with complex molecular, morphological, and functional changes. A cellular program that enables cells to undergo these multiple changes in an organized, programmed manner has been gradually uncovered. It has been termed the epithelial-mesenchymal transition (EMT), and the choice of the term "transition" reflects, among other things, the plasticity and reversibility of this program. The reversibility of EMT is documented by the fact that cells can also undergo the process in the reverse direction – the mesenchymal-epithelial transition (MET), in which mesenchymal cells convert to epithelial cells. The MET process is clearly associated with the final stages of the metastatic cascade, e.g. through the formation of micrometastases and colonization (Fig. 39), but also with physiological processes, such as the formation of a secondary epithelium during embryogenesis and with healing processes.

EMT and MET programs are mainly activated, regulated, and modulated by factors and signals from the microenvironment of cells. In other words, a cell has an inherent potential to shift in phenotype from a fully differentiated epithelial cell to a fully differentiated mesenchymal cell, and it has the ability to transition from one phenotype

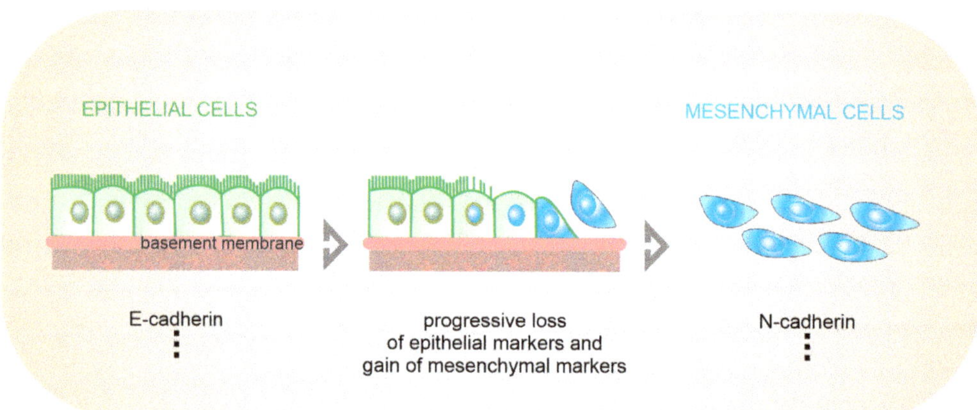

Fig. 38 Epithelial and mesenchymal cells
EMT involves the functional transformation of polarized epithelial cells into motile mesenchymal cells that secrete ECM components. Both cell types have their own characteristic molecular features that can be used to detect different transitional stages of cells that are more or less epithelial or mesenchymal in character (adapted from Kalluri, Weinberg, 2009, modified).

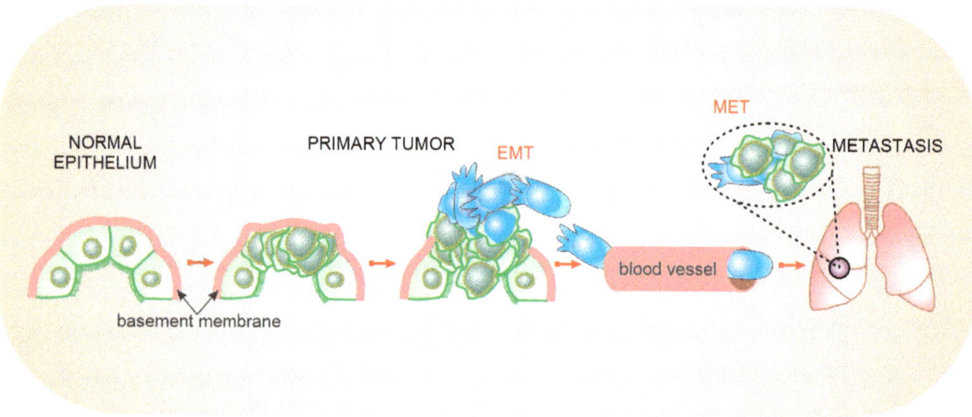

Fig. 39 Contribution of the epithelial-mesenchymal transition to tumor progression
The development from normal epithelium to progressive carcinoma occurs in several steps. The invasive carcinoma stage involves epithelial cells that have lost their polarity and are no longer connected to the basement membrane. The composition of the basement membrane also changes, as does the interaction of the cell with the ECM and signaling network. In the next step, EMT and angiogenesis are initiated, facilitating the onset of the malignant phase of tumor development. Progression from this stage to the metastatic phase involves EMT, which facilitates invasion of the vasculature by tumor cells, leaving them at a distant site where they can form micro- and macrometastases. The development of metastases may include MET and a return to an epithelial phenotype (adapted from Kalluri, Weinberg, 2009, modified).

to another (called transdifferentiation), with the degree of its epithelial or mesenchymal character depending substantially on the environment in which it is located. Loss of E-cadherin expression appears to be critical for entering the EMT program. Only E-cadherin is key factor determining epithelial phenotype. Any change that leads to a decrease in E-cadherin expression increases the likelihood that the cell will enter EMT and promote its mesenchymal character. Thus, the epithelial-mesenchymal plasticity of cells is closely related to the plasticity of E-cadherin expression.

Both the EMT and MET programs are physiologically active especially during early embryogenesis. Tumor cells exhibit their pathological reactivation. The main difference between EMT and MET processes during embryogenesis and carcinogenesis is that in carcinogenesis this process involves cells that are genetically damaged and gradually lose their ability to respond appropriately to signals from their microenvironment. Epithelial-mesenchymal plasticity allows tumor cells to dynamically adapt during the metastatic cascade. In response to EMT-stimulating signals, a subpopulation of epithelial cells at the invasive edge of the tumor may lose their epithelial properties and detach from the tumor. As the cells move away from the tumor, they are exposed to fewer epithelial signals, and in the presence of EMT signals delivered by stromal cells, their mesenchymal properties deepen. Mesenchymal cells can invade surrounding tissues, and the mesenchymal phenotype facilitates intravasation into blood or lymphatic capillaries. These spreading tumor cells are more resistant to genotoxic and environmental stresses, which is critical for their survival in the vascular system.

Once the cells reach a distant organ, the mesenchymal phenotype facilitates extravasation and invasion into foreign tissue. Here, the spreading cells are exposed to signals different from those of the primary tumor, and the mesenchymal state may give them a survival advantage or keep them in a long-term dormant state. Once appropriate environmental signals are available, the spreading cells can undergo MET and gradually regain epithelial properties, which include the ability to proliferate rapidly and form strong intercellular adhesions. The epithelial signals are amplified by autocrine and paracrine signals, leading to stabilization of the epithelial phenotype. This favors the development of macrometastases composed mainly of epithelial cells. The metastatic cascade is thus completed (Fig. 39) (Kalluri, Weinberg, 2009; Pole, Weinberg, 2009; There, Weinberg, 2013).

Organ-specific metastasis

Circulating tumor cells can theoretically spread in all directions and form metastases in any part of the organism. In reality, however, certain carcinomas metastasize preferentially to a narrow range of target organs. For example, breast carcinomas most commonly metastasize to the bones, liver, brain, and lungs; prostate carcinomas to the bones; and colon carcinomas to the liver.

Organ-specific metastasis formation or organotropism could simply result from a passive process in which circulating tumor cells are carried away in the bloodstream and subsequently become lodged in the capillary network of the first potential organ "in their path." This mechanistic approach provides a good explanation for organ-specific metastasis of a number of tumor types. For example, blood flowing from colorectal carcinomas through the portal vein to the liver contains tumor cells that are most likely to settle there. However, this explanation does not apply for many other tumors and their typical metastases.

Another possibility is that the metastatic cells are predisposed to survive in a particular tissue, for example, because of the expression of specific adhesion molecules or because of specific growth factors available in the new environment. More than 100 years ago, the English surgeon Stephen Paget formulated the "seed and soil" hypothesis, postulating that tumor cells (seeds) survive only in the environment (soil) to which they can efficiently adapt. Later, the link between some features of tumor cells (expression of certain genes) and their propensity to metastasize to specific organs was even demonstrated at the molecular level (Chambers et al., 2002; Valastyn, Weinberg, 2011).

According to the "seed and soil" model, metastases selectively colonize specific organs based on their mutual compatibility: the target tissue is a "fertile soil" for immigrating tumor cells, the so-called metastatic niche, in which they can survive and which is remodeled and even adapted to them after their arrival. Later, a model of pre-metastatic niches was proposed and approved. According to this model, primary tumor cells send signals to the body that influence distant tissues to become more helpful and hospitable to the survival and growth of circulating tumor cells, even before these cells reach the site of potential metastases. The mechanism of this phenomenon is based on the production of signaling molecules and exosomes. Exosomes are membrane-enveloped bodies released by tumor cells that contain proteins, nucleic acids, and other metabolites. They fuse with target cells, transferring their contents to them and thereby influencing and reprogramming them. The entire process of creating premetastatic niches begins with local changes, such as the induction of vascular permeability, remodeling of the extracellular matrix, and reprogramming of local stromal cells. It continues through the acquisition of cells from other tissues, such as cells from the bone marrow and immune system, and, of course, through the attraction of circulating tumor cells (Kaplan et al., 2006; Cox et al., 2012; Peinado et al., 2017).

The formation of metastases is usually considered a unidirectional, linear process. In the "seed and soil" model, circulating tumor cells are referred to as seeds through which the tumor spreads from the primary site via a metastatic cascade to distant sites in the body where the soil is suitable for the formation of secondary deposits. However, the entire process can take an unexpectedly different course. Circulating tumor cells, in fact, have the ability to migrate in all directions and "seed" the tumor in various distant parts of the body. Thus, they can also return to their place of origin – the original primary tumor. This process is called "self-seeding" and shows that tumors can grow not only from the inside out, but also from the outside in. Through this process, cells that have overcome many obstacles and acquired a number of new properties during the metastatic cascade return to the primary tumor. By returning, they significantly alter the primary tumor: they increase its heterogeneity, stimulate its development, and increase its aggressiveness. The "self-seeding" model provides insights into the complicated and hidden process of tumor spread in the organism and confirms that the dynamics and progression of the metastatic cascade can be very different in different tumors (Kim et al., 2009; Hahnfeldt, 2010; Comen, Norton, 2012).

Colonization and dormant metastases

The process of forming metastases is sometimes divided into two main phases: (1) the physical spread of tumor cells from the primary tumor to distant tissues and (2) the adaptation of these cells to a foreign tissue microenvironment leading to successful colonization, i.e. the ability to transform micrometastases into macroscopic tumor deposits. Colonization is the least efficient step of the entire metastatic cascade, so dissemination and colonization need not and often do not occur simultaneously. Many patients have been found to have numerous micrometastases that successfully disseminate throughout the body but never progress to macroscopic metastasis. They represent a condition sometimes referred to as dormant or occult metastases.

There may be other reasons for such a dormant state, and some have already been mentioned. For example, in chapter 7A on angiogenesis, (1) the mechanism of so-called concomitant resistance was mentioned, in which some primary tumors produce and release factors – for example, angiogenesis inhibitors – that prevent the effective growth and development of macroscopic metastases, thus keeping micrometastases in a quiescent state. Only after removal of the primary tumor – the source of the inhibitor – can angiogenesis be induced in the micrometastases, resulting in successful colonization. (2) In the introduction to this chapter, we mentioned that cells from some tumors can migrate very early during the tumor's development. Thus, micrometastases can arise that lack some of the aggressive properties (including the ability to activate angiogenesis) that are responsible for effective tumor growth. Micrometastases then remain in a dormant state until they develop these properties, which can take many years. (3) So-called metastasis suppressor genes have been described, the products of which are able to inhibit the growth of metastases without affecting growth of the primary tumor. More than 20 metastatic suppressors have been described that act at different phases of the metastatic cascade (Horak et al., 2008; Hurst, Welch, 2011; Wang et al., 2017). (4) Another mechanism that keeps micrometastases in a quiescent state was mentioned in chapter 4A: cells can respond to stress caused by nutrient or oxygen deprivation and oxidative stress, for example, by inducing autophagy. In other words, the tumor cells degrade some of their own components, shrink significantly, and enter a quiescent state. This allows them to withstand the adverse conditions and survive until the microenvironment becomes favorable again. (5) The next form is so-called immunological dormancy, which results from the interaction between tumor cells and the immune system, as discussed in more detail below in chapter 10A (Schreiber, 2010, p. 76; Hanahan, Weinberg, 2011; Hesketh, 2012, p. 131; Peinado et al., 2017; Aguirre-Ghiso, 2007; Narendra et al., 2013).

9B. Underestimation of the value of relationships and home

Malignant tumor cells form fewer adhesions and leave the site of their origin

Benign tumors develop and remain at their site of origin. In contrast, the cells of a malignant tumor invade surrounding tissues from their site of origin and may form distant metastases in other parts of the body. An essential part of this process is that tumor cells lose or change their connections to the other cells and non-cellular structures of the primary tumor; they leave the original common tumor site and migrate to other, "non-native" parts of the organism. For healthy cells, such a loss of attachment and leaving "home" would be fatal and would immediately trigger a form of programmed cell death known as anoikis or homelessness. Thus, the "sense of home" is essential for healthy cells to survive. And they "feel at home" because they are able to establish "proper relationships" in that place. In contrast to healthy cells, malignant tumor cells are much less dependent on relationships, i.e. connections to surrounding cellular and non-cellular structures. They can survive far from home without them. Underestimating the value of relationships and home is another, the eighth, of the non-deadly sins.

Konrad Lorenz' individualization and the sixth deadly sin

Since the last quarter of the 20th century, economically developed countries have been in transition from the industrial to the post-industrial phase. As mentioned in the previous chapters on angiogenesis and resource misuse (chapters 7A and 7B), this transition is accompanied, among other things, by an increasing distancing of individual social groups from each other and by growing inequality between them. According to sociologist Jan Keller, this transition is also accompanied by the disappearance of classes, the decline of social strata, the ongoing individualization of society, and increasing competition for jobs and life opportunities. Historically, traditional status groups moderated competition among their members, guaranteed them a certain way of life and offered common protection through shared rights and privileges. Jan Keller considers the pressure on the middle classes to be intentional and individualization to be an imposed strategy that excludes the middle and lower classes from the possibility of defending their positions collectively. The struggle for jobs and pervasive competition further disintegrate society and exclude individuals and entire groups from it (Keller, 2011). People are gradually becoming more and more detached from their formerly entrenched social positions and the fixed expectations of personal life that go with them. Of course, human life is still shaped by a set of social norms and institutions, but the earlier emphasis on conformity has loosened and transformed into a pressure for personal autonomy. Individualization, then, frees us from often restrictive statuses, social roles, and prescribed life paths, but it also brings with it a loss of support from traditional institutions. This places high demands on the individual in the area of self-awareness, finding one's own life path, setting personal goals, and taking responsibility for one's own life choices in all areas of life (Vacková et al., 2014). Individualization can thus be viewed from many different angles, at many levels, and in many different manifestations.

It is clear that relations between people have changed and loosened dramatically. As early as the 1970s, Konrad Lorenz named "breaking with tradition" as the sixth of his eight deadly sins. He was aware of the vital importance of habits and customs that are passed down from generation to generation in order to preserve the human experience gained earlier in development. He was convinced that the preservation of existing experiences is even more important than the acquisition of new ones. Part of, and at the same time a prerequisite for, the preservation of historical experience is also the preservation of the method of its transmission from generation to generation. In the course of evolution, a mechanism has been built into human behavior which ensures that earlier experiences are first examined, controlled, i.e. temporarily rejected (in puberty in the form of so-called physiological neophilia – i.e. attraction to everything new), but then a revival of love for old traditions, so-called late obedience, takes place. The healthy course of this process presupposes functioning relationships and communication between generations. And Konrad Lorenz feared that a tipping point had been reached at which the younger generation could no longer understand the older one or even identify with it, and could therefore become hostile to it. He saw the causes of this identification disorder in the rapid development of technology, which accelerates the divergence between the generations, and above all in the lack of contact of children with their parents from early childhood. He predicted that this would have pathological consequences in the future (Lorenz, 1974). Although he later toned down the sharp formulation of this "sin," subsequent history shows that his assessment was largely correct. Traditional relationships are being altered and often dissolved altogether. Even Konrad Lorenz would probably be surprised at the amazing forms of new relationships, or rather non-relationships, we can create.

Sologamy becomes more and more popular

Reading contemporary newspaper headlines like "Weddings with only the bride? Nothing unusual in Japan" or "An Italian woman marries herself: Sologamy is becoming more and more popular," Konrad Lorenz would probably be surprised, if he could even guess what is the words mean. Sologamy is marriage to oneself, marriage without a groom or without a bride, a symbolic promise of fidelity and love to oneself. Sologamy is preferred by people who are alone and without a partner, either because they have not managed to enter into a partner relationship, or because being alone, being single, is their choice. Sologamy is already possible in several countries and the number of interested people is increasing worldwide. A Japanese travel agency has been offering ceremonies for this type of marriage since 2014. Another Canadian agency writes in its advertisement for sologamy: "Singleness is a new form: Celebrate being single."

The emergence and spread of sologamy reflects the fact that more and more young people prefer to remain single, are not overly interested in sex, and are happy alone. And if solitude does not suit them, there are agencies (at least in Japan) where people can hire, for example, "caring mothers for grown daughters," "professional listeners," or "best friends who share their joys and sorrows with clients." And this solitude business is reportedly working better and better. After all, the number of people living alone is also on the rise worldwide, and the curve of the percentage of single-person households is rising steeply. In the Czech Republic, for example, only one person lives in every third household, while the share of single-person households was only 7.6% in 1930 (Vránková, Jaroš, 2016).

The age of loneliness

According to some authors, we live in an age of loneliness. Loneliness is a modern epidemic and one of the main causes of unhappiness in life. In addition to young people for whom it is difficult to establish a partner relationship, many seniors also feel lonely, as do, for example, people who live in dysfunctional marriages or in dysfunctional relationships in general. Loneliness and isolation (i.e. both the subjectively perceived feeling of loneliness and the objectively measurable social isolation) slowly kill us. Long-term loneliness causes cardiovascular disease, increases the risk of dementia, depression, anxiety, raises blood pressure and cholesterol levels, impairs the immune system, causes sleep disturbances, and accelerates the aging process. Loneliness is more harmful than obesity and takes a similar toll on our health as smoking fifteen cigarettes a day for an extending period of time. Thus, loneliness is not only unpleasant, but harmful to health and even life-threatening (Holt-Lunstad et al., 2010, 2015).

Cooperation and living in a larger group were essential to the survival of the species *Homo sapiens*. Being alone was a death sentence. That is why we have associated the feeling of loneliness with so much pain and fear. This fear and longing for others is evolutionarily ingrained in us and serves as an incentive to enter into relationships

(Trešňák, 2015). And this is not only true for *Homo sapiens*. A similar need has been found in other animals as well. Thus, the threat of exclusion from the herd is by far the most effective weapon used by a dominant mare to discipline mischievous and unruly young horses in herds of wild mustangs. The unruly foal is ruthlessly chased out of the herd by the mare and is allowed to return only on the condition that it stops hurting others. Staying away from the herd would mean certain death. And so the foal relents.

Uncovering and understanding this inner need for relationship and cooperation became the starting point and foundation for the nonviolent training of horses by the world-famous "horse wizard" Monty Roberts (2010a). He not only inspired horse lovers all over the world with his method. He successfully transferred his many years of experience working with horses and his clear understanding of their deep need for cooperation and relationship back to the human world and used it in working with troubled youth (Roberts, 2010b). The basic principles of his "method" are nonviolence, respect, and choice. Roberts assumes that the need to build relationships – both in humans and horses – is so basic and strong that all one has to do is create an environment of trust, i.e. nonviolence and respect, and no human – just like a horse – can resist and always chooses to relate and cooperate voluntarily. So there is no question that the need for relationships, the fear of exclusion and loneliness, is deeply rooted in horses and humans, shaped by evolution and innate. So what's wrong with our relationships these days? What are the causes of their deterioration, our relative apathy to this situation, and disregard for the value and importance of companionship?

Autistic society

Konrad Lorenz saw the problem in the lack of contact between parents and children in the earliest years of childhood. And many agree with him. For example, child psychologist Jiřina Prekopová believes that we are experiencing "perhaps the most critical turning point in the history of humanity, when interest in love and humanity is taking a back seat to interest in the comforts of money and technology" (Koucká et al., 2012, p. 14). She believes that the capacity for love in technological civilization is diminishing year by year. According to her, phenomena such as high divorce rates, the increasing number of single parents, and restless and aggressive children document the loosening of social ties and the dissolution of the natural social network. Life in a relationship seems to be more difficult than life without a relationship. From early childhood, personal contact with parents is replaced by technology. Instead of supporting and facilitating communication between people, the computer becomes the object and the human partner falls out of communication. This also paves the way to loneliness. Prekopová sadly notes that communication with the computer, if we can master it technically, is much more pleasant than with another human being. The computer does not argue with us, does not express disagreements and dissatisfactions, does not demand our attention when we do not want to give it. On the contrary, it works reliably with a single click whenever we want it to. Such convenience and reliable response without negotiation and self-restraint is not guaranteed in real interpersonal contacts. Prekopová speaks of an "autistic society where everyone has their own space and the ability to retreat and hide behind devices, computers, televisions and video games. Self-interest is more important than the interest of the family or the community" (Prekopová, 2007, pp. 88–89).

Another reason for the erosion of interpersonal relationships is the wealth and abundance that characterize our contemporary civilization. When everything is enough, or more than enough, one needs the other less and less. Being alone is no longer a death sentence; apparently we can survive even without the "herd." So the reflex imprinted by evolution that forces us to enter into relationships is slowly fading. And finally – according to Jiřina Prekopová – our pride betrays us. "We are blinded by the human intellect, which has produced technically perfect devices and ensured material prosperity. We have created the idea that everything, health and disease, life and death, is in our hands, that everything can be created or owned by science or money. We have succumbed to the illusion that we are masters of all beings and can rise above everything. We will fly to Mars and beyond, we do not need the love of parents to conceive a human being" (Koucká et al., 2012, p. 173).

Motherhood on ice

Do we really not need the love of parents to conceive a human being? Artificial insemination, *in vitro* fertilization (IVF), was developed as a method to treat infertility and was successfully used for the first time back in 1978. At that time, no one imagined how intensively the method would be used in a short time and how quickly the birth of "test-tube babies" would become common, how quickly it would become necessary and how quickly it would become almost standard. After IVF, egg freezing, or cryopreservation, was later developed. This method was also developed for medical reasons, for example as an option for women with cancer whose fertility might be threatened by antitumor therapy. The first child from a frozen egg was born in 1986.

Like artificial insemination, egg freezing is growing in popularity and, as with artificial insemination, its use has rapidly expanded beyond its original target population of female clients. It is being used by healthy but childless women to prevent age-related deterioration of egg quality and to preserve the possibility of becoming pregnant at an older age. In this way, women buy more time to find the right life partner or when they postpone motherhood for professional or other reasons. This phenomenon is sometimes referred to as "social freezing." The woman appears to have the opportunity to buy time to decide whether or when she wants to become pregnant. Despite the still relatively low efficiency and high price of this procedure, interest in it is growing, and the number of women undergoing it is increasing. This symptomatic and cautionary phenomenon reflects the illusions of our time mentioned by Jiřina Prekopová: dazzled by technology, many believe that science and money will solve all our problems. There is also a kind of contemptuous feeling that a person does not even need the love of parents to be conceived (Prekopová, 2007, 2012; Koucká et al., 2012).

It would be cheap, and probably unfair, to get upset and moralize about this trend in general if companies (led by Facebook and Apple) had not turned over a whole new leaf in this area by including the option of egg removal and freezing in their employee benefits packages. As more and more women postpone motherhood "for the sake of their careers" – or rather, because it is still very difficult to balance childcare and demanding work – "understanding and accommodating" companies (which take maximum advantage of the performance and creativity of their female employees literally "in their prime"), under the guise of "freeing" women from the pressures of the biological clock and declining fertility, contribute to raising false hopes of successful parenthood among postponers. With these policies, they pressure their female employees to put their careers before their children, rather than helping them find a way to have children while working reasonably.

In this way, companies and corporations are pushing the limits of work culture and corporate law and, above all, contributing mainly to normalize this pathological and relatively extreme solution, to support the tightening atmosphere and attitude of society that "you must first graduate, create something, enjoy something and then take care of your family." But as one of the women trying to preserve the hope of future motherhood says: "You cannot get everything from hard work. Especially when it comes to relationships" (Lauder, 2014). It's hard not to think of Konrad Lorenz's warning. It almost looks like we have a problematic relationship with our offspring, not just in their early childhood, but long before they are conceived and born!

Society of insecurity and home

Sociologist Jan Keller introduces his book *Feast of the Homeless* with a chapter titled "The Diagnosis of Time," in which he presents the social themes most frequently addressed in contemporary sociological literature. One of these themes is insecurity. People today face insecurity mainly in the fields of work and family relationships, but they also have problems dealing with their own position in society, security in old age, and questions of future offspring. Insecurity and the feeling of threat are increasing in society, and people are looking for ways to reduce them.

Another important issue today is the question of appropriate stimulation. How can one stimulate the desire to work if the work serves no higher purpose? Why have children if they lower your standard of living and limit the possibility of self-realization? Why vote in elections? What is the point of saving money for old age? Keller also points out that the flip side of the problem of stimulation in a bored society is the passionate search for thrills and experiences that transgress all boundaries.

The third common theme is the issue of personal identity, its formation and maintenance. Post-industrial society is changing so rapidly and so dramatically that more and more people are having difficulty understanding where their place is in the community and where they are actually going (Keller, 2013).

Jan Keller looked for points of connection between these three themes and concluded that territory is the common denominator. In general, a territory is defined as the area that an individual (humans, but also many animals) stakes out and defends against other individuals. Territory, that is, home in its broadest sense, enables humans to satisfy three important needs. First and foremost, the home provides safe shelter. Jan Keller is convinced that even the most itinerant person needs a familiar place where he or she feels safe and secure. The everyday feeling of a homeless person is one of pervasive insecurity and constant threat. Secondly, in addition to peace and security, home is a rich source of stimulation. Only in the company of loved ones and with their support can one overcome challenges that would otherwise be difficult to endure. When you do not have a home, you may lack challenges since you have nowhere to take them and face them calmly. On the contrary, homelessness brings a pervasive, deadly boredom.

And third, housing helps solve the problem of one's identity. The home extends each individual inhabitant, it is more durable, temporally and spatially larger than the space of each of its inhabitants. Thus, whoever feels at home in a particular place can identify and connect with the people who lived there in the past and with whom he shares it in the present. He discovers who he is and who he is not, he becomes himself. The fate of a homeless person is a life in nowhere land, a life in boundless anonymity (Keller, 2013).

Nostalgia, homesickness

The three benefits that the home brings to its inhabitants, namely safety, stimulation, and help in finding one's identity, have fascinating parallels in the world of cells. When cells are in the right "place" in the body ("at home"), they are surrounded by many cellular and non-cellular structures that enable them to form a rich network of contacts and communication. They have a reliable source and supply of survival signals that protect them from programmed death. They have an adequate supply of signals and resources that stimulate them to perform versatile, purposeful, and appropriate functions related to metabolism, respiration, growth, response, and adaptation to various types of stimuli. All this also leads to the development of cells, to their differentiation, i.e. to the acquisition of a distinctive, appropriate identity.

Cells that lose their "home" die by the process of anoikis. Cellular homelessness leads to death. What happens to a homeless person? The disease that arises from immense homesickness and longing for home is called nostalgia. It is incurable and in extreme cases can even end in death. Is death by homesickness a human version of anoikis? And how do we react to the ever-pervasive nostalgia in our society? Jan Keller notes that the market society is taking over and exploiting this area. Thus, the wave of homesickness was quickly recognized by businessmen, who launched a new sector – the nostalgia industry. Nostalgia fashion was born: nostalgic items, nostalgic style, nostalgic TV series… Of course, this industry has no intention of curing homesickness. On the contrary, the idea is to feed it with more and more nostalgic props in order to be economically successful (Keller, 2013).

What is home?

We use and understand the word "home" so naturally that it is difficult to grasp and clearly define. Moreover, it has several meanings. For example, it is a sociological, psychological, cultural and political term. In everyday language, we understand home to mean a place of residence, a private, intimate space separate from the public sphere. A place where we belong, understand and where others understand us. It is a place that shapes our perception of the world, gives us practical experience, is a source of personal memory and a place of dreaming.

From a political perspective, home can be a place of personal dignity and potential resistance. From the sociologist's point of view, as mentioned earlier, it is a place that provides security, that stimulates and helps people to find personal identity. At the same time, it is always a place of work, restriction, rules, hierarchy and unequal power relations. Some authors therefore speak of tyranny in the context of home: the tyranny of hierarchies, of rules, of repetition. The experience of home also includes the need to negotiate one's position and to find and take one's place. Thus, home is a multi-layered concept that refers not only to physical space, but also to ideas and visions, to feelings and solidarity (Keller, 2013; Vacková et al., 2014; Třešňák, 2015).

How are we losing our home?

How is home currently changing, where is it disappearing, and why? Just as the concept of home is multi-layered and the levels at which we perceive and define it are diverse, so too are there multiple mechanisms and sources for the "deconstruction of the modern home," as Jan Keller puts it. The process of loss of home can be understood on two parallel levels: on a general level as a loss of social anchorage, of belonging to society, in a narrower sense as a problem with a particular physical home (Keller, 2013).

The increase in homelessness is related to the onset of the aforementioned postmodern phase and the associated individualization of society. While modern society gradually granted basic rights to more and more population groups and integrated them as full citizens, with the onset of the postmodern phase the historical process of integration came to a halt and entire groups of previously seemingly reliably integrated individuals were excluded from society. This social exclusion was driven by the increasing difficulty of finding meaningful work, the weakening of social and family ties, and the problems of finding and keeping good housing. The originally unfriendly and hostile industrial society was gradually cultivated and made more comfortable for modern man. A sense of security was provided by functioning insurance systems. The revival took many forms: rising wages and the welfare system increased people's purchasing power, the space for leisure activities was expanded, and the flourishing media filled this space with new content easily accessible to all. Modern man began to build his identity primarily through the occupation he took and the professional career he achieved, possibly also through extra-occupational activities he freely chose (Keller, 2013).

It would appear that modern society has taken over some of the responsibilities of the traditional home, thereby weakening pressure to maintain a home. Moreover, a society that values economic growth, material prosperity, and technological conveniences is losing interest in the spatial and social proximity of individuals, in traditions and a fixed cultural order, as well as in affective ties between members of the same group and the sense of belonging to the human community. Current intimate ties and relationships are increasingly fragile and temporary, and it is difficult or almost impossible to find a stable sense of fulfillment, happiness, and security in them (Keller, 2013; Vacková et al., 2014). The value of home and strong family relationships is attacked and threatened "from the outside" by, and as a result of, social development, and at the same time it has been eroded "from the inside" by our decreasing ability to form and live relationships and our increasing disinterest in them, apathy, indifference, or orientation toward other values.

Home alone

This trend is documented by the sharp increase in the number of single-person households mentioned earlier. At the same time, "single life" is a new and unfamiliar experience for mankind. Never before have so many people lived independently and alone. Only in rich, socially secure societies can people afford to buy their own homes and run their households independently. And the market society helpfully responds. Just as there is a "nostalgia industry" that does not cure homesickness, but successfully makes money from it, "single life" is economically profitable, and people living alone are an attractive, growing target group for developers, bankers, architects and designers. And they are also of interest to sociologists and other experts who are mapping and researching this new phenomenon.

Czech experts dealing with the lives of people living alone were looking, among other things, for an apt Czech term for them. The term "people living alone" is too long, similar to the term "people living in one-person households." Terms like "one-person household" or "living in one" do not sound good and are not accurate in Czech, and the word "singles" has a slightly different meaning. In fact, even people who have a partner relationship often live on their own, i.e. in a one-person household. The authors who study the households of those living alone eventually chose the term "soloists."

While they acknowledge that a soloist can be someone who is not a team player, who focuses on his own success and has no regard for the needs of others, i.e. is a ruthless egocentric, the soloist is also the person who stands alone in front of the orchestra, perfectly masters his own tools, and supports everyone else's performance with his diligence, artistry, and professionalism. He is someone who ventures into the unknown, who relies on himself and knows that he himself will pay for his mistakes. Thus, he can be a pioneer and selfless helper of others (Vacková et al., 2014; Vránková, Jaroš, 2016). The very fact that a stable, universal term for solo householders is still being sought shows how new this phenomenon is, and to start the journey even in this sense means to enter an unknown space. But where is the journey going?

What will come next?

The increasing number of soloists can be perceived and interpreted as a final consequence of individualization. Taken to the extreme, one can envision the dystopian idea of completely atomized societies with childless individuals living alone (Vacková et al., 2014). The concern of Konrad Lorenz, who saw a deadly danger for civilized humanity in the decline of evolutionarily embedded social behavior and the dissolution of traditions, has already been mentioned.

According to one hypothesis, the human brain includes a social brain, that is, a functional system or specialized neural network for processing social information. Some neuroscientists believe that the social lifestyle and the associated development of certain skills such as linguistic communication, empathy, and mentalization are behind the development of the human brain and are largely what make humans human and distinguish them from other animals. It is possible that the social brain evolved in proportion to the number of members of the group in which humans lived, and that living in a group, i.e. social life, could be behind the humanization of humans (Koukolík, 2016; Koubská, Koukolík 2019).

From this point of view, individualization actually seems to be a trend against the "direction of evolution," against the flow of human history. Relationships with other people, coexistence and communication are crucial and absolutely necessary not only for the personal development and growth of man, but literally for his survival. Finally, tumors eloquently and vividly show us that the loss of ties and contacts, the loss of a secure anchorage in a place where we belong, where we are at home, is a fundamental and almost irreversible and uncontrollable leap in the malignant and lethal process of tumor development. So the loss of relationship and home may be a non-deadly sin in itself, but in the process of tumor development it is virtually a predetermined breaking point.

10A. The ability to evade the immune surveillance

Does the immune system protect against tumor development?

The presence of immune cells in tumor tissue has been known for a long time. Since the beginning of the 20th century, it has been known that the immune system influences the development of tumors and protects the host from them. In the middle of the century, the tumor immune surveillance hypothesis was formulated. It states that the immune system can recognize and eliminate tumor cells. Without this ability, tumors would be much more common in large, long-lived animals (i.e. most warm-blooded vertebrates).

From its inception, this hypothesis has had both its proponents and its opponents; however, there is much indirect evidence to support it. For example, patients with some form of immunodeficiency, i.e. a dysfunctional immune system, have an increased risk of developing cancer. Similarly, patients who undergo organ transplantation and therefore take immunosuppressive drugs, substances suppressing the immune system that function to reduce the risk of rejection of the transplanted organ or tissue, have a higher risk of tumors than people who do not take these drugs. Similarly, people infected with HIV-1 and AIDS, the acquired immunodeficiency syndrome associated with the selective elimination of a specific type of immune cell, $CD4^+$ T lymphocytes, have an increased risk of developing several types of tumors, including Kaposi's sarcoma, lymphoma, and liver cancer.

All solid tumors are penetrated by various cells of the immune system. In many tumor types, a relationship has been described between the type, density, and localization of immune cells in the tumor and the clinical course of the disease. Thus, the response of the immune system to the appearance of a tumor is different in each patient and, at the same time, fundamentally affects both tumor development and patient survival. Although all this evidence is indirect, it clearly shows that the immune system responds to the tumor in the body, and it has gradually become clear that almost all components of the complicated human immune system are involved in this response (Schreiber et al., 2011; Corthay, 2014).

Functions of the immune system

The immune system plays a fundamental role in maintaining the integrity of the organism, it is a basic homeostatic mechanism. It performs defense work by recognizing foreign pathogens (such as viruses, bacteria, parasites) and their harmful products and protecting the body from them. The second task is self-tolerance, i.e. recognizing the body's own cells and tissues and providing tolerance to them. Finally, the third task of the immune system is the aforementioned immune surveillance, which enables it to recognize and eliminate non-functional, harmful components of the body, such as damaged, infected or dead cells. Substances that are recognized by the immune system are called antigens. Virtually all chemical structures can act as antigens, but usually they must be in the form of macromolecules, soluble or bound to the cell surface (Jílek, 2014; Hořejší et al., 2017).

Components of the immune system

The human immune system consists of many cell types and signaling molecules that interact with each other as well as with non-immune cells and structures, forming a very complex and dynamic network. The immune system consists of two basic components. Innate, natural immunity, sometimes called nonspecific immunity, responds very quickly, on the order of minutes, when a problem occurs in the body. It represents the body's first line of defense. It is nonspecific in the sense that it does not respond specifically to a particular antigen, but to structural or functional features common to several or many different pathogens. The basic cellular components of innate immunity include macrophages, dendritic cells, granulocytes, NK (natural killer) cells, and mast cells.

In contrast, activation of acquired or adaptive or antigen-specific immunity requires a prior response of some components of innate immunity. This means that this type of immunity is slower to act and serves as a second line of defense. B and T lymphocytes are the key factors: B lymphocytes produce soluble antibodies that are highly antigen-specific, and T lymphocytes are equipped with similarly specific receptors on their surface. The components of specific immunity have memory and can therefore respond much more quickly and efficiently when re-exposed to the same stimulus. In addition to cellular components, the immune system also includes a variety of cytokines, signaling molecules. They are produced by the cells of the immune system and can trigger, for example, rapid proliferation and differentiation of cells that are part of the immune response (de Visser et al., 2006; Krejsek, Kopecký, 2004; Jílek, 2014; Hořejší et al., 2017).

Tumor antigens

The immune system can only respond to tumor cells if they carry tumor antigens on their surface. These antigens act as tumor-specific markers that are recognizable to the immune system. The question of whether tumor cells can be recognized by the immune system with sufficient efficacy at all, whether they are sufficiently different from healthy cells, i.e. whether they are sufficiently immunogenic, has long been debated. After all, tolerance to endogenous components is a necessary condition for a functioning immune system, and tumor cells are derived from endogenous, "normal" cells. The existence of tumor antigens has finally been demonstrated. In general, tumor antigens are defined as more or less specific molecules expressed by tumor cells but not by normal cells.

There are two categories of antigens: tumor-specific antigens and tumor-associated antigens. Tumor-specific antigens include altered proteins, products of mutated genes resulting from mutations of proto-oncogenes or tumor suppressor genes that are not found in healthy cells. In tumors associated with viral infection (e.g. cervical cancer, hepatocellular carcinoma), viral proteins present in infected tumor cells may also serve as tumor-specific antigens. Tumor-associated antigens are normal, i.e. unmodified, physiological molecules that are expressed in an unphysiological manner: either at abnormally high levels or by cells that do not normally produce these proteins at all. A seminal paper published in 1991 describes an antigen that is recognized by the immune system in melanoma patients (MAGE-1). The authors find that this antigen is not a product of a mutated gene, but a normal protein. The immune system recognizes it because it is produced exclusively by tumor cells in the adult body. Naturally, this protein is found only in the germ cells of the testis. Subsequently, a number of other tumor antigens of this type have been described (van der Bruggen et al., 1991; Schreiber et al., 2011; Finn, 2012; Hořejší et al., 2017; Higher, 2019; Feola et al., 2020).

The interaction between immune system and tumors

The immune response to tumor cells can be described very simply as recognition of tumor-specific antigens on tumor-transformed cells either directly by T lymphocytes or by NK cells (natural killers), followed by destruction of some tumor cells. The resulting tumor cell debris is then processed by macrophages and dendritic cells. These cells then begin producing proinflammatory cytokines and present tumor antigens to T and B lymphocytes. Activation of T and B lymphocytes accelerates the production of additional cytokines that further enhance innate immunity and stimulate the components of adaptive immunity – primarily the expansion of specific T lymphocytes and the production of specific antibodies by B lymphocytes. Adaptive immune mechanisms play an important role in the elimination of tumor cells, on the one hand, and in the formation of a memory that can prevent tumor recurrence, on the other. A key role in the recognition and elimination of tumor cells is played mainly by T lymphocytes, but successful immune surveillance of the tumor requires the proper functioning of all components of the immune response, their appropriate interaction and mutual communication (Finn, 2012; Šťastný, Říhová, 2015; Hořejší et al., 2017, 2019).

From immune surveillance to immunoediting

On the one hand, with time experimental and epidemiological data provided increasingly convincing support for the immune surveillance hypothesis; on the other hand, the relationship between the tumor and the host immune system proved to be more complicated and immune surveillance only a part of it. The immune system does indeed protect the host from tumors, as explained by the immune surveillance hypothesis. However, its activity also contributes to the formation and transformation of the developing tumor. In chapter 6, we described how tumor cells are constantly evolving and transforming, in part due to increased genetic instability. As a result, tumor heterogeneity increases (Fig. 23). Tumor cells or cell subclones within the tumor may also differ in their immunogenicity, i.e. their ability to elicit an immune response. Highly immunogenic cells are clearly recognized and eliminated by the immune system, whereas less immunogenic cells may escape an effective immune response. By selecting tumor cells with lower immunogenicity, i.e. cells that are more resistant, the immune system stimulates tumor development. The recognition of this dual effect of the immune system on tumor development led to the formulation of the immunoediting hypothesis. This hypothesis considers both sides of the immune system's activities – both its ability to protect the host from tumor development and its ability to fortify tumors, and thus stimulate their development. Immunoediting is a dynamic process consisting of three phases: elimination, equilibrium, and escape (Fig. 40). In English, this process is sometimes referred to as the "three E's" (Schreiber et al., 2011; Narendra, 2013).

The elimination phase reflects the original concept of immune surveillance. The immune system recognizes and eliminates tumor cells. The elimination is either complete, when all tumor cells are successfully eliminated, or incomplete, when some tumor cells remain. The efficiency and success of elimination depends on the degree of immunogenicity of the tumor and the immunocompetence of the host organism. When elimination is complete, the entire immunoediting process is complete.

In the case of incomplete partial elimination, a dynamic equilibrium is established between the host immune system and the developing tumor. Immune cells and their cytokines exert constant pressure on tumor cells sufficient to control and limit tumor growth. However, they are unable to completely eliminate the tumor. The equilibrium phase, which represents a dormant state, can last for a very long time, in extreme cases for the rest of the host's life. A dormant depot can form from many genetically unstable tumor cells. Although many older variants of tumor cells are removed during the elimination and equilibrium phases, new variants with different mutations may

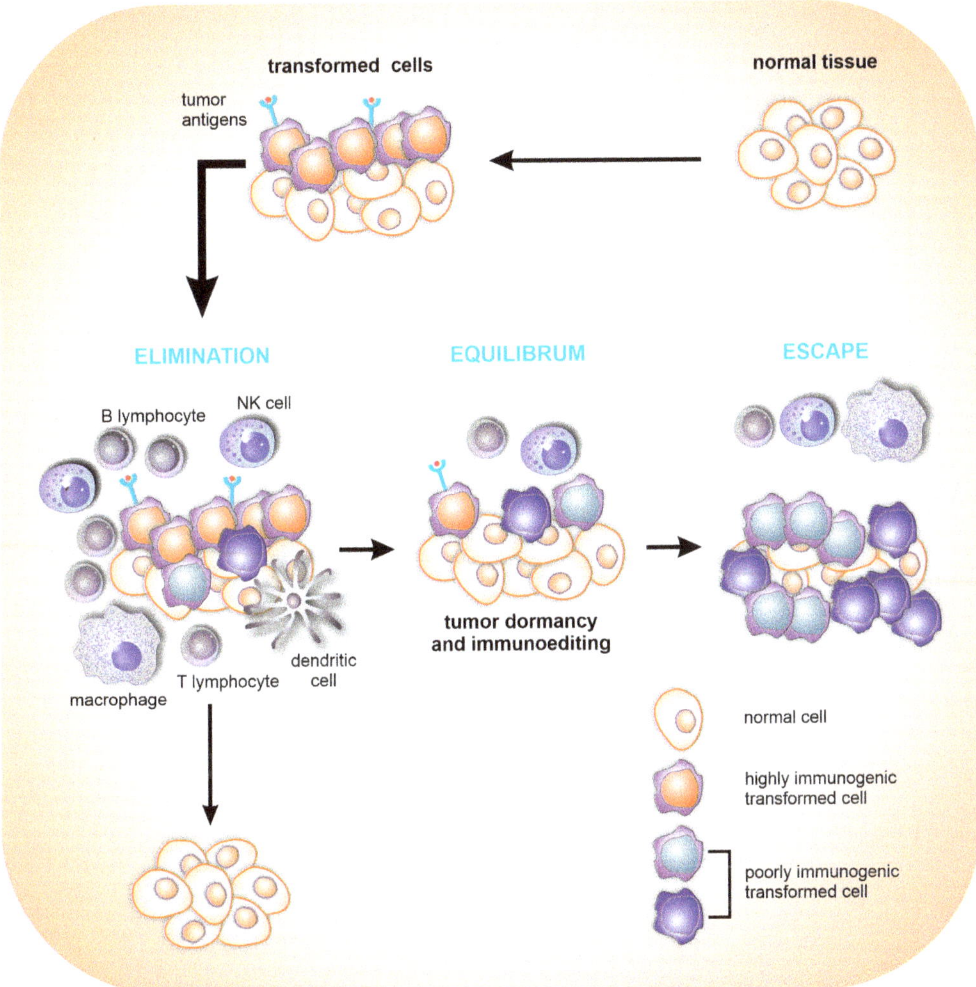

Fig. 40 Concept of tumor immunoediting

Tumor immunoediting proceeds in three phases: elimination, equilibrium, and escape. In the elimination phase, innate and acquired immunity together eliminate the developing tumor long before it can manifest clinically. When elimination is complete, the organism remains tumor-free and the process is thus finished. In the case of incomplete, partial elimination, a dynamic equilibrium is established between the host immune system and the developing tumor. The phase of equilibrium may also be the final stage of immunoediting, and the organism may contain a dormant occult tumor for the rest of its life. The constant selection pressure of the immune system on genetically unstable tumor cells in the phase of dynamic equilibrium may result in variants of tumor cells that are not recognized by the adaptive immune system and/or are insensitive to and escape the effector components and immune surveillance of the system. The immune system is no longer able to control tumor growth, and the tumor continues to develop and grow (adapted from Schreiber et al., 2011, modified).

constantly arise, some of which may have increased resistance to immune surveillance. Tumor genetic instability and heterogeneity may be a critical force that allows tumors to escape immune system "surveillance" when a new population of tumor cells with lower immunogenicity emerges. The opposite source of the imbalance may be a weakening of the immune system, which thereby loses its ability to keep the tumor under control. As a result of the imbalance, tumor development enters the escape phase, where the immune system is no longer able to control tumor growth, so the tumor continues to develop and grow (Fig. 40) (Schreiber et al., 2011; Finn, 2012; Narendra et al., 2013).

Escape mechanisms

The difference between tumor cells and normal cells is small and may be further reduced during tumor development, for example, due to loss or alteration of tumor antigens. Tumor cells can therefore be ignored by the immune system. This is their passive defense and their passive way of evading immune surveillance. Tumor cells can also actively defend themselves through mechanisms based on neutralization of some components of the immune system. For example, the activity of T lymphocytes, which play a key role in the recognition and elimination of pathogens and tumor cells, must be very tightly regulated. T lymphocytes must be activated and stimulated to respond effectively, but the duration and intensity of the immune response must be controlled by inhibitory signals called immune checkpoints to prevent unwanted autoimmune responses or excessive damage to the patient's own tissues. Some tumors violate the physiological balance between excitatory and inhibitory signals by producing high levels of immune checkpoint factors, effectively inhibiting T lymphocyte function (Šťastný, Říhová, 2015; Hořejší et al., 2017; Thommen, Schumacher, 2018; Higher, 2019; Toor et al., 2020; Tang et al., 2020).

As mentioned earlier, successful tumor immune surveillance requires the proper functioning of all the components of the immune system, not just T lymphocytes. Tumor strategies that lead to the elimination or weakening of immune surveillance may also target other components of the immune system (Sharma et al., 2017; Tang et al., 2020). Dendritic cells play an important role in the tumor-specific immune response of T lymphocytes. They are the key link between rapid innate immunity and slowly developing adaptive antigen-specific immunity. In the absence of pathogens, dendritic cells remain immature, take up the dead cells of healthy tissue and present processed fragments of endogenous molecules on their surface. These are recognized by T lymphocytes but do not activate them. Immature dendritic cells are thus instrumental in maintaining tolerance to endogenous tissue. Mature dendritic cells are formed after contact with a pathogen. They are the main cells that present antigens to T lymphocytes, activating and stimulating them and instructing them to migrate to specific sites. Dendritic cells are sensitive to various cytokines, growth factors and other signaling molecules. In tumor tissue, in the so-called tumor microenvironment, the structure and intensity of signaling molecules is greatly altered compared with physiologic tissue (see chapter 12A), and as a result, dendritic cells are also altered or defective. Therefore, instead of activating the immune response, they often suppress it (Narendra et al., 2013; Hořejší et al., 2017; Fu, Jiang, 2018).

NK cells, natural killers involved in innate immunity, along with T lymphocytes, are the key cells that provide immune surveillance of tumor cells. They are similar to T lymphocytes in many ways, but because they do not have antigen-specific receptors, they do not require antigen presentation for their activity. They respond to specific structures on the surface of tumor cells, such of those formed in the context of DNA damage or cell cycle deregulation, and can kill these cells directly. The surface receptors of NK cells are important for their function. Some of them are activating, others inhibitory, and the type of the effect depends on how strongly they are stimulated by the target cells. Tumor cells that preferentially stimulate activating receptors are strongly immunogenic to NK cells and are eliminated by them, while cells that preferentially stimulate inhibitory receptors on the surface of NK cells are resistant to them. This mechanism is the essence of NK cell-mediated immunoediting: only cells resistant to natural killers remain in the tumor (Narendra et al., 2013; Huntington et al., 2020).

Among the immune cells found in tumors, macrophages are the most abundant. The number of macrophages in tumor tissue correlates with poorer prognosis and faster tumor growth. Macrophages are extremely plastic cells. Depending on the signals from the microenvironment, they can take on many subtypes of different phenotypes, from strongly proinflammatory M1 macrophages to M2-type macrophages, which are inversely involved in the termination of the inflammatory response and in the healing processes of damaged tissue (see the next chapter, chapter 11A). Tumor-associated macrophages also consist of multiple cell populations that exhibit characteristics of both M1 and M2 macrophages. However, some factors in the tumor microenvironment, including hypoxia and some signaling molecules, stimulate the development of M2 macrophages. For example, they produce growth factors (e.g. VEGF) and metalloproteinases (MMPs), which paradoxically stimulate tumor growth and promote angiogenesis and metastasis (Narendra et al., 2013; Sawa-Wejksza, Kandefer-Szerszeń, 2018).

The immune system is a double-edged sword

Detailed analyses of the actions of individual cellular components of the immune system have shown that some immune cells and their typical cytokines tend to have an antitumor effect, while others are more involved in stimulating tumors in the escape phase. However, it is difficult to draw a sharp line between them because most components of the immune system are heterogeneous and highly plastic, and the behavior of immune cells is highly dependent on signals from their environment.

The tumor microenvironment (as will be shown in chapter 12A) is specific, different from physiological conditions, and changes during tumor development. This also has implications for the individual components of the immune system. At different stages of tumor development, they may be involved either in eliminating tumors or in stimulating their progress. In this sense, the immune system appears to act as a double-edged sword in tumor development (Narendra et al., 2013).

11A. Chronic inflammation

Inflammation

Inflammation is a basic defense mechanism, a stereotyped response of the organism to infection or tissue damage that serves to maintain tissue homeostasis. It is usually triggered by infectious pathogens, signals from endogenous cells and tissues associated with tumor development, and also stimuli related to cell death, particularly necrosis. The immune system plays a key role in inflammation, but neuroendocrine signals are also involved in its regulation. During inflammation, the entire organism adapts to the stressful situation. The body responds with changes in metabolism (mobilization of energy reserves) and behavior (fatigue and loss of appetite). This suggests that both the immune system and the neuroendocrine system are involved in the process. The latter influences the overall state of the organism and also significantly modulates the immune response itself. In addition to the components of the immune system, other cells such as endothelial cells, fibroblasts, and smooth muscle cells, as well as the components of the extracellular matrix, are also significantly involved in inflammation. The inflammatory process is associated with the activation, proliferation and migration of the cells involved. The precise sequence of the individual, highly ordered and interconnected steps is ensured by adhesive cell interactions and is controlled and regulated by the signaling molecules (cytokines, mitogens, enzymes and other molecules) that are released by the cells or the extracellular matrix in a precise sequence to the environment.

The inflammatory response can be divided into three phases, although the division is approximate because the phases are intertwined and their dynamics are a little different each time. The initiation phase occurs immediately after the stimulus and usually lasts hours. Of the components of the immune system, innate immune mechanisms clearly dominate, and a characteristic feature of this phase is the rapid amplification of the immune response and the pronounced accumulation of cells of the immune system at the site of injury. The cellular components of innate immunity are activated by intracellular signaling pathways and signaling molecules. The transcription factor NF-κB is the most important of them. Activation of NF-κB is associated with stimulation of inflammation, but also significantly stimulates proliferation, cell survival, and dedifferentiation (Elinav et al., 2013).

The most intense phase of inflammation is characterized by heightened activity of specific immunity and usually lasts several days. This phase is crucial for the completion of the body's defense response and creates the conditions for the development of an immune memory. The next encounter with the same stimulus – an antigen – triggers a much more effective response. Clonal expansion of T and B lymphocytes specific for the antigens that caused the inflammation is the typical feature of specific immunity. The intense proliferation of the corresponding lymphocytes continues as long as the triggering antigens are present. If they are eliminated, the proliferation of the corresponding lymphocytes ceases. All cells (lymphocytes and components of innate immunity) whose function is no longer needed in the ongoing inflammation die by apoptosis.

The late phase of inflammation is primarily determined by repair processes aimed at restoring the integrity of the tissue. In addition to cells of the immune system, endothelial cells, fibroblasts, and other cell types are involved. Fibroblasts and related cells are formed by the epithelial mesenchymal transition process mentioned in chapter 9A (Kalluri, Weinberg, 2009). Regeneration of the extracellular matrix and vasculature are also part of tissue healing. The repair phase may take several weeks. Almost every healing process involves some degree of fibrotization, i.e. replacement of tissue by ligaments with consequent limitation of its functionality (Krejsek, Kopecký, 2004).

Acute and chronic inflammation

The inflammatory reaction usually subsides quickly after the damaged tissue has healed. This is called acute inflammation. However, if the inflammation persists, it is referred to as chronic inflammation, which, unlike acute inflammation, can damage the tissue or lead to marked fibrotization. The difference between acute and chronic inflammation lies not only in the duration, but also in which components of innate and specific immunity are involved, the interactions between them, and how this occurs. While in acute inflammation the innate component of immunity forms the first line of immune response and subsequently stimulates specific immunity, the opposite may be true in chronic inflammation. Sustained activity of components of specific immunity may permanently hyperactivate the innate immune response and lead to damage of tissues exposed to chronic inflammation.

Persistent stimulation of the immune system and thus persistent, i.e. chronic, inflammation may be caused by certain genetic predispositions associated with, for example, the development of ulcerative colitis, pancreatitis, or rheumatic diseases; by infectious pathogens such as the bacterium *Helicobacter pylori*, which causes chronic gastritis, or the hepatitis B and C viruses, which cause chronic hepatitis; or by toxic and irritating factors such as smoke, asbestos, and long-term exposure to ultraviolet light (de Visser et al., 2006; Keibel et al., 2009).

Chronic inflammation and tumor development

While acute inflammation is part of an effective immune system response to the presence of a tumor and may contribute to its elimination, chronic inflammation is more likely to contribute to cancer development. People who are prone to chronic inflammation have a higher risk of developing a tumor. The same is true for the association between persistent infection with the bacterium *Helicobacter pylori* and the development of stomach cancer, or hepatitis virus infection and the development of liver carcinoma. These pathogens are thought to cause cancer indirectly by triggering chronic inflammation. Perhaps the most convincing evidence for a causal relationship between chronic inflammation and tumorigenesis is provided by epidemiological studies showing that drugs that inhibit chronic inflammation significantly reduce the risk of cancer (de Visser et al., 2006). Infections are generally considered to be the most common cause of chronic inflammation that initiates and promotes tumor development. In this context, it is interesting to note the possibility that commensal microorganisms, i.e. microorganisms that are quite abundant in the human body, also influence the variability of the inflammatory response and the potential development of tumors (Elinav et al., 2013).

Chronic inflammation favors the development of tumors in all their stages, from the initial stage to their progression, and contributes to the development of almost all the typical signs of tumors. On the one hand, it supplies the environment with bioactive molecules, including growth factors, enzymes, and signaling molecules, which subsequently influence cell properties and behavior, and on the other hand, through the production of chemical substances (mainly oxygen radicals) that have mutagenic effects, it stimulates the generation of additional mutations in the developing tumor cells, thus accelerating their transformation. Indirectly, inflammation increases genetic instability by suppressing DNA repair mechanisms and causing damage to cell cycle checkpoints. In addition to genetic instability, inflammation significantly stimulates cell proliferation, cell survival, supports angiogenesis and cell invasiveness, stimulates epithelial-mesenchymal transition and other cell properties and behavior of tissues associated with tumor development (Schreiber et al., 2011; Elinav et al., 2013).

Immunoediting and inflammation

The observations made at the start of this chapter suggested that the inflammation that stimulates tumor development and the immune response directed against the tumor are mutually exclusive processes. However, it is much more likely that they are dynamically interconnected processes that compete for dominance and victory as tumor cells evolve and undergo a process of immunoediting (Schreiber et al., 2011). The balance between the tumor-protective and tumor-promoting effects of inflammation is itself similarly fragile and dynamic. Moreover, the relationship between chronic inflammation and the developing tumor is not one-sided. Chronic inflammation comprehensively affects the entire course of tumor development, but on the other hand, tumor cells influence almost every step of the antitumor inflammatory response through their own signaling molecules that block effector function and stimulate the immunosuppressive function of immune cells. Thus, the tumor itself actively modulates the immune response. Chronic inflammation and tumor development are intertwined processes. In a sense, tumor development can be viewed as a form of tissue repair and healing that has spun out of control, as "wounds that do not heal" (Elinav et al., 2013).

Interactions between the immune system and the neuroendocrine system

If there is such a close relationship between the function and state of the immune system and the development of tumors, then it is clear that anything that affects the immune system will also have an impact on tumor development. The intuitive association of some psychological problems, particularly stress and depression, with an increased risk of tumor development (as well as a number of other diseases) is thus given a rational basis and allows for a more conscious understanding (Reiche et al., 2004).

The immune system and the neuroendocrine system are the two major regulatory systems of the body. Together they are involved in maintaining a constant internal environment, homeostasis.

They are functionally and anatomically interconnected. Nerve fibers are found in all organs of the body, including the organs of the immune system. On the other hand, the cells of the immune system are located near nerve fibers and are part of the nervous tissue. Mutual interactions between the neuroendocrine system and the immune system are provided mainly by the sharing of humoral components of both systems. Hormones and neurotransmitters effectively affect the cells of the immune system, and conversely, cytokines and other signaling molecules produced by cells of the immune system influence the functions of the nervous tissue (Krejsek, Kopecký, 2004).

Thus, there is constant two-way communication between the neuroendocrine system and the immune system. Various physical, psychological, and psychosocial stressors can modulate the immune response through this connection. The body's adaptation to stress is controlled by the hypothalamic-pituitary-adrenal glands and the sympathetic nervous system. The body's adaptive responses to stress therefore include changes in the immune system. Not all types of stress are harmful. Short-term stress can activate and stimulate the immune system, while the effects of long-term stress are harmful. The function of the immune system is dampened by long-term stress. The body must respond to stress in a very specific way and in a hierarchical order according to its actual importance. Because the immune response is extremely demanding from a metabolic point of view and almost always involves limited damage to the body's own structures, it must be postponed to the later stages of the adaptive response to stress. However, the limitation of immunological reactivity during long-term stress also has its downsides: among others, an increased risk of tumor development (Krejsek, Kopecký, 2004; Reiche et al., 2004).

10B. Circumvention of laws

The immune system in tumor development: A double-edged sword, the Janus, Jekyll and Hyde principle

The relationship between tumors and the immune system is complex, multifaceted, and anything but simple. Of the eleven typical features of tumors, two relate to the role of the immune system. One characteristic is referred to as "the ability of tumors to evade immune surveillance" and the second is referred to as "the presence of chronic inflammation supporting the tumor" (Hanahan, Weinberg, 2011). Therefore, it is impossible to describe the immune system's relationship with tumors with a single sign. Moreover, both signs illustrate the exact opposite effects of the immune system on tumors. While "the ability to evade immune surveillance" demonstrates, albeit indirectly, that the immune system "oversees" tumor development, i.e. controls, regulates, restricts, and ideally prevents it, the second sign, on the contrary, says quite directly that inflammation – in its chronic form – promotes tumor development. This alone clearly shows the ambiguity of the immune system's relationship to tumors. First, the developing tumor is recognized, attacked, and possibly even destroyed by the immune system, and acute inflammation can also be part of this effective defense phase. In contrast to acute inflammation, which is undoubtedly one of the protective barriers of the organism against the tumor, chronic inflammation supports the tumor in its ability to overcome those protective barriers. Thus, the inflammatory response itself is a typical double-edged sword. It helps fight the tumor, but it can also significantly support it (Narendra et al., 2013). Monitoring of the tumor by the immune system itself can also be double-edged. If the monitoring is not consistent and does not lead to complete removal, it contributes to the formation of the tumor – by modifying it to a less immunogenic state and thus supporting its further development.

The "double-edged sword" analogy (mentioned earlier in chapter 6B) is used frequently in biology today and in a variety of contexts. This probably reflects the fact that contemporary biology is increasingly concerned with complicated, convoluted processes whose individual components have neither simple, unambiguous relationships with each other nor clear, straightforward effects. To capture the ambiguity of the immune system's role in tumor development, some authors use the metaphor of the god Janus. In Roman mythology, Janus is the god of gates, doors, entrances, beginnings, and endings. He is usually depicted with two different faces, turned inward and outward, bearded and beardless, and representing also the sun and the moon. He is a symbol of change and transformation, of passing from one state to another, such as passing from the past to the future, from childhood to adulthood, and the like. The "Janus principle," as authors have called it, was used precisely to explain the dual, "two-faced" role of the immune system in the development of tumors: an immediate, effective, and ultimately protective response of the organism, including acute inflammation, which transforms over time into a long-term, protracted, chronic condition that is not only ineffective in controlling tumor development and growth but, on the contrary, begins to support and stimulate the tumor (Finn, 2012). Some authors have chosen the apt name "Dr. Jekyll and Mr. Hyde" for these two Janus-faces of the immune system (Theo-Harides, Conti, 2004).

Do all cells in a multicellular organism pull together?

In a well-functioning multicellular system, cells do not compete with each other but, on the contrary, cooperate. Their genome most likely contains the genes that code for factors maintaining stability and harmony between the somatic cells within the organism. These genes are inherited in the germ line and are positively selected, because only a coherent and functioning system guarantees that its genes will be successfully passed on to the next generation. However, this does not mean that, at the somatic cell level, the ability to grow and divide rapidly cannot also be favored, as is typical of cells that live independently as unicellular organisms. In terms of "competition," the cells of multicellular organisms cultured *in vitro*, i.e. outside the context of an organism, are not very different from unicellular organisms (Breivik, 2005, 2006). We have already reflected on this conflicting internal setting of cells within a multicellular organism in chapter 6B and also quoted Nick Lane, who writes: "An individual is an organism composed of genetically identical cells, which are specialized to perform diverse tasks for the good of the organism as a whole. From an evolutionary point of view, the question is: why did these cells subordinate their selfish interests to collaborate so altruistically in the body? Inevitably there were conflicts between the various levels of organization in the body. Such conflicts spurred the evolution of molecular 'police force', which curbs selfish interests much as the legal system enforces acceptable behavior in society. In the body, programmed cell death, apoptosis, is central to the policing of conflicts" (2005, p. 191). Lane compares the function of apoptosis, whose role in maintaining the integrity of a multicellular organism was discussed in chapter 4, to that of the police.

Genetic stability as a system of rules for decent behavior, and adherence to them (chapter 6), is another part of this control. However, when cells do not follow the rules and begin to violate them, they begin to change. This transformation is characterized by the appearance of new tumor antigens on the cell surface. Such cells must be stopped by "the law." They are recognized by the immune system. In the body, the immune system plays the same role as the law enforcement system in human society: the thin blue line that protects the body/society from any cell/person that does not obey the law and thus threatens the integrity of the system it belongs to.

The immune system contributes to maintaining the integrity of the organism

Organisms have an internal environment that is different from the external environment, and they must maintain stasis despite significant fluctuations in the external environment. For example, organisms maintain a very constant temperature, a constant pH, a constant concentration of various substances in the blood, a constant blood pressure, and the like. They have the capacity for homeostasis. Homeostasis is a self-regulatory mechanism based on the feedback principle that enables organisms to maintain a state of dynamic equilibrium when the values of individual factors fluctuate only within a narrow range of permissible tolerances. General regulatory systems are involved in maintaining the homeostasis of multicellular organisms, particularly the nervous, endocrine, and immune systems. The nervous system serves to perceive and process stimuli and provides the appropriate response. The endocrine system is a network of endocrine glands that secrete hormones circulating through the body via the bloodstream that act on both nearby and distant target organs. They regulate metabolism, the body's response to stress, and act as primary regulators of growth and reproduction. Unlike the nervous system, which serves preferentially to immediate response to stimuli, the endocrine system acts more slowly and over a longer period of time.

The immune system is the body's defense system and also controls the healing of injuries and the regeneration of tissues. On the one hand, the mechanisms of the immune system ensure that the organism is reliably defended against all disturbing or threatening agents; on the other hand, they must have an incorruptible self-tolerance. They must recognize and eliminate all foreign factors, such as antigens, pathogens, toxins, and even altered structures of its own, that could potentially harm the health of the organism, but must not turn against its own physiological structures. The immune system, along with the nervous and endocrine systems, forms a regulatory and coordinating system that oversees the integrity of the organism. Traditionally, the three have been considered separate, self-contained systems that were and continue to be studied in specific, well-defined disciplines such as immunology, neurology, and endocrinology. However, the systematic study of neuropeptides and their receptors led to the discovery of the structural and functional interconnection of all three systems, which should therefore be considered as a single dynamic information and regulatory network of psycho-immuno-endocrinology (Pert et al., 1998; Pert, 2016).

Do all people in the human community pull together?

Cells that are part of a multicellular system seem to be in constant internal conflict. Evolutionarily, they have the need and the ability to assert themselves individually, to grow and divide as fast as possible, to act faster than other cells. At the same time, when they are an integral part of the organism, they accept subordination to serve a common need – the success of the whole. This shared success of the community of cells in the organism results at least partly from the subordination of individual cells to the greater interest of the organism and their willingness to suppress the "desire to emerge individually." This internal conflict must be kept under control, and a successful organism must be equipped with tools and mechanisms for this function. One of these tools is the immune system and its ability to recognize and neutralize those cells that go their own way, reject subordination, and threaten to harm or destroy the entire organism.

When people live in a society that they want to function well in the long run, they also have to follow the rules. In every society there are rules that smooth relations between people and help to avoid conflicts. The rules, or more precisely the widespread following of the rules, keep people from getting into quarrels and prevent the fight of all against all. People cannot pursue only their own interests and enforce them ruthlessly. If they try to do so despite the rules, the law must stop them. If they are to form a functioning community, fulfill their lives meaningfully and achieve something, they need order. "Lack of order means chaos, poverty, death and destruction" (Ringen, 2017, p. 199). Order enables people to live lives that are beneficial both to themselves and to society as a whole. The state is the instrument of order. It has the power to govern, judge, and create laws for society.

The legal system contributes to the maintenance of the integrity of the state

The legal system summarizes all the legal norms that apply within the territory of a given state at a given time. It is a system of commands, prohibitions or permissions recognized or directly established by the state and followed by people and institutions. It is secured and enforced by state authority. To be incontestable, universal and valid, legal regulations must express natural law. The nature and meaning of law must be consistent with the natural world and common life, and contain some general principles that ensure this. The most important principle is justice, which is one of the foundations for the organization of human relations. It includes equality before the law. Equality before the law means that the law applies equally to everyone, including state actors. Compliance with this principle must be ensured in some way to prevent its circumvention.

In a constitutional state, therefore, the principle of separation of powers applies – power in the state is divided into three distinct branches: legislative, executive, and judicial. None can exercise unlimited power, and none may have the power to weaken or abolish the other parts. A system of checks and balances, in which neither the will of the legislature, nor the will of the government and the president, nor any decision of the Constitutional Court or the Supreme Court has unlimited validity, is considered the main characteristic of a democratic constitutional state and democratic politics. The search for a balance between the different types of power – executive, judicial and legislative – is the main condition for seeking and establishing the public interest (Přibáň, 2013). This proves that the legal system alone is not sufficient for the functioning of the state, but it is absolutely necessary and irreplaceable for the functioning of the state and the guarantee of order. In fact, it is comparable to the function of the immune system in the body. The immune system does not itself perform all the regulating, coordinating and integrating functions, but shares this role with, for example, the nervous and endocrine systems. But the weakening of the immune system fatally threatens the function and life of the organism.

Balancing the interests of the individual and the whole

Another principle of good law is balancing the interests of the individual and the whole. This is comparable to maintaining healthy relations between cells in a multicellular organism. On the one hand, the cells must subordinate themselves to the interests of the organism; on the other hand, in a well-functioning organism, each cell must thrive. Similarly, a delicate balance must be maintained between the interests of society and the interests of the individual members of society. The extent to which priority is given to the interests of the individual or the whole may vary from society to society, as attitudes always depend on the nature of the community and the worldview of the legislator. However, the effort to balance competing needs always affects the functioning of the community, its success and efficiency, and the satisfaction of its members. If only the whole is valued and the individual is obliged to submit and conform in everything and always, dictatorship arises, which leads to egalitarianism and suppression of individuality.

The status of citizens' rights can serve as a yardstick to measure the level of development of human society. Respect for human rights determines the actual level of democratization of society and the rule of law, as well as the well-being of each individual. However, if only the freedom of the individual were taken into account, chaos would result. The rights of the individual end where the rights of others begin. If these limits are not respected by the people involved, the state must intervene with its own means of enforcement. The state must not allow an individual or a group of individuals to pursue their selfish interests and use the system only for their own benefit at the expense of other members of society. Or at the expense of the common interest of the entire system.

How to suffocate democracy

We are currently witnessing a gradual process of suppression and waning of democracy in many parts of the world. Democracy as a "government of the people," that is, a form of organizing society in which political decisions are made according to the will of the majority, remains as a facade, but its inner pillars – independent institutions, supervisory authorities, civil society, free, public media, and also components of the legal system – are gradually being weakened and crippled. The so-called illiberal democracies are booming. These systems preserve procedural, formal features of democracy (such as regular elections), but real politics and the actual exercise of power are permeated with populist and authoritarian tendencies. The will to balance the various components of constitutional power is disappearing, the control of politics by law is weakening, and the power of capital is growing stronger. Politicians are losing respect for constitutional principles and rules, even though they are the essence and source of legitimacy of their power. They trample on and disregard these rules and principles: they question and disregard judicial rulings, they sweepingly replace senior officials according to party interests, they fuel citizens' distrust of these institutions through their emotive, disparaging labeling ("judocracy") when rulings or verdicts do not suit them. Questioning and undermining the independence of legal institutions reflects the intention to weaken the control of compliance with rules and laws so that individuals and interest groups (e.g. political groups, corporations, holdings, etc.) can advance their narrow private or group interests at the expense of other parts of society (Přibáň, 2013; Pehe, 2018). Once these organizations manage to escape the control of the legal system, they are actually in the "escape phase." And the legal system that previously kept them in check often comes under their control and influence and begins to support them in their actions.

Suffocated justice

The immune system must reliably and carefully distinguish between the foreign and its own. It must clearly recognize what can or must be tolerated and what must be removed immediately. Tumor cells gradually develop from healthy, physiological, i.e. "own", cells. Their otherness, their "strangeness" increases only gradually, step by step, and may continue to change, but may also decrease, as they develop. Therefore, their early and reliable detection can be a very difficult task for the immune system. To accomplish such a difficult task, the immune system must be in excellent condition. The "success" of cancer cells, as evidenced by their evasion of immune system surveillance, may result not only from the development of cells with increasingly pronounced resistance to the immune system due to reduced immunogenicity, but also from the weakness of the immune system. A weakened immune system may lose the ability to monitor the developing tumor. Such hesitation may have fatal consequences for the organism. The weakening of the legal system can have similar consequences for the state. What may seem "insignificant" in times of relatively stable political conditions, a harmless, expedient concession to political power that brings immediate benefits, may in the longer term result in inability to deal with the processes that threaten the state and may even point to its future collapse.

The fifth deadly sin according to Konrad Lorenz: Genetic decay

In his argumentation on the fifth deadly sin, which he called *genetic decay*, Konrad Lorenz raises questions about the human sense of justice and social behavior. He does not come to a clear conclusion about its genetic or cultural-historical origin, but considers its loss or weakening a deadly threat to humanity. He observes the great similarities in the legal systems of different countries and their consensus view that the exclusion of all "socially dangerous parasites," i.e. selfish people who do not respect the good of the whole, is crucial for the survival of any society. By the way, even Konrad Lorenz metaphorically compares these social parasites with tumor cells and the necessity of legal norms with the necessity of the creation of specific antibodies that effectively protect the body from malignant tumors: "We may accept it is a scientific fact that *Homo sapiens* has a highly differentiated system of behavioral patterns at his disposal, which, in a way analogous to that of the system of antibody formation in the cell state, serves to eliminate socially dangerous elements" (1974, p. 46). "We must learn to combine judicious understanding of the individual with considerations for the rights of the community. The individual, deficient in certain social-behavior patterns and the feelings that go with them, is a sick man deserving our pity, but the deficiency itself is unmitigated evil." Lorenz warns that "through the decay of social behavior we are threatened by apocalypse in a particularly horrible form" (1974, pp. 58–59).

11B. Constant mobilization

Inflammation is both protective and self-destructive

Inflammation is a form of immune system response to various types of tissue damage and the presence of foreign antigens. It is an aspect of protective adaptation and one of the homeostasis mechanisms. Inflammation is a routine, complex process aimed at containing the damaged tissue, preventing further spread of damage, and healing the injury. It involves both the removal of damaged cells and structures and regeneration and restoration of tissue. Inflammation can restore tissue nearly to its original state. Healing may be accompanied by fibrotization, when the regenerated tissue is more or less permeated by ligaments, or the inflammatory process may turn into a chronic state. The development of chronic inflammation may be due to a prolonged infection, a recurrent infection, the coincidence of multiple lesions that permanently cause inflammation, or autoimmune diseases. In addition, inflammation can also be primarily chronic. Thus, inflammation is at the same time the most important defense, but also a self-damaging reaction of the organism. And it is often difficult to distinguish when the inflammatory reaction is mobilized for acute defense of the organism and when it is causing damage unnecessarily and chronically.

Mobilization: To set in motion

The word "mobilization" is usually associated with preparation for a warlike conflict. A state that declares mobilization is put on military alert. In anticipation of military conflict or other extraordinary events, the government, in conjunction with the military authorities, seeks to increase the number of armed forces. Reservists are called up into the army and new units are formed. To provide for the needs of the military, logistics and the economy are activated. Everything is set in motion.

The word "mobilization" comes from Latin and actually means "to set someone or something in motion." So, it has a much more general meaning and is used in many contexts other than military emergency. For example, a person preparing for a particularly difficult task may also need to "mobilize all his forces." In a non-military sense, the word mobilize is probably most commonly used in politics and the public sphere. Politicians and political parties "mobilize" voters to win their votes in elections; governments then "mobilize" citizens to fulfill their own electoral programs. But are there also "programs" and goals of a different kind?

Let us roll up our sleeves!

Politics and political struggle, as well as the exercise of power, by their very nature involve conflict and the permanent work of dealing with differing or opposing opinions, attitudes, and aspirations. "The public space must be fed by conflict and protest" (Přibáň, 2013, p. 38). This is the essence of democracy. Democracy is a form of political institution that allows all citizens to participate in the administration of the state. Political decisions are made according to the will of the majority. The minority does not lose its rights and freedoms, but submits to the majority. In a democracy, everyone can have their say, can express different opinions, options, plans and proposals to solve problems, and all kinds of political programs can be proposed. Through public debates, argumentation and agitation, and using mechanisms established by law, a political program is chosen that convinces the largest part of the citizenry – the voters. The winning political alliance can form a government, which then seeks to implement its political program. A government has two tasks: to make decisions and to implement them. Implementation is difficult because the government faces what veteran Norwegian sociologist and political scientist Stein Ringen calls a "nation of devils." The "nation of devils" is an amorphous mass of people who should accept the government's decisions but often do not want to (Ringen, 2017). Humans, like cells, experience internal conflict. The conflict between the pursuit of personal well-being and the need to work for the common good.

The art of governing, then, is to draw unwilling and sometimes even hostile people to the government's side so that they do not thwart the government's intentions but, on the contrary, support them and are involved in the pursuit of the common good. One of the last options available to the government in its implementation efforts is coercion: the threat of use, or use of, power to require obedience through force, through the pressure of executive institutions. But for a society to function truly efficiently, the government cannot rule only through commands and controls. It must also encourage, persuade, and motivate people, to influence their mindsets and preferences to gain their willingness and voluntary support for its goals. The effort to achieve this ability to influence society (by the action of public officials, experts, entrepreneurs, workers, voluntary organizations, etc.) to a willingness to participate in achieving a shared vision is what Ringen calls the art of mobilization, which he considers a necessary component of a good and successful government. The starting point for successful mobilization is, of course, an appealing and inspiring vision, the realization of which is in the interest of society and has the potential to bring benefits to all. Mobilization then means winning support from below through a wide range of different incentives, including salaries, working conditions, pensions, titles, and even shares in power. This constructive mobilization, then, is a way to effectively achieve specific goals, but it is also a strategy with certain long-term consequences. Mobilized people stay, live, thrive, and also gain influence. This changes and builds up the whole society. It develops and becomes rich, not only economically, but also institutionally and humanly (Ringen, 2017). Social mobilization therefore has a similar positive, creative effect as acute inflammation, which not only protects and fights, but also heals, repairs, and restores tissue.

Up and at 'em! Critical polarization and destructive mobilization

Democracy presupposes diversity of opinion, i.e. multiple and opposing views, and thus also constant disputes and conflicts. The constant struggle, the so-called agon, is permanent and a perfectly legitimate part of democracy, of public life in general. But there are two types of conflict. Agonism is a relationship in which rival parties respect each other and recognize their mutual legitimacy. They compete for the opportunity to impose their opinions and proposals in carrying out tasks and solving problems and crises, and so vie for the support and votes of citizens.

One important mechanism by which citizens select governing representatives in democratic systems is through elections. Voters decide which of the competing politicians and which of the competing parties will govern for a set period of time. And so, before the elections, parties fight for the voters' favor with all kinds of weapons. That's mobilization! And sometimes with indiscriminate, even dangerous weapons. To win elections, for example, a legitimate political opponent turns to face an implacable enemy. And that enemy, whether internal or external, is portrayed in such a way as to create a sense of existential threat, and voters are mobilized into a fictional battle of good against evil. As a result, agonism is replaced by antagonism, which intensifies into an irreconcilable relationship between "us" and "them" in which there is no possibility of agreement, compromise, or mutual respect. The danger of such political mobilization lies in the fact that instead of a constructive, objective tone and a struggle within the rules and by legal means, politicians, in the name of saving the country from the abuse of a "dangerous and threatening enemy", twist and adapt the laws to their needs in every way.

The lawyer and sociologist Jiří Přibáň says: "If a political opponent becomes an existential enemy, then whoever has power can do with him what he wants. And that is the end of democracy" (Přibáň, 2013, p. 114). When antagonism replaces agonism, the result is a divided, polarized society that loses the willingness and the possibility to search for and find common solutions. An ideologically irreconcilably polarized society tends to stick to and follow a leader, even if it is a crazy, dangerous, malignant leader, who leads the mobilized masses and uses his power and the party or state apparatus to adapt the laws to his needs with impunity. Such destructive mobilization, which resembles chronic inflammation, does not protect society but, on the contrary, diverts it onto dark, dangerous trajectories (Macháček, 2011; Přibáň, 2013).

Permanent mobilization, permanent revolution

Permanent mobilization of the masses is one of the typical characteristics of fascist states, along with nationalism, messianism, and the leader principle. Fascist regimes require permanent mobilization of the masses as well as permanent expressions of public support, for instance through demonstrations, marches, and participation in certain rituals. A common mobilizing element is the theory of a conspiracy whose victim is the nation and whom those labeled "traitors," i.e. irreconcilable enemies – either groups excluded from the nation or certain members of the nation – are the perpetrators. The mobilization and enthusiasm of the masses can reach such proportions (and this is desirable for fascist movements) that many supporters and activists begin to profess fascism as the meaning and fulfillment of their own lives, and the distinction between private and public space becomes blurred. Even privacy becomes political and thus ceases to be private. Perhaps in this there is even a certain similarity with the development of tumors, when mechanisms of the affected organism (immune system, angiogenesis) work in favor of tumors and can no longer differentiate and thus defend or prioritize the interests and health of the whole system. The emphasis on permanent mobilization of the masses, sometimes referred to as "permanent revolution," is a characteristic that fascism shares with other totalitarian ideologies (Duraj, 2013).

Chronic inflammation promotes tumors; who and what promotes permanent mobilization?

Fascism, permanent revolution. It may seem that these dangers are distant in time, or at least in space. But it turns out that even in today's world, including Europe, nationalist and populist movements easily and ruthlessly stir up emotions and create information bubbles in which everything seems clear and simple, and as a result they tear down existing systems without bringing constructive, rational proposals and solutions. And just as the inflammatory response is typically a double-edged sword that helps to effectively fight cancer but can also significantly support it, mass mobilization can also be a positive tool in the hands of good politicians who can recruit the "devil's nation" for the effective fulfillment of certain constructive visions and goals, but it can also be an instrument of destruction and subjugation of masses of people and entire nations for base personal or group interests. And similarly as it is difficult to distinguish when the inflammatory response is mobilized for the acute defense of the organism and when the organism is already unnecessarily, chronically damaged, it can be difficult to distinguish between constructive and reasonable mobilization and fraudulent, devious, and destructive mobilization.

12A. The tumor microenvironment: Two perspectives of tumor development

Robert Alan Weinberg is one of the best known biologists in the field of cancer research. Among other things, he is the discoverer of the first human oncogene (*ras*) as well as the first tumor suppressor (*RB*), and the author or co-author of a number of key discoveries in the field of tumor biology. In 1991, he contributed to the book on the origin of tumors, *Origin of Human Cancer*, with the chapter "Oncogenes, tumor suppressor genes, and cell transformation: Trying to put in all together." He introduces it with the words of the French poet and philosopher Paul Valéry: "That which is simple is always wrong: that which is complicated is useless" (Weinberg, 1991).

In his work he summarizes the most important discoveries of the last 15 years. The discovery of the first oncogenes and proto-oncogenes, their functional diversity and the model of damaged cell signaling pathways. The realization that activation of a single oncogene is not sufficient for tumor transformation, that tumor development is a long-term process and involves multiple steps. He comments in detail on the discovery of the first tumor suppressor and the link between its damage and retinoblastoma development. And even then, in 1991, given the increasing and rather chaotic amount of new data and evidence, he wonders whether it would be possible to formulate some central – albeit perhaps simplistic – principles that would allow a clearer understanding of tumorigenesis and its detection, or whether this field would take the form of an increasingly complex, boundaryless phenomenology (Weinberg, 1991).

In the 1980s and 1990s, the list of recognized oncogenes and cancer suppressors grew explosively, and it became increasingly clear that the complexity and chaos in the field weren't diminishing. On the contrary, each tumor seemed to be "a unique experiment of nature" that gradually acquired a unique set of mutated genes in an unpredictable chronological order (Weinberg, 2014). Even then, Weinberg believed that there must be some order behind this chaos reflecting the genetic and biochemical changes in the various signaling pathways. As mentioned in chapter 2A, he and his colleague Douglas Hanahan moved away from the biochemical and genetic levels and focused on the general biology of tumors – the phenotypes of tumor cells and tumors that arise from them. They believed that there they'd find the order hidden beneath the surface. Their efforts resulted in the naming of seven typical features that at the time reflected the general biology of all tumors. In 2000, they summarized their conclusions in a seminal work with the brief, apt title "Hallmarks of Cancer" (Hanahan, Weinberg, 2000). Weinberg later recalled that he and Hanahan were almost certain that their oversimplified view, as much as it reflected their own thinking, would disappear without provoking a strong reaction. But it didn't disappear! It turned out that the need to identify unifying key themes to describe tumor development was profound, as evidenced by the thousands of authors and articles that later cited the paper. In 2011, they wrote a sequel to their already iconic work, this time titled "Hallmarks of Cancer: The Next Generation" (Hanahan, Weinberg, 2011), in which they expanded the list of typical features of tumors (Fig. 8). This work also elicited an overwhelming response, the only difference being that this time no one was surprised.

Somatic Mutation Theory (SMT)

The concept of Weinberg and Hanahan is now widely accepted to explain the origin and development of tumors. It is the mainstream of contemporary tumor biology, and also serves as the axis and outline of this book. It describes carcinogenesis as a process of gradual accumulation of mutations in proto-oncogenes and tumor suppressors. The entire process has a monoclonal origin, starting with a single "renegade" cell that, through the gradual accumulation of changes, transforms step by step from a healthy cell into a fully malignant cell (Figs. 6 and 7). Therefore, this concept of tumor development is sometimes referred to as somatic mutation theory (SMT). Somatic mutation theory focuses on unraveling the genetic and molecular context of cancer development, explaining tumors by mutations and genetic causes, the expression of damaged genes, and thus the cellular level of the entire process (Malaterre, 2007). In addition to numerous scientific articles, a popular science book on the origin of tumors and on important discoveries in the field of tumor biology was written by Robert Weinberg in 1998. He gave the book an apt title: *One Renegade Cell: How Cancer Begins* (Weinberg, 1998).

In formulating the somatic mutation theory, the findings of modern biology, especially molecular biology, were largely taken into account. In 1953, the structure of DNA was elucidated and the central dogma of molecular biology was formulated, defining the flow of genetic information between DNA, RNA and proteins (Fig. 41). In principle, it allows the transfer of information between nucleic acids and their translation from RNA to proteins. This means that the flow of information from

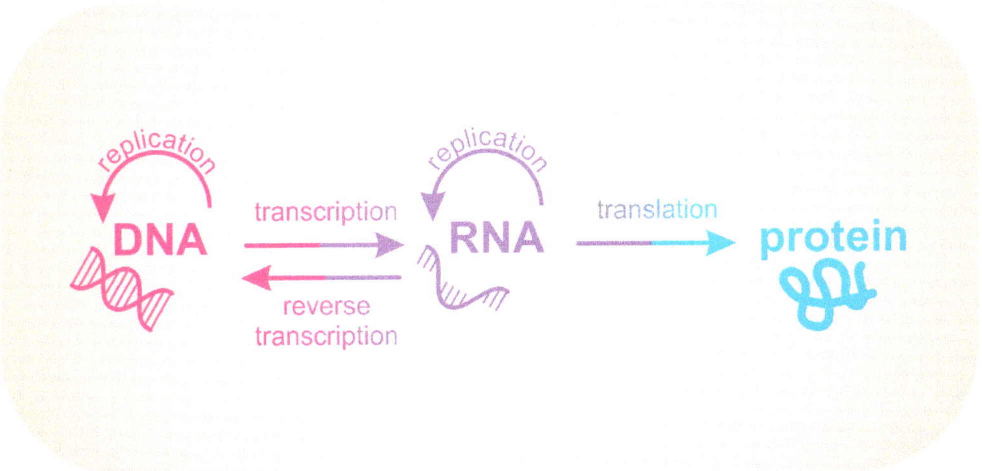

Fig. 41 The central dogma of molecular biology describes the flow of genetic information between DNA, RNA and proteins. The transfer of information between nucleic acids is possible. Commonly, the pathway involves DNA replication, transcription of DNA into RNA, and translation of RNA into proteins. Less common is RNA replication and reverse transcription of RNA into DNA. Information flow from proteins to nucleic acids is not possible.

proteins to nucleic acids and thus a "Lamarckian" contamination of genetic information by the resulting changes in the organism is not possible. The term "central dogma" was coined by Francis Crick, one of the co-discoverers of the DNA structure, and he chose the term "dogma" to describe something obvious rather than something that cannot be questioned. But the term dogma seems symptomatic. It has gradually reached its fulfillment even in the belief that the phenotype is a direct and accurate reflection of the genotype, and that, by analogy, the development of the organism is only the gradual unfolding of the genetic program and the influence of the environment is almost marginal.

However, the somatic mutation theory is not only a reflection of the *zeitgeist* and the prevailing way of thinking (so-called genocentrism). It stands on the solid foundation of a number of indisputable findings in tumor biology. For example, it has been demonstrated that a number of carcinogens, i.e. agents (physical, chemical, biological) that initiate or promote tumor formation and development, damage nucleic acids and cause mutations. The ability of some oncogenic viruses to cause cancer is related to their ability to carry active oncogenes or to deactivate key tumor suppressors in host cells. Tumor development has been shown to be associated with increased genome instability (see chapter 6A). All these discoveries found their logical interpretation in the somatic mutation theory. Moreover, the fields of molecular biology and molecular genetics have not only been vehicles of general progress by bringing new perspectives, deep insights, and better understanding of biological processes in general, but they have also demonstrated their tremendous power, efficacy, and productivity directly in oncology. They have contributed concrete advances in tumor prevention, diagnosis, and therapy, and opened the door to what is called targeted tumor therapy. The merit of somatic mutation theory as a productive concept guiding research is undeniable in this context.

Tissue organization field theory, Carlos Sonnenschein and Ana Soto

The somatic mutation theory is an absolutely dominant concept in cancer biology, but it is not the only concept. There is a significant alternative to it. It is the so-called tissue organization field theory (TOFT), first proposed in 1999 by Carlos Sonnenschein and Ana Soto in their book *The Society of Cells: Cancer and Control of Cell Proliferation* (Sonnenschein, Soto, 1999). They rely on two central claims that distinguish their theory from the somatic mutation theory.

First, they claim that the usual, given, i.e. natural state, of cells that are part of a multicellular organism is a state of proliferation. It is regulated and kept in check by inhibitory signals from the external, extracellular microenvironment. In contrast, SMT is based on the assumption that cells are in a quiescent state and only enter a state of deregulated proliferation through mutations (Alberts, 2010). Proponents of TOFT argue that bacteria and even unicellular eukaryotic organisms naturally grow and divide continuously when suitable conditions exist, such as in the presence of adequate nutrients and the absence of toxic substances, and conclude that this state is maintained even in the cells of a multicellular organism. They consider the statement that proliferation is a natural state of cells to be a fundamental biological postulate.

They also argue that the ability to move is a basic, natural property of all cells and therefore, like cell division, can only be inhibited. Thus, according to the proponents of TOFT, the mobility of metastatic tumor cells in the formation of metastases is not a newly acquired property, but a restoration of the original, natural state (Sonnenschein, Soto, 2005, 2011, 2016, 2018).

While SMT postulates that carcinogens induce mutations in cells that then lead to loss of control over proliferation, TOFT is based on the fact that the primary effect of carcinogens on tissues is the disruption of normal tissue organization, the disruption of physiological interactions between tumor and stroma, i.e. non-tumor cells and the non-cellular environment – the extracellular matrix. This loss of adequate structure, and thus loss of adequate physiological signaling between cells within the tissue, then leads to deregulation, i.e. loss of inhibition of cell proliferation and movement, and is also a major cause of tumor development. Thus, proponents of TOFT understand carcinogenesis as a "relationship problem." According to this view, tumor development is the result of the loss of constraints imposed by the interactions between cells and the physical forces created by the cellular arrangement within tissues, allowing cells to return to their natural state of proliferation and movement (Sonnenschein, Soto, 2016, 2018).

From this point of view, proponents of somatic mutation theory are sometimes called reductionists because they seem to reduce the cause and nature of tumors to the level of genes, i.e. they take a genocentric view, whereas proponents of TOFT are often called organicists because of their systemic view (Gilbert, Sarkar, 2000; Malaterre, 2007; Saetzler et al., 2011). As mentioned above, SMT is based mainly on the findings of molecular and cell biology. TOFT is rather based on the findings of developmental biology and embryology. As early as the 19th century, it was suspected that there is a parallel between embryonic development and cancer development, and that the essence of tumor development is altered tissue organization. The developmental plasticity of human fetuses was already known at that time, and the role of the environment in fetal development was well understood, as well. A number of recent observations and experimental data support and confirm the view of organicists that what is essential for tumor development is the context in which the cells – cancerous and non-cancerous – exist. For example, if a normal early mouse embryo is transplanted to an ectopic site, i.e. a place where it does not normally belong, such as the kidney or the abdominal cavity, it will behave like a malignant tumor – a teratocarcinoma. Conversely, teratocarcinoma cells injected into an early embryo (at the blastocyst stage) become part of normal tissues and organs (Stewart, Mintz, 1981). These experiments suggest that a tumor cell need not always be a tumor cell under all circumstances. Rather, the tumor phenotype appears to be plastic, adaptable, and controllable by the environment. This would imply that tumors do not arise primarily at the cellular level, but rather at the level of tissue organization (Soto, Sonnenschein, 2004, 2005, 2014; Sonnenschein, Soto, 2005, 2016, 2018).

Mina Bissell, beyond the cell boundaries

The main difference between the somatic mutation theory and the reductionists on the one hand, and tissue organization field theory and the organicists on the other, seems to be well expressed by the book titles of their representatives. On one side is Robert Weinberg's "single renegade cell," on the other Carlos Sonnenschein and Ana Soto's "society of cells" (Weinberg, 1998; Sonnenschein, Soto, 1999).

In December 2007, the California weekly newspaper *East Bay Express* published a long article about the life and work of Mina Bissell and her views on tumor development. The article was prefaced with the information that her work had been ignored by the scientific establishment for decades, even though her unique approach could fundamentally change the way we think about tumors and their treatment. The article was titled "Thinking Outside the Cell" (Platoni, 2007). The title of the article is apt and provides an accurate picture of the position from which Bissell views tumors and, at the same time, the position she takes or would take in a virtual battle between reductionists and organicists. She has a radical but increasingly accepted view, justified primarily by her life's work, that in both normal development and some pathological phenomena, "phenotype dominates genotype" (Bissell, Hines, 2011).

Early in her career in the 1970s, Mina Bissell accepted the above findings indicating that the tumor phenotype of cells is plastic and adaptive and is influenced by the environment (Stewart, Mintz, 1981). She further developed these findings in her laboratory. For example, she used the Rous sarcoma virus, which carries the *src* oncogene. Injection of the virus into chickens caused sarcomas. Injection of the same virus into chicken embryos failed to induce sarcomas. However, when the cells of this embryo exhibiting no signs of tumor transformation were transferred to tissue culture, they resumed a transformed character. Bissell's conclusion was clear: it is the context or microenvironment outside the infected cell that determines whether cells carrying an activated oncogene will form a tumor (Dolberg, Bissell, 1984).

Among the literally "revealing" results of Mina Bissell's laboratory are the conclusions drawn from countless experiments with breast cancer cells, showing how strongly the phenotype of the cells depends on the type of cell cultivation. In standard cultivation of tissue cultures on the bottom of Petri dishes, breast epithelial cells significantly change their morphology, lose their polarity and stop producing milk. They completely "forget" their original form and function in the organism. When cultured on surfaces that simulate the three-dimensional (3D) environment of real tissue, the cells reorganize and form structures that look physiological again. Bissell showed that tumor cells can resume normal characteristics in a 3D environment, even though they still have mutations in their genes. She showed that the environment has a crucial influence on the phenotype of the cells and that adequate signaling from the microenvironment keeps the transformation process under control (Weaver et al., 1997; Bissell, 2017). Further work extended this knowledge and examined in detail the role of tissue architecture, stromal cells, and the extracellular matrix during development, homeostasis, and also in tumorigenesis (Nelson, Bissell, 2006; LaBarge et al., 2007; Ghajar, Bissell, 2008; Bissell, Inman, 2008; Weigelt, Bissell, 2008; Spencer et al., 2010).

Her findings also helped Bissell clarify a phenomenon that was one of the starting points for all of her work. These observations concerned the occurrence of occult tumors, that is, small, undetected, hidden tumors. They are probably much more common than we realize, and Bissell addressed the question of why people do not have tumors much more often. She started with considering the trillions of cells in the human body that are created by repeated division from a single initial cell, the fertilized egg. During the human lifetime, they are repeatedly exposed to a range of harmful substances, both exogenous and endogenous, that cause mutations, including mutations in oncogenes and tumor suppressors. According to existing dogma (i.e. somatic mutation theory), these mutant cells should lose control of their proliferation and form the basis of new tumors. In reality, however, the number of detectable tumors in humans is much smaller than the number of such deleterious mutations (and occult tumors).

Based on many of her findings, Bissell suggests that in these cases the microenvironment surrounding the tumor provides inhibitory signals for its further development and keeps it in an "occult" state. In other words, tumor initiation cannot be prevented, but as long as tissue architecture and homeostasis are maintained, its further progression remains under control (Bissell, Hines, 2011). In this context, it should also be noted that Bissell, in contrast to Soto and Sonnenschein, does not deny the involvement and contribution of oncogenic mutations in tumorigenesis, but, in agreement with them, points to the central contribution of the microenvironment, the context, in controlling tumor progression. And it is precisely this knowledge that in a different, almost heretical, way (in relation to the central dogma of molecular biology) supports her statement that "phenotype dominates genotype." The physiological, properly structured environment inhibits tumor development, whereas the pathological environment, altered, for example, by a developing tumor, can support its further progression (Bissell, Hines, 2011) (Fig. 42).

Mina Bissell: The dialectic of micro- and macroenvironment and the organ phenotype

How does a single cell, a fertilized egg, develop into a multicellular organism composed of tissues and organs with different morphologies and functions? We asked this question in chapter 1A (Fig. 1) and briefly outlined the answer. Why only briefly? Primarily because it seemed sufficient for an introduction to "our" questions about tumors and their formation. Not everyone in the large community of scientists involved in cancer research has a thorough knowledge of developmental biology, morphogenesis, and organogenesis. Many of them study tumors according to the somatic mutation theory where the absence of this knowledge is not crucial. Another reason for the lack of a deeper explanation of the evolution of organisms is that many connections in and laws of this process are still unclear and incompletely understood. On the contrary, the findings of many tumor biologists – in the spirit of Konrad Lorenz's conviction that "far from being an insurmountable obstacle to the analysis of an organic system, a pathological disorder is often the key to understanding it" – have contributed significantly to the understanding of these processes (see Introduction; Lorenz, 1974, p. 13).

A place of honor among them is occupied by Mina Bissell, her students and colleagues. She studied tumor development and the factors influencing it until she gradually understood the physiology of tissue, organ, and organism development, and the cell differentiation involved. She shows that the differentiation of cells is influenced by signals from their immediate environment, but also by signals from more distant parts of the organism. Proper development depends on the ability of cells in developing tissues and organs to perceive and respond appropriately to all incoming signals. They must perceive the context (and this is more important for proper development than dependence on possible mutations of their own genome). Bissell speaks of a "dialectic," a kind of conversation, a constant dispute between the microenvironment, the macroenvironment, and the phenotype of organs, referring to the phenotypic plasticity of cells, to the ability of their genotype to "produce" different phenotypes depending

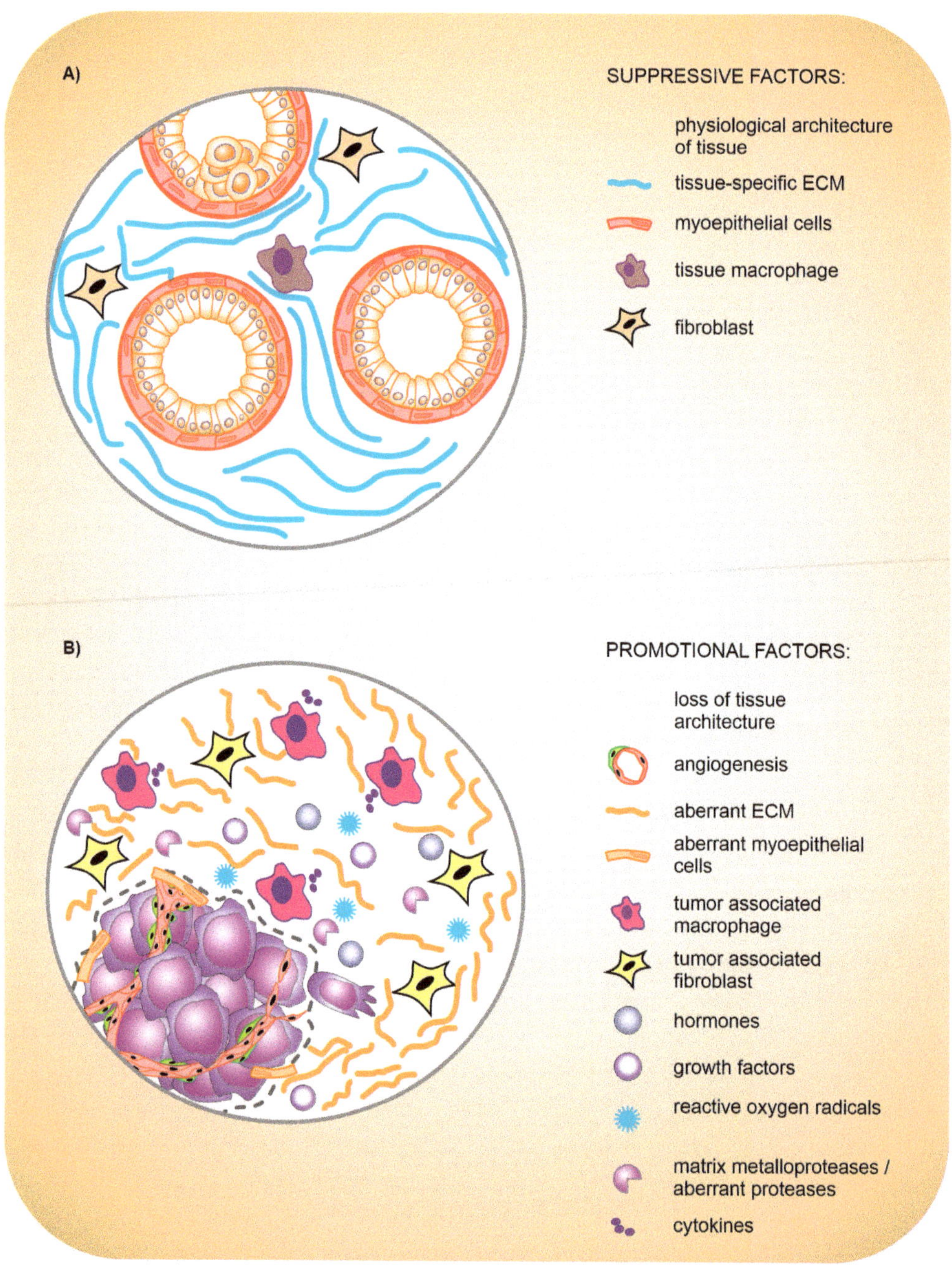

Fig. 42 Tissue Organization Field Theory (TOFT)
(A) The normal structure of the tissue microenvironment acts as a barrier to carcinogenesis: a healthy microenvironment surrounding the tumor provides signals that inhibit its further development and keep it in a quiescent state. (B) In contrast, a damaged microenvironment can significantly promote tumor development (adapted from Bissell, Hines, 2011, modified).

on extracellular signals. However, she also emphasizes that once organogenesis is complete, the fully-formed organs must be quite resistant to stimuli from the external environment that would disrupt their structure, although at the same time they must remain sufficiently receptive to specific signals related to their function. When this recognition of various external stimuli is weakened and the architecture of the tissue is damaged, the organ is "open" to developmental and evolutionary forces, and tumor formation is one of the possible consequences (Bhat, Bissell, 2014).

Convergence of SMT and TOFT views? Any hybrid theory?

Are somatic mutation theory and tissue organization field theory incompatible? Or is it just a matter of time and evolution and they will gradually converge? Are there hybrid concepts?

Mina Bissell's concept assumes that carcinogenesis involves substantial contributions both from the microenvironment and oncogenic factors such as mutations (Bissell, Hines, 2011), and this example suggests that the boundary between reductionists and organicists is not impenetrable. On the side of the proponents of somatic mutation theory, there are also more ideas than those of Weinberg and Hanahan, and some seem closer to TOFT. For example, Richard Klausner (director of the National Cancer Institute from 1995 to 2001) believes that tumors have three main characteristics: (1) genome instability, (2) altered cell behavior, and (3) altered tissue behavior. With his third point, he clearly names and emphasizes the importance of relationships and signaling between cells in highly organized tissues (Klausner, 2002).

However, Hanahan and Weinberg themselves have repeatedly emphasized that to understand tumorigenesis, the study of tumor cells alone is not sufficient. As early as their 2000 review article, they clearly point out that the tumor is a complex tissue containing many different cell types and noncellular structures that cooperate, interact, and are collectively involved in tumor development (Hanahan, Weinberg, 2000). They argue that the complex relationships between tumor and stromal cells should be considered when trying to understand processes such as tumor angiogenesis and tumor invasion and metastasis. In their 2011 follow-up paper, they expand the list of hallmarks of tumors to include the presence of chronic inflammation, communication with the immune system, and the role of the microenvironment, precisely the features that underscore the importance of the microenvironment in tumor development (Hanahan, Weinberg, 2011) (Fig. 8).

Douglas Hanahan – a longtime collaborator of Robert Weinberg and co-author of their iconic articles – provides another clue to the importance of the tumor microenvironment for tumor development in a review article with Lisa Coussens emotionally titled "Accessories to the Crime: Functions of Cells Recruited to the Tumor Microenvironment" (Hanahan, Coussens, 2012). They show that almost without exception, all typical features of tumors are fundamentally influenced by non-tumor cells, particularly infiltrating cells of the immune system, tumor-associated fibroblasts, and endothelial cells and pericytes (Fig. 43). Although the paper is published in a prestigious scientific journal (*Cancer Cell*), we find there a very literary, almost emotional statement: "Cancer cells do not manifest the disease alone, but rather conscript and corrupt resident and recruited normal cell types to serve as contributing members to the outlaw society of cells" (Hanahan, Coussens, 2012). This sentence, with its sentimentality and a certain poeticism, is really not a typical sentence for a scientific text, and it even sounds like an "overlap." However, in the context of the topic of this chapter, what is most remarkable is that it is clearly to the tumor cells that the authors attribute the controlling, decisive role in the process of tumor development.

What results from the "dispute" between SMT and TOFT?

How compatible are SMT and TOFT? Is it even meaningful to ask the question about these rival concepts of tumor development? Is it not just a historicizing view that will dissolve by itself with time (compare Figs. 42 and 43)? If we admit the importance and usefulness of such a question, what is the reason? Soto and Sonnenschein, in their tissue organization field theory (Sonnenschein, Soto, 1999), adamantly reject somatic mutation theory, reductionism, and genocentrism. They repeatedly emphasize and declare that SMT and TOFT are two incompatible and irreconcilable concepts of carcinogenesis. They believe that carcinogenesis can be understood either as a process that begins with a single renegade cell and manifests as a disease of cells with unregulated proliferation, or one can focus on a community of cells and view the tumor as a problem of tissue organization and cell communication. They question not the data and observations, but their interpretation, which they believe affects the proper understanding of the whole process. For example, do inherited mutations of certain genes (tumor suppressors, according to SMT) primarily deregulate cell proliferation, or do they disrupt tissue structure, which then cause inadequate signaling and communication problems that translate into altered proliferation of certain cell types? While TOFT is primarily concerned with the context, the tissue arrangement, SMT considers cells within tissues as quasi-autonomous entities controlled by their genes (Soto, Sonnenschein, 2005).

One might think that the steady shift of mainstream thinking in cancer biology (SMT) to the microenvironment and the complex relationships within tumors would prove TOFT right and satisfy its proponents. However, this is clearly not the case. In one of their earlier papers, the authors of TOFT reject shifts in the thinking of SMT advocates as mere corrections to an inherently wrong concept: "Lack of fit stemming from the adoption of the somatic mutation theory of carcinogenesis has been dealt with by supporters of the theory in a manner comparable to those adopted by pre-Copernican astronomers who dealt with the lack of fit between theory and data on the motion of the planets. They added ever-increasing numbers of epicycles to the planets'

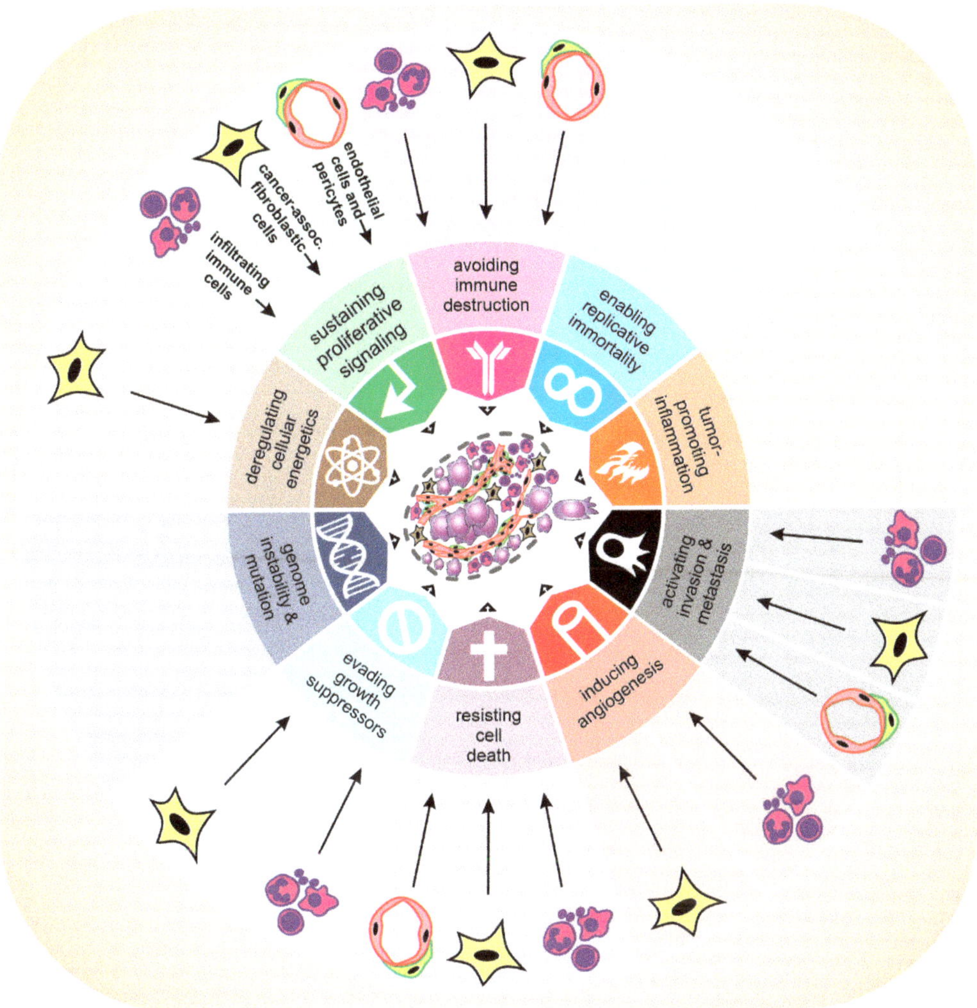

Fig. 43 Tumor microenvironment influences the typical features of tumors
Most features of tumors are influenced by non-tumor cells present in the tumor microenvironment: tumor-associated fibroblasts, immune cells, endothelial cells, and pericytes (adapted from Hanahan, Coussens, 2012, modified).

orbits. The resolution of those inconsistencies was to remove the Earth from the center of what was the Ptolemaic universe and place the Sun there instead" (Sonnenschein, Soto, 2000). Soto and Sonnenschein insist that a fundamental paradigm shift in the way we view the processes of tumorigenesis is indeed necessary. For them, SMT and TOFT are incompatible (Soto, Sonnenschein, 2005, 2008).

Their long search for the "correct" concept of carcinogenesis and the real causes of tumorigenesis gradually led Sonnenschein and Soto to the realization that the whole of contemporary biology is in a certain crisis. The reason for this is, on the one hand, the explosive development of technologies that can be used in biological research, providing huge amounts of new data, and, on the other hand, a kind of inability to adequately interpret and integrate the obtained results and, most importantly, to provide breakthrough insights and understanding of complex biological phenomena and problems. Gradually, they came to believe that contemporary biology lacks an up-to-date global theory and an appropriate paradigm. In the past, this role was successfully played by cell theory and evolutionary theory, which provided suitable starting points and frameworks for the interpretation and integration of the new findings at that time.

The organicists believe that we currently lack a general and unifying "theory of organisms" and that the prevailing reductionism at all levels (ontological: organisms consist of molecules and their interactions; epistemic: that systems and phenomena of a higher level in the hierarchy can be reduced to a lower level; methodological: the best way to study biological systems is to use the lowest possible level), although it has increased our knowledge of many aspects of living systems, is an obstacle to a deeper understanding of both the physiological functioning of organisms and their pathological disorders, including tumors (Soto et al., 2016; Sonnenschein, Soto, 2016, 2018).

Robert Alan Weinberg: A reductionist?

Robert Weinberg is considered the main proponent of somatic mutation theory (Malaterre, 2007). Is there any merit to this claim? And is it fair? Weinberg, along with Douglas Hanahan, is the undisputed co-inventor of the concept of the "hallmarks of cancer." He is the author of the book *One Renegade Cell: How Cancer Begins.* Undoubtedly, his views are based on and most clearly represented by the somatic mutation theory. On the other hand, in many papers published recently with his team, he emphasizes the synergy between the microenvironment and tumor cells during tumor development. This is true, for example, in the study of epithelial mesenchymal transition, its involvement in tumor metastasis, its links to tumor stem cell development, tumor immune microenvironment, and other phenomena (e.g. Polyak, Weinberg, 2009; Tam, Weinberg, 2013; Pattabiraman, Weinberg, 2014; McAllister, Weinberg, 2014; Chaffer et al., 2015; Lambert et al., 2017; Binnewies et al., 2018; Zhang, Weinberg, 2018; Dongre, Weinberg, 2019). But it is not really necessary to look for Robert Weinberg's attitudes in his individual contributions to cancer research. He himself has clearly expressed his own views on several occasions.

As early as 1991, when he was still searching for a holistic concept of tumor biology, he chose the aforementioned quote from Paul Valéry: "That which is simple is always wrong: that which is complicated is useless" as the motto for his article, mentioned above, "Oncogenes, Tumor Suppressor Genes, and Cell Transformation: Trying to Put it all Together." This documents that he is well aware of the simplification involved in his approach, but he also explains why he does so (Weinberg, 1991).

A few years later, in his unsurpassed textbook *The Biology of Cancer*, he writes quite explicitly:

"A simple and powerful conceptual paradigm pervades most of the previous discussions in this book: cancer is a disease of cells, and the phenotypes of cancer cells can be understood by examining the genes and proteins within them. The origins of this idea are clear, being traceable directly back to bacterial and yeast genetics… Applying this concept to metazoans and their tissues has had obvious advantages for biologists. Metazoan organisms are complex almost beyond measure, and this complexity has frequently prevented researchers from extracting simple and irrefutable truths about organismic function. In response, many researchers embraced the credo of reductionist science: when working with complex systems, the best way to arrive at solid and rigorous conclusions is to take apart these systems into simpler, more tractable components and study each separately. While the lessons learned may relate only to small parts of very large systems, at least these lessons are solid and definitive and will not require substantial revision each time a new generation of researchers revisits these complex systems and their component parts… Such reductionism has allowed many areas of cancer research to thrive… By the end of the twentieth century, it had become increasingly clear that many of the traits of tumors could not in fact be traced directly back to individual cancer cells and the genes they carry. The grand simplifications agreed upon a generation earlier by cancer researchers had begun to lose their utility. Increasingly, evidence turned up that cancer is actually a disease of tissues, in particular, the complex tissues that we call tumors (Weinberg, 2007, pp. 527–528; 2014, pp. 577–578)."

In the opinion of Robert Weinberg, reductionism represents only the first, initial, and temporary stage in cancer research, to be followed, of course, by a much more comprehensive approach.

Does it matter which paradigm prevails?

The question is to what extent Weinberg is correct in his certainty that the foundations of our understanding of cancer, based on reductionist ideas, will undergo smooth transition into much more complex and contextual ideas and concepts that seek the true nature of cancer. Or, are Soto and Sonnenschein correct in believing that a gradual, incremental change in the approach to cancer research is not possible and that a major paradigm shift such as "replacing the sun with the earth at the center of the universe" is necessary?

Regardless of whether we consider the two different concepts to be completely incompatible or convergent, it is certain that simply knowing about them, being aware of their existence, can have a significant impact on tumor biology. At the theoretical level, the prevailing paradigm always affects the design of experiments (e.g. awareness of the limitations of the informative value of 2D tissue cultures, the need for different cell types to cooperate, etc.) and the interpretation of their results. The prevailing paradigm also affects the objects of study and funding. In general, then, the prevailing paradigm has a direct impact on our understanding of the process of carcinogenesis and its further study (Kuhn, 1989; Lazebnik, 2004; Kolata, 2009; Soussi, 2010). In practice, for example, it can significantly influence the strategy of cancer therapy (Kenny et al., 2007; Hanahan, 2014; Kim et al., 2017).

Recognition of the importance of context, i.e. tissue structure and, in particular, its functional contribution to tumorigenesis, will expand the targets of cancer therapy from primary tumor cells to fibroblasts, endothelial cells, leukocytes, and other cells of the immune system and other cell types, as well as noncellular structures (the extracellular matrix and its components). A number of such therapeutics are already in use and more are sure to follow.

The question of the meaning and implications of the prevailing paradigm recalls the commentary by Judah Folkman, the tumor angiogenesis researcher (chapter 7A). In 2000, when explaining the benefits of antiangiogenic therapy, he rejected the prevailing paradigm of somatic mutation theory, which he repeatedly referred to as gene-centric, and wittily called his dissenting view the "endothelium-centric paradigm." This might seem like a funny bon mot from a respected titan of tumor biology if his expression had not been preceded by years of rejection, even ostracism, by the proponents of somatic mutation theory. Not only have mainstream researchers long ignored his work and labeled him a charlatan, but they have also hampered his efforts to work effectively on his discovery (Folkman et al., 2000; Schreiber, 2010). Mina Bissel has experienced a similar story (see above; Platoni, 2007). An awareness of the existence of the two concepts and a deeper knowledge of tumorigenesis could lead not only to a more cautious interpretation of the results of cancer research, taking into account more possibilities and points of view, but also to a greater openness and respect for different, seemingly insignificant, approaches in the search for the nature of cancer. And, last but not least, to greater openness and respect for the researchers themselves.

A third theory?

As described in the previous section, antiangiogenic therapy of tumors was not the result of a fundamental paradigmatic reassessment of somatic mutation theory, but rather the result of evolution of science and a quite natural shift in the understanding of carcinogenesis toward higher complexity. This is consistent with the work of philosopher Christophe Malaterre, who in 2007 wrote a detailed, insightful study of organicism and reductionism in tumor research, concluding that the positions of the SMT and TOFT are likely compatible and have a tendency to converge. Malaterre predicts, or at least acknowledges the possibility of, the creation of a third theory – a systemic theory of carcinogenesis – that will combine the two original views (Malaterre, 2007).

In a sense, the idea of a systemic theory of carcinogenesis is a reflection of Andrew Feinberg's stochastic epigenetic model of tumor development. According to this model, epigenetic changes, rather than mutations, are the cause of tumor development. Epigenetic changes are heritable alterations in gene expression that leave the DNA sequence intact, primarily imprinting changes – DNA methylation associated with covalent modifications of histones. Based on his findings, Feinberg suggests that some degree of spontaneous stochastic heterogeneity of the epigenome is natural and inherent in all cells, including healthy cells, and likely serves as a source of adaptability and evolution. A substantial increase in epigenetic heterogeneity may result from stress, for example repeated cycles of damage and healing. That is, from the influence of the microenvironment or even the broader environment of the tumor. Increased epigenetic changes can then lead to variability in the expression of genes, including those involved in the regulation of important cellular properties and functions that we associate with tumor development (Feinberg, 2014; Timp, Feinberg, 2013).

And how does the epigenetic model of tumor development relate to SMT? Feinberg literally says that the genetic model is good, only incomplete, in that it explains only 50% of the causes of tumor development. In his opinion, epigenetics is at the heart of tumor evolution. He sees the alteration of the epigenome as the first stage of tumor evolution, followed first by the phase of generation and selection of gene mutations and then by the escalation of genetic and epigenetic instability (Feinberg, 2014). The evolution of genetic and epigenetic alterations in tumors is closely linked. In the context of the topic of this chapter, it is interesting to note that the door to recognizing the contribution of epigenetic mechanisms to tumor development was opened by the discovery of mutations in genes encoding components of the epigenetic machinery. In other words, it was a pure SMT approach that led to the recognition and appreciation of the involvement of epigenetic mechanisms in tumor development (Rodríguez-Paredes, Esteller, 2011; You, Jones, 2012; Chatterjee et al., 2018).

12B. Pride and introspective individualism

SMT or TOFT? Genes or environment? An individual or a community? Individualism or collectivism? Liberal or social humanism?

Is it useful to consider SMT and TOFT and their "clash"? What is the point? As for the answer to the immediate question raised by both theories – namely, how tumors develop and what is the key point of their formation – we can assume from all experience to date that it will be found sooner or later, depending more on the amount of time and experimental effort spent than on the theoretical question of whether the truth is held by the reductionists or the organicists. Moreover, they are already in broad agreement that the nature of the origin and development (as well as the diagnosis, treatment, and prevention) of cancer cannot be found at the level of the tumor cells themselves, but that it is imperative to study their relationships with all the structures of the affected organism. The systemic approach is essential.

It turns out that the life of cells in the body is not controlled only by their genes. Of course genes are important, but they represent only a kind of storehouse of all potentialities, a collection of all possible cellular plans and destinies. But which specific life path the cell will take depends on the signals coming from its external environment and how the cell responds. Thus, it is not only the "single renegade cell" but also the character of the "society of cells surrounding it" that will determine whether and how a tumor will form, and more generally, how the whole community and each individual within it will continue to live. If we look at the conflict of SMT and TOFT from a distance, from the overlap perspective, we can find in it a lot of insights into the completely non-biological sides of our life.

Is individualism, with its preference for the individual, his independence, self-sufficiency and minimal restriction by the surrounding society, correct? Or collectivism, which sees the individual primarily as part of a whole and subordinates his interests to that whole? What is the relationship between the individual and the community? How important is the influence of the environment and our specific circumstances on who we are, how we develop, and what direction we take? How influential are the environment and our circumstances on our decisions and life choices?

The greater the influence of the environment we admit, the more fundamental it seems to be to maintain a proper and healthy environment – a properly structured and functioning community. And, on the other hand, what is the role of the individual in society and its history? To what extent can the individual influence the state of society and history?

And what does the term "humanity" mean? Is humanity an attribute of the individual, and the freedom of each individual is therefore inviolable, as liberals believe? Or is humanity not a matter of individuals but of the collective, and the highest value is not the freedom of the individual but the equality of all people, as socialists believe?

How much do the rules and structure of the community influence the behavior of the individual? To what extent are the rules of the community superior to the rules and rights of the individual? Are we even aware of the power of the community? Are we even aware of how much we are influenced by our environment? And are we aware of how much responsibility we have for creating and maintaining the (proper) structure of the community?

Genes or environment? Lessons from cells

When James Watson and Francis Crick proposed their model of DNA structure in 1953, its general acceptance depended on its ability to explain the mechanisms that provide the basic functions of DNA. That is, the ability to store and encode genetic information, replicate it accurately, and transcribe it into an RNA structure so that it can be translated into the structure of proteins. Proteins are responsible for carrying out most vital cellular functions. In addition to DNA, RNA can also store and accurately replicate genetic information, and RNA-encoded information can be transcribed back into DNA. In contrast, the flow of genetic information from nucleic acids to proteins is unidirectional. There is no known mechanism for the transfer of information from the structure of proteins to the structure of DNA. These possibilities and limitations of the flow of genetic information summarize the central dogma of molecular biology (Fig. 41).

And it is the central dogma of molecular biology that has often been interpreted and understood in a simplistic, superficial and literal, even dogmatic way that leads to the belief that genes determine and direct our lives. That is, the belief known as genetic determinism. Genes carry accurate information about the structure of the proteins they encode, but they cannot turn themselves on or off to determine the rate of self-expression, that is, the rate at which the encoded proteins are produced. In this respect, genes depend on the function of proteins and signals, including those coming from the extracellular environment (as briefly discussed in chapter 1A). It is the proteins that decide what part of the genetic information is used and how in the cell. In a sense, the proteins as well as extracellular signals, i.e. signals from the environment in which the cells are located, are even superior to the genes.

The set of all the genes of a cell, the genome, represents "only" the blueprint, the plan, the full potential of the cell, but which part of this plan is actually used is decided by the "Master," and that is the environment. Of course, even signals from the environment cannot cause the cell to exceed the limits of the blueprint and force the expression of information that is not included in the "offer." If a gene is damaged by mutation and codes for a protein that does not function properly or at all, the cell will lack the appropriate function even though extracellular signals are present to activate it. This is the basis of the somatic mutation theory. But at the level of the individual cell it also becomes clear that the influence of the environment is significant and largely determining, which in turn confirms the tissue organization field theory.

"Cytopomorphism" and Bruce Lipton's idea that we can learn something from cells was mentioned earlier in this book in the *Overlap* chapter. In his book *Biology of Belief*, Lipton very strongly rejects genocentrism and proves his position with a witty, somewhat provocative, but illustrative question. If genes control the life of cells, just as a person's life is controlled by his brain, where is the brain of the cell located? Is it in the cell nucleus which contains its genome? Seeking an answer to this question, he recalls that if the nucleus were the cell's brain, its removal would result in immediate cell death. However, experiments have repeatedly shown that the cell can live successfully for an extended period of time without a nucleus. Without a nucleus, the cell does not lose its "brain" – what it loses is the ability to reproduce. So the cell nucleus is not the brain, but the sex gland of the cell! So what actually is the brain of the cell? Lipton claims it is the cytoplasmic membrane with its receptor proteins and associated effector proteins that convert the signal into an appropriate cellular response (see chapter 1A).

Genes alone cannot preprogram the life of a cell. Survival depends on the cell's ability to respond dynamically to signals from the extracellular environment (as well as from the intracellular space) and to adapt constantly to changing conditions. For cells to exhibit "intelligent" behavior, they must have a functioning membrane with protein receptors and corresponding effectors. Thus, it is proteins and protein complexes that are the basic units of "cellular intelligence" (Lipton, 2005). This suggests, in great agreement with the theory of field tissue organization, that communication and appropriate responses to influences from the extracellular environment are key to the survival and healthy life of the cell.

The individual or the environment? The theory of broken windows

To what extent do our environment and the state of society influence us as individuals, and to what extent are we independent and strong in our personal attitudes? It turns out that certain circumstances can significantly influence our behavior both positively and negatively. In 1982, political scientist James Wilson and criminologist George

Kelling published the Broken Windows Theory (Wilson, Kelling, 2007). This theory states that a neglected environment (with broken windows, trash, spray-painted walls) can promote antisocial behavior, including petty and serious crimes. In an environment where no one seems to care, people lose the scruples they would have in a clean and tidy environment to avoid littering or destroying the property of others. An unethical environment leads to unethical behavior by citizens.

This hypothesis has been criticized, but also repeatedly tested experimentally. For example, two used cars were parked without license plates and with their hoods open, one in the busy and poorly maintained Bronx, the other in a quiet, affluent neighborhood of Palo Alto. In the Bronx, a family removed the battery just minutes the car was left. Within 24 hours, the car turned into a wreck with broken windows where children played cops and robbers. Nothing happened to the car in Palo Alto, and when the experimenters removed it five days later, people called the police and asked if the car had been stolen. In another experiment, the authors left an envelope, apparently containing money, hanging from a mailbox. When the box was clean and the area tidy, only one person out of eight came by and stole the envelope. One in five such envelopes was stolen from a mailbox painted with graffiti, as well as when trash was strewn around the mailbox.

The context of the situation, the environment, has been shown to actually influence the extent of immoral behavior. In another experiment, scientists erected a fence along the street that forced users of the adjacent parking lot to take a detour. A police officer posted no trespassing and no bicycle parking signs to the gate in the fence. The gate remained open. Only about one in five violated the prohibition and walked through the gate. However, when the experimenters locked bicycles to the fence (thus violating the norm), almost no one complied with the prohibition. The conclusion of this study, as well as many other similarly designed experiments, was clear. Violation of one norm leads to easier violation of other norms as well (Wilson, Kelling, 2007; Keizer et al., 2008; Houdek, 2010; Valenta, 2015). The results of such experiments clearly show that we are anything but independent in our actions. The environment and the behavior of other people influence us significantly.

Milgram's experiment: Obedience or conscience?

In 1963, inspired by the trial of Nazi criminal Adolf Eichmann, American social psychologist Stanley Milgram, who was of Jewish descent, tried to find out how far people can go in obeying authority, even if it violates their conscience. He was interested in the psychological causes of the Holocaust and wanted to know if the concentration camps were a purely "German" affair or if they could recur anywhere. Through an advertisement, he recruited volunteers and offered them four dollars to participate in an experiment designed to test human learning and memory.

Volunteers in the role of "teacher" asked a "student" in the next room questions over a microphone. The student answered by pushing buttons. The "teacher" sat next to an electrical device that he used to punish the "student" with an electric shock for each mistake. The generator had 30 switches with a gradation of 15 V, labeled from 15 V to 450 V. The "teacher" was told that the experiment would study the effect of punishment on learning. In case of a wrong answer, the "teacher" would electrocute the student. After each wrong answer, the "teacher" was to increase the electric shock by one step, or 15 V. Milgram told the "teacher" that the shocks could be very painful but would not leave any lasting effects. Although the "students" indicated that they were in pain and pleaded for mercy, the "teachers" gradually increased the voltage after each incorrect response. At 300 V, the "student" stopped answering, but Milgram told the "teacher" that no answer would be counted as a wrong answer and the test must continue. All participants continued to 300 V. Sixty-five percent of the "teachers" administered shocks up to the very end of the scale, although they often hesitated and worried about the "student's" health. A full 65% of the "teachers" used the highest possible voltage of 450 V, which was labeled "dangerous – severe shock"! The experiment was repeated many times in different variations, always with the same result. The conclusion is that, being directed by someone with sufficient authority, 65% of people, in contradiction to their own conscience, will torture their neighbor with a lethally dangerous dose of electricity, collect four dollars, and go home feeling they have done their job well (Urban, 2011; Milgram, 2017).

Only a few years after Milgram's experiment, psychiatrist Charles Hofling came to the same conclusions. In his experiment, an unknown drug was secretly slipped in amongst other drugs in 22 different hospital wards. Then a doctor called the nurse, who knew the doctor only by name, and asked her to administer the drug to a patient at a dose higher than the maximum recommended daily dose listed on the drug's package leaflet. The physician told the nurse he would come later and sign the appropriate form. Only one of 22 nurses refused to give the patient the medication; the other 21 nurses followed the doctor's advice, although this was in stark contradiction to hospital policy. In another study, nearly half of the nurses admitted that they had ever followed a doctor's order, even when they felt it would harm the patient. Hofling thus confirmed Milgram's conclusions. Obedience to authority and learned, "normal" behavior in an organization can cause even good, decent people to do evil (Houdek, 2010).

The Stanford Prison Experiment: The Lucifer Effect

The world-famous, brutal and controversial Stanford Prison Experiment came to the same conclusions as Milgram's experiment and was directed by his contemporary and high school classmate Philip Zimbardo. For his experiment, he selected 24 mentally balanced, healthy individuals from 75 student volunteers, randomly divided them into 12 prisoners and 12 guards, locked them in an artificial prison, and observed their behavior. He wanted to find out whether aggressiveness and brutality depended more on the psychological predispositions of individuals or were caused by circumstances. He wanted to determine if and how people's behavior changed in an authoritarian environment. Although the experiment was planned for two weeks, he had to end it prematurely after only six days. The prisoners were rioting, depressed, and the guards resorted to increasingly frequent and cruel punishments and humiliations.

It has been found that even a healthy person, when subjected to extreme pressure of circumstances, can radically change his behavior in a very short time. The phenomenon when a decent and honest person becomes a criminal under the influence of a system that gives him great power over anonymous people, Zimbardo called the Lucifer Effect. Zimbardo, Milgram, and others studied the corrupting influence of situations on individuals and the power of institutions. In an institutional setting, no one explicitly asks anyone to do anything bad. But institutions have their fixed rules and habits, and a person plays an assigned role in them: teacher, student, doctor, prisoner, jailer, father or mother. And when a person is confronted with a situation in which he is assigned a certain role, he can easily discard some personality traits. It has been shown that quite "normal" people can commit ugly acts, because the potential to do good and evil is in every person and our physical and social environments affect us much more than we can imagine and admit (Houdek, 2010; Urban, 2011; Zimbardo, 2014).

Theory of the correction factor and the system of locks

How independent are we and how much are we influenced by circumstances and the environment?

Economist and psychologist Dan Ariely (mentioned earlier in chapter 6B) has contributed to answering this question, but from a slightly different angle. Ariely has long been interested in the causes of dishonesty and fraud. Does the perpetrator simply and rationally choose to deceive based on a calculation of potential gain, risk of getting caught, and magnitude of potential punishment? How many people commit robberies and dishonest acts? Just a few "bad apples" or a large number of more or less "decent" people?

To answer these questions, Ariely and his colleagues conducted a simple experiment. They asked volunteers to solve twenty mathematical problems and rewarded each correct solution. Participants were divided into two groups: one, after completing the task, put their answer sheets in the shredder so they could not be checked; the other gave them to the organizers. Then everyone reported how many tasks they had successfully completed. Those who had their answers shredded reported an average of six correct answers; those who had to give the answer sheets to the organizers reported an average of only four answers. Without the opportunity for verification, people lied to get a higher reward. At the same time, the experiment showed that the higher average number for the first group was not due to a few "immoral" people reporting a high number of correct solutions, thus raising the average. On the contrary, almost everyone reported a slightly higher score, about six. The follow-up experiments showed that the frequency of cheating was influenced neither by the amount of the reward for each correct answer nor by the level of risk of disclosure.

Thus, Ariely discovered that everyone has the potential to be dishonest, and that almost everyone cheats a little. He formulated the theory of the correction factor, according to which human behavior is driven by two opposing motivations. On the one hand, we want the gain that comes from cheating; on the other hand, we want to be perceived and respected as honest and sincere. The correction factor represents a delicate balance between these two motivations. We can cheat, but only a little, so we can excuse it and still feel good about ourselves.

Ariely illustrates the theory of the correction factor with the story of his student Peter, who one time locked himself out of his apartment and had to call a locksmith to open the door. Peter was very surprised that the locksmith took less than a minute to open it, and he doubted the security of such a lock. In response, the locksmith explained to him that locks are only for decent people to stay decent. According to the locksmith, one percent of people will always steal and open the door even if it is locked. Another one percent will always be honest and never steal anything, not even from an apartment with the door open. But the other 98 percent of people would like to open the door if it is not locked. If these 98 percent of people are exposed to temptation, their honesty will disappear. It is a system of locks – external constraints – not internal moral equipment, or conscience, which keeps most "decent" people in check (Ariely, 2012; Ryska, Pruša, 2013).

These interesting results reveal a lot about us. But perhaps even more interesting are the results of other modified follow-up experiments aimed at other sub-aspects of human deception and dishonesty and the circumstances that favor or, on the contrary, diminish them. For example, studies show that deception is contagious and its frequency increases when we see other people acting deceptively around us. When we meet other members of our social group behaving inappropriately, we are likely to reconsider our own standards and use their behavior as a model for our own. If it's an authority figure – a parent, boss, teacher, or other person we value or who holds a higher position – they are even more likely to drag us down with them. On the other hand, Ariely and his team have also successfully defined the circumstances that reduce dishonest practices. These include reminders of moral rules, making a personal commitment, and, of course, effective supervision (Ariely, 2012; Ryska, Průša, 2013).

The individual or the environment? What is the cause of corruption? A politician or a political system?

What is the cause of corruption and failure of politicians? And where can we look for solutions to the problems of corruption? What is the key to political recovery? Is it about finding an ideal, decent individual who is dependable under all circumstances, or is it more about establishing laws and rules, i.e. creating a "healthy" environment?

These questions are perceived very sensitively by the whole society. The above-mentioned experiments (and many others like them) generally aimed at the same questions. Are their results consistent with the somatic mutation theory, according to which the key is to find decent, resilient and reliable individuals, or rather with the tissue organization field theory, according to which the key to a healthy body politic is to create a social environment and organizations that prevent individuals from falling?

The results are quite clear: what determines people's behavior is the system. The degree of personal decency or moral strength of the individual is secondary. So, these results are very relevant to understanding and combating corruption in politics. Once a politician is elected, he has the same opportunities as the characters in the stories above. Like the prison guard, he is given great power over the anonymous mass of people who are just numbers to him. Like the test-taker whose performance cannot be assessed because of shredding, a politician tries to make as much money as possible in his position because, being in the public sector, the effectiveness of his policies cannot be fully attributed to his efforts alone. And just like the person who encounters a door without a lock, politicians can use their position to obtain money and property because they have the power to redistribute wealth in society. Just as the majority of "decent" and normal participants in these experiments failed, originally "decent" politicians will always fail. And this will happen to the extent permitted by the system settings, which may be better or worse at preventing corruption. The results of the above experiments not only offer insights into our problems with politics, but also clearly show where we need to look for effective tools to solve these problems. Looking for and trusting ideal individuals will not help. What we need is to establish and maintain a functioning and restrictive oversight system (Ryska, Průša, 2013).

Czech lawyer and sociologist Jiří Přibáň sums this up in his book *Tyranized Justice*: "The rules are more important than the person who wins the election" (Přibáň, 2013, p. 156). And in another passage of the same book, he encourages us to pay much more attention to general norms than to specific individuals. He acknowledges that politics is essentially about gaining and exercising power, and for this reason many things are permissible in political struggle. But he urges making a careful distinction. While attacking and weakening one's opponent is a hard but legitimate part of political struggle, in it impermissible and dangerous to challenge the very rules of the game. Deliberately bending and rewriting laws, even the constitution, to meet one's immediate needs sets a dangerous precedent, destroys the architecture of the legal environment, and in the long run disrupts and destabilizes the entire system.

Three levels of responsibility

Among the "overlaps" discussed in this book, we have offered a concept of cancer not only as a physical disease of a multicellular organism, but as a general disorder of a complex living system, including human society. And we have asked what lessons we can learn from tumors and whether we can draw inspiration from them. This chapter concludes our review of the eleven typical hallmarks of cancer listed by Douglas Hanahan and Robert Weinberg in 2011 (Hanahan, Weinberg, 2011). While one group of hallmarks (increasing self-sufficiency in growth signals, resistance to cell cycle arrest signals, damage to apoptosis, acquisition of unlimited replication potential, increased genetic instability, reprogramming of energy metabolism) is directly related to tumor cells, the development of other typical features (induction of angiogenesis, acquisition of metastatic potential, ability to evade immune system surveillance, presence of chronic inflammation) requires tumor cells to cooperate and interact with non-cancerous cells.

In order to prevent tumor-like processes in human society, it is necessary for us as individuals to strive to avoid behaviors that we might, with exaggeration, call "tumorous" or "tumor-promoting." This is a clear responsibility that each of us has. This is what I call the first level of responsibility: not being a "tumor" person!

But this is not enough!!! We have seen that "successful" tumor development heavily depends on non-tumor cells and their cooperation with tumor cells. This leads to the second level of responsibility: never cooperate with "tumor" people!

But even this is not enough!!! The last typical hallmark of cancer discussed in this chapter is the involvement of the tumor microenvironment. We have seen that the environmental conditions, and therefore the entire system, are essential to the successful development of the tumor. We have documented how the correct, physiological structure of the tissue can contribute significantly to stopping tumor development. From this realization, the third level of our personal responsibility and an urgent imperative inevitably arises: active involvement in the maintaining an appropriate, healthy environment! It is not enough for us to strive to prevent the onset of cancer development by joining and collaborating with the wrong people. Passive resistance, no matter how strong, is not enough. As citizens, we must consciously, actively, passionately and tirelessly insist on the proper organization and functioning of things in our immediate and more distant environment, we must insist on the observance of all conventions, tasks, rules and orders.

Premature epilogue, or Mina Bissell's "Think outside the box!"

Chapter 12A mentioned an article about the life and work of Mina Bissell and her view of tumor development that was published in the California weekly newspaper *East Bay Express* in December 2007 (Platoni, 2007). It was entitled "Thinking Outside the Cell." This title is symptomatic and apt for the way Bissell views tumors. Compare it to books written by the main proponents of the two other views of tumors, Robert Weinberg's *Single Renegade Cell* and Carlos Sonnenschein and Ana Soto's *Society of Cells*. That made sense! A few years after it was printed, I realized to my surprise that the article's headline probably referred to something else that made much deeper sense, and with even more "overlap" than I thought. In 2012, Mina Bissell gave a great talk on tumors as part of the TED (Technology, Entertainment, Design) series and introduced it with her favorite joke (Bissell, 2012).

The joke (Fig. 44A) paraphrases the saying "think outside the box." The phrase refers to the "nine-dot puzzle" (Fig. 44B) and is a metaphor for unconventional, creative thinking that uses a different angle and perspective. And Bissell used this joke to briefly explain that her greatest contribution to tumor biology – thanks to her training in chemistry and microbiology, which was unusual for oncologists – was that she brought a unique perspective to the study of tumors and, indeed, thought "outside the box."

"Never, ever, think outside the box."

Fig. 44 Mina Bissell's favorite joke about nonconformist thinking, published in The New Yorker in 1988 (A). The phrase "thinking outside the box" is related to the "riddle of nine dots" (B). The task is to connect nine dots (arranged in three rows of three) in one move with a maximum of four straight lines. The brainteaser can be solved only by going beyond the usual conventions and drawing the lines outside the boundary of the "virtual box" surrounding the dots.

And she persisted in this position despite being overlooked and ostracized by more conformist colleagues.

This point is perhaps also one of the messages of this chapter. Perhaps the determination of whether SMT and TOFT are two incompatible and mutually exclusive theories or whether they are just two temporarily separate streams of a future systemic theory of carcinogenesis is not crucial. Perhaps the most important benefit of this scientific dispute is the opportunity to see and recognize the value of not insisting on one's own point of view, on one's own truth, and, conversely, the value of making one's own way outside the mainstream, the value of thinking differently – even at the cost of being a minority. Or at least to be tolerant of someone who thinks differently. Perhaps, at least for a short time in our lives, we should have the courage to think "outside the box." Or at least be "out of bowl," as the Czech version of that phrase goes. Like the author of this text.

13A. The tumor suppressor p53

We have already encountered the tumor suppressor p53 several times in the previous chapters. The p53 protein is involved in the regulation of almost all cellular functions affected by the process of carcinogenesis. Thus, it is an important tumor suppressor, a crucial guardian of the organism against the development of tumors. The *TP53* gene, which encodes the p53 protein, is one of the most frequently mutated genes in human tumors. Therefore, it is also one of the most studied. This is illustrated by the fact that every two years an international "workshop" is organized that focuses exclusively on p53 and is attended by hundreds of scientists from around the world. This is a privilege that has hardly been granted to any other gene or protein.

The history of the p53 discovery

The p53 protein is literally one of the "evergreens" of modern oncology. It was discovered independently in 1979 in five different laboratories around the world as a cellular protein about 53 kDa in size that interacts with viral proteins that are critical to the ability of a virus to cause malignant transformation and that are produced in infected cells. It was soon recognized that this protein is present at high levels not only in virus-infected cells, but also in many types of tumor cells that have formed without the involvement of the virus, whereas little or no protein is present in normal cells. The p53 protein has been the subject of intense research since its discovery, but for several years it was mistaken for an oncoprotein encoded by an oncogene. Later studies showed that this error was due to work with an originally unnoticed mutation in the p53 gene.

It was not until 1989 that it was clearly demonstrated that p53 is not an oncogene but, on the contrary, a tumor suppressor. In 1992, one of the discoverers, Sir David Lane, coined the somewhat literary but apt title "guardian of the genome" (Lane, 1992). In 1993, p53 was named Molecule of the Year. This reflects its popularity, the great interest it has attracted, and its central importance in protecting the cell from tumor transformation. And the importance, popularity and intensity with which this small protein is studied has not changed to this day. Because it was discovered independently by several research teams, it was given several names, for example, p53, the 54K protein, the 55K protein, the 48-55K protein. All of these names reflect the molecular weight of the protein, which was one of the few pieces of information known about the protein at the time of its discovery. At the very first international workshop devoted to this gene and protein in 1983, a nomenclature was discussed. After a long debate, the simple name "p53" finally prevailed. Paradoxically, it later turned out that the molecular weight of the p53 protein is quite different – only 43.7 kDa (Levine, Oren, 2009).

How p53 works: A molecular mechanism

The major mechanism by which the p53 protein exercises its function is by controlling the transcription of its target genes. The p53 protein forms a tetramer (the functional molecule consists of four subunits – the p53 monomers) and binds to DNA in the so-called consensus sequence, which it specifically recognizes and which is located in the regulatory regions of its target genes. Thus, the p53 protein functions as a sequence-specific transcription factor. This is its most important and crucial mode of function in the cell and this mechanism is also the key to its tumor suppressive effect.

In addition to activating transcription of many of its target genes, p53 can also effectively repress transcription of other genes. This function of p53 is not associated with binding to its consensus sequence. Control of transcription by p53 depends on its direct interaction with DNA in the nucleus. However, p53 also has quite different, "non-transcriptional" activities that occur both in the nucleus and in the cytoplasm of the cell. These include direct interactions with other proteins, for example with proteins of the Bcl-2 family in mitochondria (see chapter 4A).

What triggers p53 activity and what is the role of the MDM2 protein

One of the target genes whose transcription is activated by the p53 protein is the *MDM2* gene. The product of this gene, the MDM2 protein, can interact with p53, inhibiting it and causing its effective degradation. MDM2 is an important cellular regulator of p53 and mediates a negative feedback loop. This is responsible for maintaining the p53 protein at a low level in the cell under normal conditions: the p53 protein is constitutively produced in the cell, but because it activates the transcription of the *MDM2* gene, thereby promoting the formation of the MDM2 protein, it causes the immediate effective degradation of the p53 protein. However, when cells are exposed to stress, such as DNA damage, critical shortening of telomeres, high activity of some oncogenes, hypoxia, deficiency of certain types of nutrients, loss of normal cell contacts, etc., p53 is stabilized and activated. These stimuli lead to changes in the proteins of p53 and MDM2, which prevent their mutual interaction and thus the degradation of p53. The p53 protein level in the cell increases and p53 can fulfill its functions (Fig. 45).

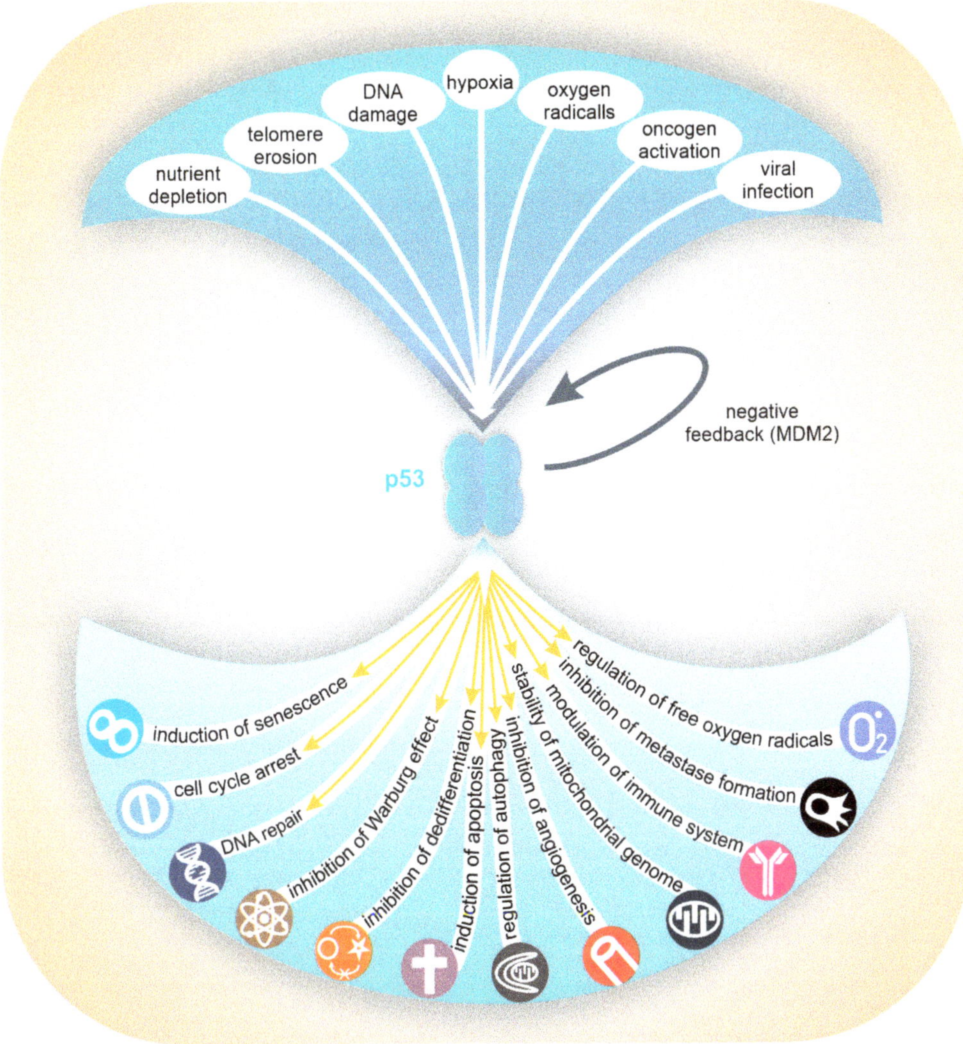

Fig. 45 Functions of the tumor suppressor p53
Various forms of stress to which the cell is exposed lead to the stabilization and activation of p53. The active p53 protein then ensures an adequate cellular response, usually by stimulating transcription of its target genes. One of these genes is the *MDM2* gene, coding for the MDM2 protein that causes efficient degradation of the p53 protein. It is thus an important regulator of p53 and forms a negative feedback loop with it.

How p53 works: Cellular functions

The first recognized cellular function of p53 was its ability to arrest the cell cycle (Baker et al., 1990). Target genes of the p53 protein include the gene coding for the p21 protein, an inhibitor of cyclin-dependent kinases. Activated p53 effectively stimulates transcription of this gene. When the level of p21 inhibitor increases in the cell, sufficient phosphorylation of the RB protein is prevented. Therefore, the cell cycle cannot pass the restriction point and stops (see chapter 3A) (Fig. 10). Thus, loss of p53 function can significantly contribute to "the release of cell cycle brakes" and contribute to the loss of control over cell division.

Another uncovered function of the tumor suppressor p53 in the cell was its ability to effectively induce apoptosis (Yonish-Rouch et al., 1991). One of the first genes recognized whose transcription is activated by p53 and whose product contributes significantly to the activation of apoptosis was the gene encoding the protein Bax, a pro-apoptotic protein of the Bcl-2 family (Chipuk et al., 2004; see chapter 4A). Gradually, it became clear that the involvement of p53 in the control of apoptosis is multifaceted. The p53 protein is involved in the regulation of both the extrinsic and intrinsic apoptotic pathways. Extrinsically, for example, by activating the transcription of genes encoding some death receptors. The intrinsic mitochondrial pathway is controlled by p53 at multiple levels and through multiple mechanisms. In addition to the Bax protein, p53 activates transcription of other genes encoding pro-apoptotic proteins of the Bcl-2 family. In contrast, it represses the expression of some anti-apoptotic factors such as survivin and, as mentioned earlier, is also involved in the regulation of apoptosis through a non-transcriptional pathway. Directly inside mitochondria, it interacts with some anti-apoptotic proteins of the Bcl-2 family and blocks their function. The ability of p53 to effectively induce apoptosis is considered key to its function as a tumor suppressor. Cells lacking functional p53 are much more resistant to programmed cell death than cells with functional p53 (Amaral et al., 2010; Kamp et al., 2016; Giorgi et al., 2016).

In addition to inducing temporal cell cycle arrest and apoptosis, p53 can (together with the tumor suppressor RB and the inhibitor p16) trigger a permanent block of the cell cycle associated with senescence. Again, the key protein mediating association with p53, p21, an inhibitor of cyclin-dependent kinases, is one of the targets transactivated by p53. And while the activity of p53 through p21 is involved in the induction of senescence, paradoxically it appears to suppress the gradual development of certain hallmarks of senescent cells, including high levels of the inhibitor p16 and SASP (production and secretion of numerous extracellular factors, the so-called secretory phenotype associated with senescence – see chapter 5A).

Thus, p53, via the inhibitor p21, can cause both reversible transient cell cycle arrest and cellular senescence associated with irreversible cell cycle arrest. How is the decision between these two options made? Cell cycle arrest is required to trigger cellular senescence, but not every cell cycle block triggers senescence. The temporality of cell cycle block and the ability to resume cell division are preserved in cells that are truly "at rest" and not stimulated by, for example, growth factors and mitogenic signals. When cell cycle block is induced in cells that are stimulated to grow, cellular senescence is induced by geroconversion. Interestingly, the p53 protein not only regulates the expression of p21, but also has a significant impact on the regulation of metabolism and growth (see below), thus substantially influencing the decision between transient cell cycle arrest and irreversible senescence. According to the proposed model, p53 activity is necessary for the induction of senescence, but its subsequent decline is required for the maintenance and development of typical signs of senescence (Blagosklonny, 2012, 2014; Johmura, Nakanishi, 2016.)

The p53 protein is also involved in the inhibition of angiogenesis in a very complex manner. In chapter 7A, p53 was described as an activator of thrombospondin-1 (TSP-1) expression. TSP-1 is one of the major inhibitors of angiogenesis. Indeed, the gene encoding this protein is under the control of p53 as its target gene (Dameron et al., 1994). In addition, p53 also stimulates transcription of other antiangiogenic factors. However, in regulating angiogenesis, p53 also acts in other ways. It is not only a transcriptional activator of several inhibitors of angiogenesis, but also a transcriptional repressor of several activators of angiogenesis, including the important pro-angiogenic VEGF. Thus, these two mechanisms act in the same direction and tip the balance of the angiogenic switch to the off position (see chapter 7A). Moreover, p53 can physically interact with the HIF-1α factor, which is involved in the effective stimulation of angiogenesis, and initiate its degradation. This mechanism is also used by p53 to inhibit angiogenesis. Thus, the ability of the p53 protein to inhibit angiogenesis, including tumor angiogenesis, is very complex and contributes significantly to its tumor suppressor function. Conversely, loss of p53 function substantially shifts the angiogenic switch to the "on" position (Teodoro et al., 2007).

The induction of tumor angiogenesis and the increase in metastatic potential are closely related. Through its ability to inhibit angiogenesis, p53 also contributes to preventing invasiveness and metastasis formation. However, in addition to this indirect mechanism, p53 is even more directly involved in suppressing metastasis formation, for example by maintaining the epithelial state of cells by actively suppressing the epithelial-mesenchymal transition program (see chapter 9A). The p53 protein is effectively involved in this process by both activating transcription of EMT inhibitors and repressing transcription of their activators. Functional p53 thus interferes with the development of the metastatic process, and its loss greatly increases the likelihood of this process (Muller et al., 2011).

The p53 protein also represents a significant obstacle to the conversion of energetic cellular metabolism to aerobic glycolysis. It effectively prevents the Warburg effect by inhibiting glycolysis and stimulating oxidative phosphorylation through coordinated mechanisms that include, for example, effective inhibition of glucose transporter expression (see chapter 8A). The ability of p53 to regulate cellular metabolism is even more general. In addition to glucose metabolism, it also regulates lipid metabolism. While most adult differentiated cells obtain lipids mainly from food, highly proliferating cells, including embryonic stem cells and also tumor cells, intensively synthesize lipids and use them to build the membranes of dividing cells and to synthesize steroid hormones that stimulate cell growth. The p53 protein inhibits the synthesis of lipids and conversely stimulates their degradation in mitochondria.

The ability of p53 to regulate glucose and lipid metabolism is closely related to its ability to control the level of free oxygen radicals and how cells respond to them. Among other things, this links p53's ability to control metabolism to its function as a "guardian of the genome." When damage to cellular structures by oxygen radicals is transient and repairable, p53 can activate a whole range of antioxidant responses, many of which are related to the metabolism of sugars and lipids. On the other hand, as part of the cellular response to intense stress, p53 can increase the amount of oxygen radicals to high levels, triggering cellular senescence or programmed cell death. This suggests that the p53 protein can either induce responses that help the cell to eliminate and/or reduce the effects of oxygen radicals on its structures or, conversely, use oxygen radicals to trigger cell death (Vousden, Ryan, 2009; Wang et al., 2012; Berkers et al., 2013).

Moreover, cellular energy metabolism is associated with the process of autophagy, which eliminates dysfunctional organelles and macromolecules and releases the nutrients bound within them, thus contributing to the maintenance of cellular energy self-sufficiency. The involvement of p53 in the regulation of autophagy is very complex, and depending on the circumstances, p53 can both stimulate and inhibit this process. It is symptomatic that this ambiguity of the role of p53 in the regulation of autophagy mirrors the ambiguity of the effects of autophagy on cell fate and thus on the process of carcinogenesis. Autophagy can lead to cell death, but it can also be a mechanism that enables cell survival (see chapter 8A). In regulating autophagy, p53 applies all the mechanisms by which it functions. It regulates the transcription of the appropriate target genes in the nucleus and is also involved in the regulation of autophagy in a nontranscriptional manner, namely through direct physical interactions with some proteins in the mitochondria and the endoplasmic reticulum. Not only the complexity with which it regulates autophagy, but more generally the versatility with which it intervenes in cellular metabolism, gives the p53 protein the ability to tune the most precise survival strategy during starvation and metabolic stress (Vousden, Ryan, 2009; Wang et al., 2012; Berkers et al., 2013; Giorgi et al., 2016).

The p53 protein also has a major influence on the activity of the immune system (chapters 10A and 11A). It regulates the course of both innate and adaptive immune responses. At least 30 genes whose expression is controlled by p53 encode proteins that directly affect immune system functions. It contributes to the fine-tuning and balance of immune responses in multiple ways. For example, the involvement of p53 in the regulation of inflammation is very complex. The p53 protein stimulates acute inflammation but suppresses the development of chronic inflammation. During inflammation, free oxygen and nitrogen radicals (ROS and NOS) are also produced, which activate p53. Active p53 stimulates the expression of enzymes involved in both the production and removal of free radicals. Therefore, p53 alone has a significant impact on the levels of ROS and NOS (see above). The relationship between p53 and the viral infections that can trigger an inflammatory response is similarly complex. Activation of p53 is part of the response to infection by many viruses, for example SV40, Epstein-Barr virus, adenoviruses, influenza viruses, HIV-1, and others. By arresting the cell cycle or inducing apoptosis, p53 can regulate or eliminate viral infection. In addition it can also directly stimulate other antiviral responses. On the other hand, there are viruses that can alter the function of p53. For this reason the p53 protein was discovered: as a protein that interacts with certain viral proteins. The first viral protein discovered to interact with p53 was the large T antigen of the SV40 virus. This type of interaction usually results in some form of degradation or inactivation of p53.

Indicative of the complex role of p53 in immune system function is its multilayered functional association with another protein – the transcription factor NF-κB, an important regulator of immune responses. NF-κB stimulates inflammation, cell proliferation, survival, and dedifferentiation. Therefore, it is primarily perceived as a p53 antagonist. However, the relationship between NF-κB and p53 is more complex than that. It is tissue-specific, depending on the type of stimulation, and there is even room for cooperation between the two proteins. This is evident, for example, in the regulation of cellular senescence, which involves cell cycle arrest mediated by p53 and induction of SASP mediated by NF-κB. The process of cellular senescence is also closely linked to immune system activity in other ways, particularly in its second phase, during the elimination of senescent cells, which depends on efficient macrophage activation (chapter 5A; Gudkov et al., 2011; Elinav et al., 2013; Lowe et al., 2013; Cooks et al., 2014).

In chapter 6, p53 was described as a protein capable of a versatile response to DNA damage, thus contributing to the maintenance of genetic stability. It is also clear from the previous sections that it fulfills its role as "guardian of the genome" (Lane, 1992) by employing its previously mentioned capabilities, such as arresting cell division, inducing senescence or apoptosis, modulating cellular metabolism, the immune system, and others. In addition, the p53 protein is undoubtedly involved in all five major DNA repairs, namely nucleotide excision repair (NER), base excision repair (BER), mismatch repair (MMR), homologous recombination (HR), and non-homologous end joining (NHEJ). It regulates DNA repair through both its transcriptional activation function and a non-transactivating mechanism. When DNA damage occurs in a cell, such as a chemical modification of a base, a base substitution, or a DNA break, signals are triggered that target p53. A rheostat model has been proposed to explain the role of p53 in DNA repair in a cell. When DNA damage is minor, the p53 protein present in a cell at normal levels activates the appropriate repair mechanisms and provides an immediate remedy. If the extent of damage is greater, stabilization of the p53 protein occurs, which not only activates DNA repair but also halts the cell cycle. This gives the cell time for more extensive repairs. If DNA damage persists or exceeds the cell's repair capacity, p53 induces programmed cell death or senescence. Thus, p53 as a cellular rheostat ensures that the cell's response to DNA damage is always proportionate to the extent of the damage (Sengupta, Harris, 2005; Menon, Povirk, 2014).

In addition to its essential role in maintaining genome stability in the nucleus, p53 is also instrumental in maintaining mitochondrial genome stability. Mitochondrial DNA (mtDNA) contains only 13 protein-coding genes, but all of these proteins are necessary for respiration and oxidative phosphorylation. In addition, mitochondrial DNA is much more frequently exposed to oxygen free radicals, ROS, which are generated during respiratory processes in mitochondria. Therefore, efficient repair of mtDNA and maintenance of its correct structure are absolutely necessary. The p53 protein is involved in the maintenance of mtDNA stability by activating and repressing transcription of its target genes, but also by a non-transcriptional mechanism. With the exception of NER, all other repair mechanisms are involved in mtDNA repair. In particular, BER is important because it repairs DNA damage caused by free oxygen radicals. Mitophagy, a form of autophagy degrading damaged mitochondria, is also involved in maintaining the stability of mtDNA. The p53 protein is physically present in mitochondria and therefore can influence the regulation of mitophagy in both transcriptional and nontranscriptional ways, both positively and negatively. Because of its complex involvement in maintaining mtDNA stability, the p53 protein is referred to by some authors as the "guardian of the mitochondrial genome" (Park et al., 2016; Kamp et al., 2016).

However, the relationship between p53 and mitochondria is even more complex. The p53 protein is involved not only in maintaining the stability of mitochondrial DNA and regulating mitophagy, but also in regulating mitochondrial biogenesis, for example by directly regulating the expression of some mitochondrial components, such as enzymes that function as parts of the respiratory chain. It also influences mitochondrial activity by controlling energy metabolism and thus the provision of "fuel" for mitochondria. In addition, it is involved in regulating the levels of oxygen radicals and mediating appropriate responses to them. One of the consequences of this complex collaboration between p53, mitochondria, and respiration is that p53 stimulates the aerobic capacity of organisms, i.e. the amount of oxygen that the organism can consume during intense physical exertion. The p53 protein is also necessary to increase oxygen consumption during aerobic exercise. At the same time, aerobic capacity is not only an indicator of physical fitness, but also a predictor of the incidence and mortality of some diseases, including cancers (Wang et al., 2012; Kamp et al., 2016).

What determines the effects of p53? How does p53 select its target genes?

An almost unbelievable conclusion emerges from the previous sections. The p53 protein mediates the cell's response to almost every stress stimulus that may occur within it.

It selects from a wide range of possible cellular responses, and it not only actually regulates all of these cellular functions, but also controls and coordinates them with each other. For example, it can initiate immediate repair in response to minor DNA damage, but it can also halt the cell cycle and allow the time-consuming major DNA repairs to take place. In the event of even greater damage and/or insufficient capacity of repair mechanisms, it can initiate cellular senescence or trigger a cell death program. And, of course, the appropriate response may be a much more complex, finely tuned cellular reaction. But how can p53 do that? How does it decide which of the possible cell fates it will set in motion?

Although p53 acts at the molecular level through several mechanisms (see above), the main mechanism – and also the key mechanism for its tumor suppressor function – is activation of transcription of its target genes. It binds to the p53 consensus sequence in the regulatory regions of its target genes and stimulates their expression. It appears that we can turn the question of how p53 selects an appropriate cellular fate into how p53 selects its target genes. However, it is not certain that this will make the answer any easier. Several hundred genes have been described that contain the p53-binding site in their regulatory region. How p53 selects from them is not entirely clear. It is difficult to clarify this point because the effects of p53 on the expression of its target genes depend on many factors. These include, for example, the exact sequence of a particular p53-binding site in a given target gene (the p53 consensus sequence is only generally defined), its localization relative to the transcription start, and the number of p53-binding sites in a given gene. The type of stimulus that activates p53, its intensity and duration are also critical. For example, high and prolonged stress triggered by irreparable DNA damage leads to cellular senescence or death, whereas mild stress triggers only a mild p53 response that allows the damage to be repaired. The type of stimulus that activates p53, and its duration, among other factors, have a significant impact on the amount of p53 protein and its post-translational modifications. These are other circumstances that influence the function of p53. The specificity of the cell type and the general functional context, such as the availability of other protein cofactors and the presence and intensity of other signals, are also important. For example, we have seen that cell cycle arrest triggered by p53 can be transient and cycling can subsequently resume, but it can also be irreversible and lead to cellular senescence. And this depends on whether the cell with activated p53 is stimulated to grow at the same time or not.

In any case, the signaling and functional network that has evolved in the cell around the p53 protein is extremely complex, and the hundreds of p53 target genes are only one of many sources of this complexity. Another source of complexity in p53 function is the wide range of stimuli that activate it. How p53 selects target genes depending on the nature of the stimulus is not yet clear. According to one of the proposed models, p53 always induces the expression of up to hundreds of its target genes when activated in the cell, and other circumstances, including cellular context and cell type, then determine cell fate (Laptenko, Prives, 2006; Halazonetis et al., 2008; Menendez et al., 2009; Berkers et al., 2013; Serrano, D'Amours, 2014; Meek, 2015; Andrysik et al., 2017b; Fischer, 2017).

The gene encoding p53: The most frequently mutated gene in human tumors

The p53 tumor suppressor gene is one of the most frequently mutated genes in human tumors. It is reported that up to half of all human tumors have a mutation in the *TP53* gene. However, the function of p53 can also be damaged by a mechanism other than direct gene mutation. Therefore, the p53 pathway appears to be damaged almost universally in all tumors. This suggests that p53 is a serious obstacle that – if the tumor is to develop "successfully" – must be eliminated. As long as cells are equipped with a functioning p53 tumor suppressor, they are very well protected against possible tumor transformation.

Further evidence of the fundamental importance of p53 in protecting against tumorigenesis is provided by Li-Fraumeni syndrome (LFS). LFS is an inherited syndrome that predisposes the development of various tumor types such as sarcomas, adrenocortical carcinomas, leukemias, and breast carcinomas from a young age. The lifetime risk of developing a tumor is nearly 100% in women with LFS and about 80% in men. The most common causes of this severe and, fortunately, rare syndrome are congenital mutations of the tumor suppressor *TP53*. This means that inborn loss of a *TP53* allele predisposes its carrier to a tremendous risk of developing a tumor later in life (Wang, 2000; Cobb et al., 2002; Hickson, 2003; Shiloh, Ziv, 2013).

The *TP53* gene is not large; its coding sequence consists of 393 codons (trinucleotides). Mutations that occur in tumors can affect virtually any *TP53* codon. However, they most commonly occur in the portion encoding the DNA-binding domain, the part of the protein responsible for binding p53 to the consensus DNA sequence. Most p53 mutations are missense mutations causing single amino acid substitutions in the resulting protein. These relatively "minor" deviations from the standard structure of the p53 protein are usually sufficient to corrupt its binding ability to the consensus sequence. However, this type of mutation is not typical of tumor suppressor genes. The extent of damage to other tumor suppressors in tumors is usually much higher and they are often completely absent in tumor cells. In contrast, the typical mutant p53 protein not only differs from its standard form by a single amino acid, but it is also present in high amounts in tumor cells, often in extremely high amounts. One of the possible explanations is that the non-functioning p53 cannot activate the transcription of its target genes and thus cannot activate the *MDM2* gene, whose product mediates the degradation of p53. Therefore, the MDM2 protein is absent in cells with a nonfunctional p53 protein, so p53 cannot be degraded and accumulates in the cells. It is likely that other mechanisms are also involved in the extreme stabilization and accumulation of mutant p53 in cells (Brosh, Rotter, 2009; Freed-Pastor, Prives, 2012).

p53: Tumor suppressor or oncogene?

It has already been mentioned that p53 was originally mistaken for an oncogene and only later was clearly shown to be a tumor suppressor. The occurrence of inactivating mutations mainly in the DNA-binding domain proves that it is the ability to activate transcription of target genes by binding to the consensus sequence in the regulatory regions of these genes that makes p53 an effective tumor suppressor. The mutants lose the ability to bind to the p53 consensus sequence and activate transcription of their target genes, and at the same time they lose tumor suppressive function. And the loss of p53 tumor suppressive function dramatically weakens the cell's ability to resist tumor transformation. But often there is more to the story than that. Many mutant forms of p53 have been found not only to lose their tumor suppressive function, but also to retain or acquire other abilities that contribute even further to cancer development. They take on the properties of oncogenes. This means that tumors with mutated p53 proteins are more aggressive and dangerous than tumors that completely lack p53. These oncogenic forms of p53 likely function either as independent transcription factors or as cofactors that work with other transcription factors to affect the expression of a range of target genes – both positively or negatively – that differ from those activated by the standard p53 variant. They can also act through direct physical interactions with other cellular proteins, such as the p53-related proteins p63 and p73 and mitochondrial proteins. High amounts of mutant p53 proteins in the cell also play an important role in this effect (Brosh, Rotter, 2009; Oren, Rotter, 2010; Freed-Pastor, Prives, 2012; Meek, 2015; Kamp et al., 2016).

p53: Tumor suppressor, oncogene… anything else?

We think about the p53 protein almost exclusively in the context of the disease in which it was discovered and studied, i.e. as a tumor suppressor or – in mutated form – as an oncogene that not only loses its original safeguarding function but even contributes to the development of tumors. The certainty of this understanding of p53 is reflected in the established name we currently use for it: the p53 tumor suppressor. This is not surprising, given what a great, versatile, and indispensable guardian against tumorigenesis p53 is. Perhaps even more surprising is that protection from tumors is not the only function of this protein. As incredible as the finding may sound, protection from tumors is not even the main function of the p53 protein. Why is this so surprising? Probably those who think that the power of p53's tumor suppressive function blinds us are right. We are blinded by its light (Vousden, Prives, 2009). Indeed, the function of p53 clearly goes beyond only carcinogenesis…

p53 beyond carcinogenesis I

We assume almost dogmatically that the p53 protein is normally more or less inactive and is activated only under severe stress conditions, such as high activity of oncogenes or significant DNA damage. But p53 activity can probably also be triggered by everyday, ordinary "suffering," minor fluctuations in normal mammalian life, and circumstances of normal growth and development. The p53 model as a "cellular rheostat," proposed in the context of DNA damage repair mentioned earlier, is based on such an assumption. According to this model, the p53 protein provides not only for the repair of DNA damage, the extent of which proportionally increases the risk of tumor transformation, but also for any, even relatively small, "ordinary" DNA damage that occurs quite physiologically in daily life (Sengupta, Harris, 2005). Similar is the involvement of p53 in the regulation of glucose metabolism. Reduced glucose activates p53. The p53 protein stimulates oxidative phosphorylation, promoting short-term survival of cells that are temporarily starved. This is positive for the physiology of the organism. For example, mice lacking functional p53 tire rapidly, probably because they are unable to efficiently obtain energy through aerobic respiration. In tumor cells, the loss of these physiological responses due to p53 inactivation facilitates an increase in genetic instability and the development of aerobic glycolysis, milestones on the path to cancer. However, the p53 protein is clearly not "only" a tumor suppressor that protects the cell from tumor transformation, but a general functional factor that ensures an appropriate response of cells even to completely physiological stimuli, physiological fluctuations (Vousden, Lane, 2007).

p53 beyond carcinogenesis II

In addition to its tumor suppressive effects, the p53 protein is important in other aspects of life. In vertebrates, there is an entire p53 protein family. In addition to p53, it includes two other members, p63 and p73. All three proteins are similar, have a similar structure and a similar molecular mechanism of action. All three genes have also evolved from a common ancestor. Compared to p53, the effects of loss of function of p63 and p73 on tumor development are not nearly as pronounced. These proteins are mainly needed during development. However, p53 has also been found to have effects on the development of the organism. For example, it inhibits teratogenesis and reduces the incidence of developmental defects. On the other hand, the activity of p53 can sometimes be harmful to the organism. For example, the induction of apoptosis in response to ischemia can damage some organs. Apoptosis induced by p53 is also likely to be involved in the development of some neurodegenerative diseases such as Parkinson's disease, Alzheimer's disease, and Huntington's disease. Similarly, p53 appears to be involved in the aging processes of the organism and in reducing longevity. Thus, it not only acts as a tumor suppressor, but also regulates numerous other cellular and developmental processes. These include processes that have both positive and negative effects on the life of the organism (Vousden, Lane, 2007; Vousden, Prives, 2009).

p53 beyond carcinogenesis III – an evolutionary perspective

Is it justified to consider the p53 protein exclusively or mainly as a tumor suppressor by its evolution? There are three members of the p53 family in human cells: the p53, p63, and p73 proteins. They have evolved from a common ancestor. The p53 protein mainly functions as a tumor suppressor, while the p63 and p73 proteins are preferentially involved in the development of the organism. The roles of these proteins are also very similar in the cells of mice and other vertebrates. In lower animals, for example flies and worms, the situation is different. They have only one gene coding for the p53 protein, which activates apoptosis in the germ cells under stress conditions. Thus, its key role is to protect the germ line (Lu et al., 2009; Belyi et al., 2010). Surprisingly, even the simplest multicellular organisms, placozoans, have the p53- and MDM2-coding genes. Their bodies, composed of four cell types, resemble a simply organized colony of cells rather than a sophisticated system (Lane et al., 2010). And the p53 gene has even been found in unicellular organisms such as flagellates and amoebae. So it is clear that the p53 gene is not only present in lower, short-lived organisms in which tumors have not been found, but existed before multicellular organisms evolved. It has also been found in choanoflagellates, the closest relatives of multicellular animals (Nedelcu, Tan, 2007; King et al., 2008; Lu et al., 2009). When we consider human evolution, we must keep in mind that the life expectancy of humans in evolution history was much shorter than today (about 29 years on average). Therefore, humans were not threatened by tumors at such a young age. They died before tumors could develop. The risk of tumor development could not create the selection pressure driving the evolution of p53. The evolution of p53 could not be linked to its tumor suppressor function. The original function of the p53 protein had to be different. Perhaps, as mentioned earlier, p53 was originally responsible for gamete protection. Later, during the development of complex multicellular organisms, this function was extended to the protection of stem and progenitor cells. The ability to interfere with tumor development was another evolutionary extension of p53 activity (Aranda-Anzaldo, Dent, 2007; Vousden, Lane, 2007; Vousden, Prives, 2009; Lu et al., 2009; Belyi et al., 2010; Lane et al., 2010).

Complex organisms and stem cells

Unlike simple organisms (e.g. Hydra, Drosophila, nematodes) whose bodies, with the exception of gametes, consist of postmitotic, non-dividing cells, the tissues of complex organisms (e.g. vertebrates) are renewable. Damaged cells can be replaced. This results from the existence of stem and progenitor cells, which are responsible for tissue regeneration (see chapter 1A). Stem cells are undifferentiated cells that can divide and transform into other cell types (Fig. 3). Therefore, the body can produce the new cells it needs and repair damaged tissue. Stem cells are also characterized by the ability to self-renew through the mechanism of asymmetric cell division. While one of the two daughter cells differentiates "normally" to give rise to progenitor cells and gradually

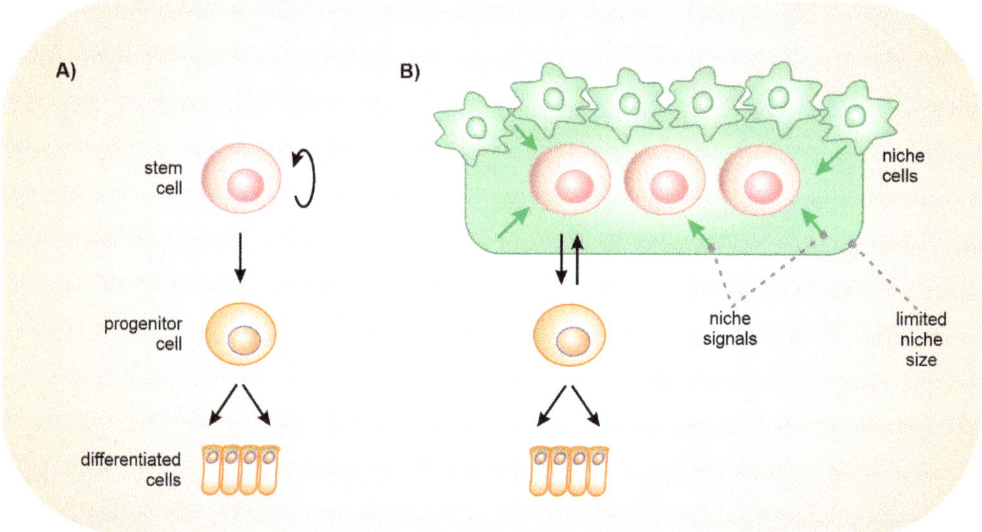

Fig. 46 Two models of stem cell function
According to the hierarchical model (A), differentiation is a one-way process with a fixed lineage of development from a rare stem cell through a progenitor cell to fully differentiated cells. A differentiated cell cannot return to a stem state. According to the neutral competition model (B), cells are highly plastic. The degree of cell "stemness" is significantly influenced by the microenvironment, called the niche, and cells follow neutral competition dynamics: they divide and differentiate, but can be reprogrammed by their niche and return to the stemness state (adapted from Batlle, Clevers, 2017, modified).

fully differentiated, specialized cells, losing the ability to divide indefinitely, the other fully retains the properties of the parental stem cell, including the ability to divide indefinitely (Fig. 46). Stem cells are equipped with an active telomerase that enables maintenance of stable telomere length during repeated cell division cycles (see chapter 5A; Campisi, 2003; Belyi et al., 2010).

Embryonic stem cells are derived from an early embryo at the blastocyst or morula stage (see chapter 1A) and are pluripotent. Adult stem cells are parts of postnatal tissue that exhibit tissue specificity, multipotency, or unipotency. For a long time, differentiation was considered a unidirectional process that cannot reverse and restore the state of stemness. This idea of a unidirectional, fixed developmental pathway leading from a rare stem cell through the progenitor cell to fully differentiated cells (Fig. 3) prevailed in the hierarchical model (Fig. 46A). Experimentally, however, induced pluripotent stem cells have been successfully produced from quite "ordinary" fully differentiated cells of the adult body. Their production is a rather complicated process requiring the activation of several key transcription factors, but this shows that cells are much more flexible in the differentiation hierarchy than originally thought. They are quite plastic. This plasticity is reflected in the model of neutral competition. According to this model, the "stemness" of cells is significantly influenced by the microenvironment, the niche, and cells follow the dynamics of neutral competition: they divide and differentiate but can be reprogrammed by the niche and return to a state of stemness (Fig. 46B) (Batlle, Clevers, 2017).

The growth of tumors, like the growth of physiological tissue, is determined and stimulated by stem cells, in this case tumor stem cells. They are poorly represented in the heterogeneous tumor tissue, but have the unlimited ability to proliferate and differentiate into any tumor cell type and also self-renew. They have some features of adult stem cells. This is one reason why adult stem cells were thought to give rise to tumor stem cells, namely though their gradual oncogenic transformation, the gradual accumulation of mutations (Fig. 47A). Thus, adult stem cells, after an initial intervention, represent "the first renegade cells." On the other hand, a low probability that rare and slowly dividing stem cells are "successful" in this process, and especially experiments showing that induced stem cells can be generated almost routinely from differentiated stem cells and that the cells are highly plastic, suggested that tumor stem cells could also arise from transformed somatic cells by dedifferentiation, as proposed by the neutral competition model (Fig. 47B) (Antoniou et al., 2013; Friedmann-Morvinski, Verma, 2014; Chaffer, Weinberg, 2015; Batlle, Clevers, 2017).

The p53 protein, gametes and stem cells

The repeated cycles of cell division that stem cells of all types undergo are associated with a high risk of mutation formation, accumulation, and spread. Therefore, they need mechanisms to reduce this risk. What is the role of the p53 protein in this process? The original role of the prehistoric p53-like protein in the simplest animals was to maintain the integrity of the gametes, the germ cells. It acted preferentially by inducing apoptosis in response to DNA damage (e.g. in Hydra, Drosophila) and later also to hypoxia and starvation (in *Caenorhabditis elegans*). A similar mechanism, the induction of apoptosis (and cell cycle arrest) in response to various types of stress, is also used in somatic cells of complex organisms.

But what about stem cells that are very resistant to apoptosis? Even in them, the p53 protein is strongly activated in response to DNA damage. It suppresses one of the key transcription factors responsible for maintenance of stemness and, conversely, induces cell differentiation. The differentiated cells are then much more willing to undergo the cell death program. In this way, cells with damaged DNA are effectively eliminated. Thus, the functional p53 protein allows stem cells to differentiate and, at the same time, by preventing dedifferentiation of differentiated cells it blocks the resumption of stemness (Fig. 48A). In contrast, differentiated cells lacking functional p53 can be easily reprogrammed into pluripotent stem cells and eventually into tumor cell stem cells (Fig. 48B). Thus, the p53 protein forms a kind of protective shield in the cells that prevents them from making a reversible transition between a less and

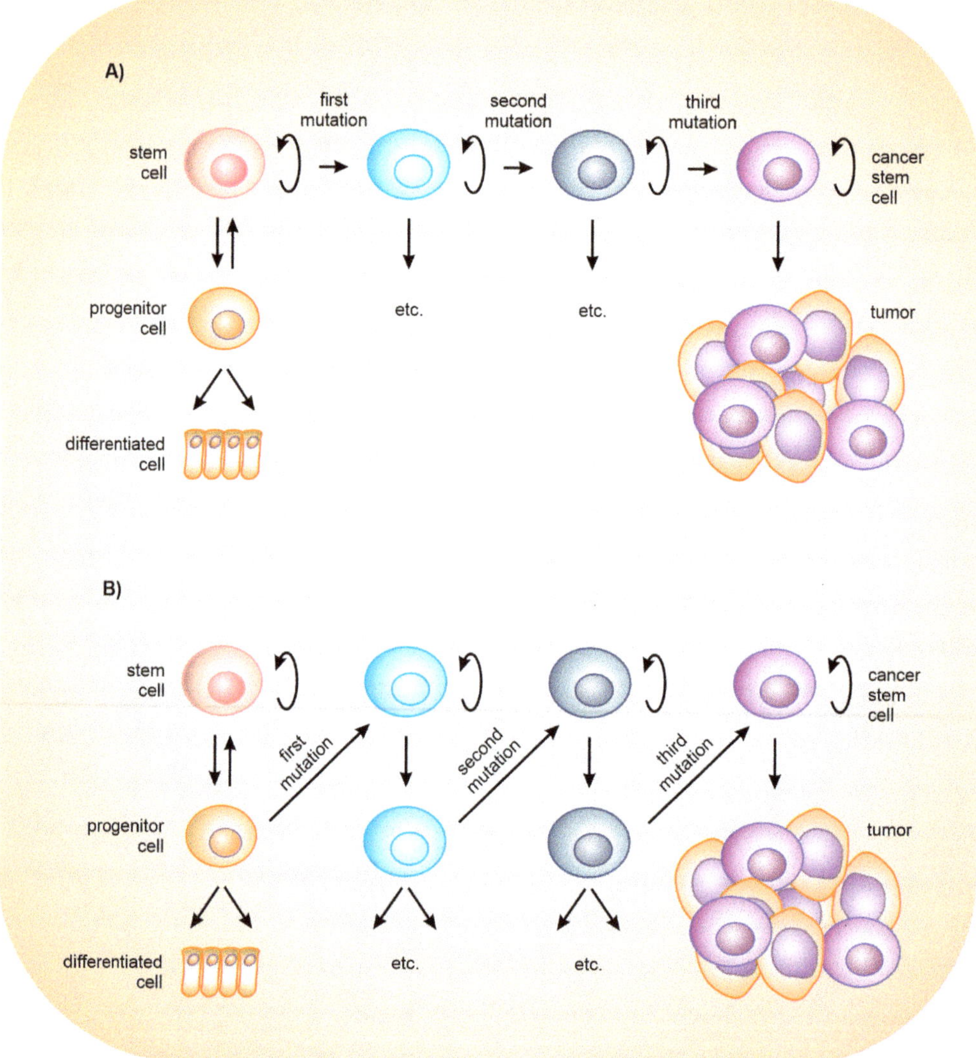

Fig. 47 The development of a tumor stem cell
(A) Tumor stem cells can arise from adult stem cells by gradual oncogenic transformation due to a gradual accumulation of mutations. (B) They can also arise by dedifferentiation of tumor-transformed mutant progenitor cells (adapted from Chaffer, Weinberg, 2015, modified).

a more differentiated state. It only allows a one-way progression toward terminal cell differentiation (Bonizzi et al., 2012; Lin, Lin, 2017) (Fig. 48). Thus, the p53 protein is a crucial inhibitor of the formation of induced pluripotent stem cells and also tumor cell stem cells. In this way, it acts as an exemplary tumor suppressor. This ability complements and extends its incredible versatility as tumor suppressor.

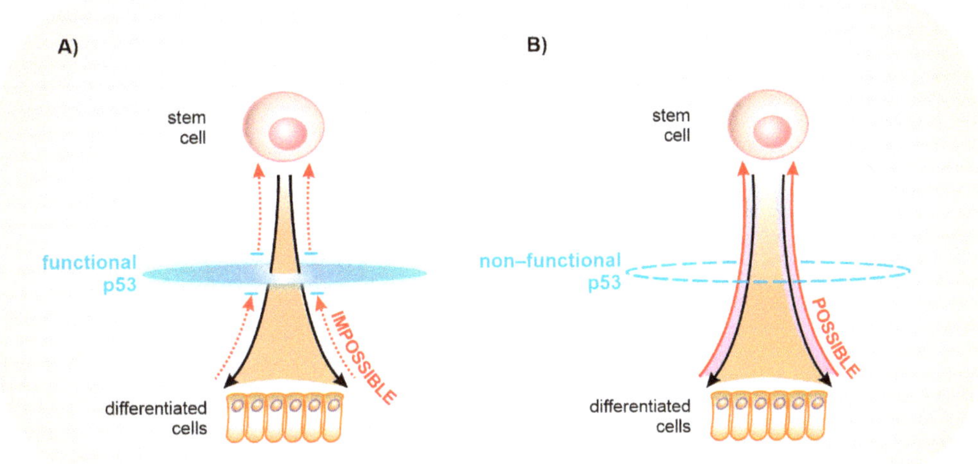

Fig. 48 The p53 protein as a stem cell switch
(A) Under normal circumstances, when the p53 signaling pathway is functional, stem cells can differentiate into specialized cells, but specialized cells cannot return to the state of stemness because of the functional barrier of p53. (B) When the protective barrier of p53 is not functional, differentiated cells can be reprogrammed into stem cells (adapted from Lin, Lin, 2017, modified).

13B. Wisdom and responsibility

The p53 tumor suppressor: A cellular Ferda the ant? Commander of the integrated rescue system?

The p53 protein effectively participates in the regulation and control of all cellular functions and properties that have been implicated in human tumorigenesis (Figs. 8 and 45). Incredibly, this single small protein can affect many different functions in a cell. Virtually all of them are important for protection against the onset and development of cancer. The p53 protein really is the most versatile tumor suppressor known. In the cell, it does work of all kinds; it is a cellular Ferda the ant (a popular character in Czech fairytales and comics). Like Ferda, an anthropomorphic ant created by Ondřej Sekora, "the ever-optimistic guy with a red polka-dotted kerchief around his neck who is fearless, smart, resourceful, and always finds his way around," p53 knows every problem that can arise in a cell. It mediates the cellular response to almost any stress stimulus (damage to DNA, hypoxia, elevated oncogene activity and many others) and chooses from a wide range of possible responses (arresting the cell cycle, inducing DNA repair, senescence, death and many others). Unlike Ferda, however, the p53 protein itself does not immediately perform major and specific work in the cell. Like a transcription factor, it activates the expression of many of its target genes, and the products of these genes then perform their own specific work. Thus, the p53 protein decides which of its target genes gets activated and how the cell will respond. It plays a role similar to the commander of the integrated rescue system, who does not take much action himself, but has almost everything under his thumb, from firefighters, maintenance workers and paramedics to the food supply, i.e. dozens of subordinates who do their work according to the commander's orders (Andrysík, 2017a).

The p53 tumor suppressor: A conscience?

Another metaphor for the cellular role of p53 was chosen by those studying it in the context of regulating the inflammatory response who compared it with the role of the transcription factor NF-κB, which, unlike p53, stimulates inflammation, cell proliferation, their survival, and de-differentiation and is therefore widely considered a p53 antagonist (chapters 10A and 13A). These authors note that p53 responds primarily to intrinsic stress signals such as DNA damage, high oncogene activity, damage to the mitotic spindle, or shortening of telomeres. According to them, the role of p53 in this sense is analogous to the human conscience. The human conscience is an internal mechanism that prevents members of the human community from acting unethically and, in extreme cases, can lead to suicide or self-sacrifice, which in a metaphorical sense corresponds to apoptosis at the cellular level. The prerequisite for unethical and antisocial behavior is the elimination or weakening of this function, whether in the human or cellular sense. In contrast, NF-κB responds primarily to external stress signals, such as infection. When there is an external threat that requires mobilization of all internal and external protective mechanisms, altruistic self-limiting or self-punishing behavior is not a rational strategy. At the cellular level, NF-κB activation "suppresses the conscience" by causing a transient inhibition of apoptosis and creating conditions that mobilize cell survival and stimulate the immune response (Gudkov et al., 2011). Before leaving this paragraph, let us summarize what properties the authors associate with the function of the p53 protein. According to them, p53 provide cells with the ability to behave ethically, socially, and altruistically.

The p53 tumor suppressor: Wisdom

Which of the above metaphors best fits the function of p53 in the cell? Is p53 a striving jack-of-all-trades like Ferda the ant? Or an experienced and energetic commander of the integrated rescue system? Or does it represent the cellular conscience that drives the cell with deep self-reflection to respond appropriately to its shortcomings, even to the point of self-destruction? For example, if a cell's DNA is damaged, is p53 strong and "brave" enough to change it? Should it stop the cell division and begin the demanding process of DNA repair? Or does the cell have no choice but to accept its inevitable fate and either initiate the process of senescence or activate the cell death program? "God, grant me the serenity to accept the things I cannot change, courage to change the things I can, and wisdom to know the difference." This prayer for the gift of discernment by Friedrich Christoph Oetinger, which is a version of the prayer of St. Francis of Assisi, considers wisdom to be the ability to make the right choices. If we accept this view, then the p53 protein, which selects the cellular response and decides the fate of cells, represents cellular wisdom.

The p53 tumor suppressor: The art of making the right choice

Discernment, i.e. making a decision, is an essential part of life and shapes our life course significantly. Every decision has an impact on our lives. The more important the decision, the more serious the consequences. Making the right decisions, making the right choices, is critical to our life's journey. But, the more choices we have, the more difficult it is to choose between them. There is no doubt that the modern and postmodern world of today, with its almost unlimited possibilities, has become very complicated. Everything, including our choices, is entwined, and even the consequences of our choices are unexpectedly and often confusingly connected and intertwined. The ability to make choices, the ability to discern, is thus both essential and difficult. Even according to the Christian tradition, the gift of discernment is one of the basic charisms, the fundamental spiritual gifts, the extraordinary abilities. According to some, it even occupies a privileged position among charisms, because without the gift of discernment, other gifts cannot be used properly. At the same time, there is neither a simple and universal recipe for the gift of discernment nor a universal "right" choice for everyone. What is right for one person may not be right for another. And what is right today may not be right tomorrow. And that is why no treatise on the ability to make decisions and the art of discernment can provide a mechanically applicable recipe or even a general rule of thumb for making the right decision. It can only provide guidance and advice on how to seek and achieve the right choices (Georges, 2011; Kiechle, 2015). "True happiness is not tied to a particular way of life, but to the fact that we are in the right place" (Georges, 2011, p. 24).

Even in a human body with more than 10^{13} cells, no two cells are exactly the same, they cannot occupy exactly the same space. And therefore the cells in the body do not have to respond equally to the same stimuli and the same situations. The response depends on their particular condition, fitness, age, and personal history, as well as their exact location in the body and the specific conditions of the environment. But if their p53 function – their ability to make the right decisions – is preserved, they will respond appropriately. "Appropriate" here means in the cell's own interest, for its personal benefit, its own good life, and at the same time best in relation to its immediate environment and therefore in the highest interest of the whole body, the whole system of which it is a part. After all, the good condition of the body, in return, provides good conditions for the life of each cell. And it is not surprising that one of the most important clues in the human search for the right life decisions is the recommendation to make decisions in one's own personal interest and at the same time in the interest of others. It is considered wise to strike a balance between egoism and altruism (Kiechle, 2015).

The p53 tumor suppressor: Responsibility

Decision making is closely related to responsibility. Chapter 1A explains that although the structure, organization and function of a multicellular organism is very complex, multicellular organisms have no central control and there is no control center anywhere in the body. Cells communicate intensively with each other and with their environment. Their activity is coordinated by the nervous, immune and hormonal systems, but actually each individual cell is responsible for its own good condition and for its harmonious integration into the whole body. Thus, it bears its share of responsibility for the condition and destiny of the entire organism. This at the same time brings us back to the idea that p53 is a kind of cellular conscience. Conscience is defined as the ability to distinguish between morally good and evil, the ability to discern what we should and should not do. And morality means a general idea of how to behave properly in society. Whether we refer to the p53 protein simply as the conscience of the cell, the wisdom, or the responsibility of the cell, we are talking precisely about its ability to make decisions about the fate of the cell and to strictly consider the interests of the entire organism in making that decision.

Does a cell lacking p53 suffer from an antisocial disorder? Is the tumor cell deprived? Cells that have lost the function of p53, i.e. the ability to consider the interests of the whole organism, resemble deprived people. This is what a certain subgroup of psychopaths is called. Psychopaths can be criminals, real human monsters, for example serial and mass murderers or rapists. In addition, there are also the deprived people, sometimes called partial, incomplete, socially successful, socially skilled, socially adapted or subcriminal psychopaths. These are people with antisocial traits. Compared to criminal psychopaths, they are much less conspicuous, i.e. harder to detect, and therefore pose a significant danger to society. Deprived people are not usually violently criminal; on the contrary, they can be powerful and economically very successful, often highly educated. They occupy top positions in power and finance and are largely responsible for "white-collar crime." However, we also find them in small businesses, in families, in short, at every level, in every social environment (Koukolík, Drtilová, 2006).

According to the psychobiological seven-dimensional model, human personality can be characterized by four temperament dimensions and three character dimensions. The temperament dimensions include (1) novelty seeking, (2) harm avoidance tendency, (3) reward dependence, and (4) perseveration, and represent mainly innate types of automatic responses that are largely due to heredity. Character dimensions include (5) self-focus, the ability to take responsibility for one's own decisions and actions in order to fulfill a meaning rather than to satisfy immediate needs. Another dimension is (6) willingness and ability to cooperate, the capacity for compassion and empathy, respect for the rights and differences of others rather than the pursuit of personal gain at any cost. And finally, the last character dimension is (7) self-transcendence, the awareness of personal unity with the whole being, perceiving oneself as part of cosmic evolution.

The character dimensions are mainly the result of learning and cultural influences, and express the extent to which a person feels like an autonomous individual and the extent to which he or she feels like a part of human society, of the whole, of the universe. A low level of self-focus, cooperativeness, and self-transcendence is the common denominator of all types of antisocial personality disorder. It is characterized by noncompliance with social norms and disregard for others. People affected by antisocial personality disorder do not even know feelings of remorse or guilt. They have no regard for other people. Their only concern is to achieve their goals and gain power. They are selfish and strictly impose their will; they do not care about the interest of the whole. Does not this description fit a cell that lacks the function of the tumor suppressor p53? Does it not also fit a tumor cell? František Koukolík and Jana Drtilová, Czech authors who have studied deprived people in depth, describe their behavior as "similar to the behavior of a malignant tumor, including the consequences for the group they influence or control" (Koukolík, Drtilová, 2006, p. 63).

In this context, let us briefly return to chapter 12B and the story of Dan Ariely's student who locked himself out of his apartment, called a locksmith for help, was surprised at how easily the lock opened, and was taught by the wise locksmith that the lock system protects against 98 percent of our "normally decent" fellow citizens. We learned a lesson from this for the problem of political corruption and (political) life in general, namely that "the rules are more important than who wins the election." Dan Ariely and his colleagues came up with Correction Factor Theory, based on the fact that most people like to make money by committing a little fraud but need to feel good about it. These people are kept in check by the system (constitution, laws, rules). But the wise locksmith also said that in addition to the one percent of our fellow citizens who are always honest and never steal anything, there is another percent of citizens who are inveterate thieves and can always somehow open the door. This one percent of deprived people without conscience is extremely dangerous! In deprived people – we know this from cells lacking the function of the p53 protein – no lock system works!

p53: Guardian and (precondition?) of multicellularity

If part of the ability of the p53 protein is to select from the possible cellular responses to various stimuli those that are not only in the interest of the cell itself, but also in the interest of the whole organism, it is quite justified to consider p53 as a guardian of multicellularity. Perhaps even as its necessary precondition. After all, one of the major problems facing multicellular organisms is the risk of the emergence of "renegade" cells, which, through a random mutation, may gain a growth advantage over other cells and become the starting point of tumor development.

"Renegade cells" inevitably arise because of the genetic instability inherent in all living cells, as explained in chapter 6A. At this point we also referred to the estimated rate of about 10^{-6} at which mutations in a gene occur during human cell division. The more cells there are in the body, the greater the likelihood that one of them will sooner or later take on characteristics of a "renegade" cell and threaten the entire system by developing into a tumor. When the p53 function is lost, the development of tumors is almost inevitable. Therefore, one might think that p53 is a gene that evolved primarily to solve this problem. If the p53 protein did not exist, another would have to evolve to provide protection for multicellular organisms from renegade cells and tumors (Assaily, 2005).

An elegant proof of this statement is the observation that elephants, whose bodies consist of many times more cells than the human body, should therefore also be much more susceptible to tumor formation – but this is not the case. They have twenty copies of the gene p53 in their genome. The high number of genes for p53 explains an otherwise surprising observation – called Peto's Paradox after its author – that elephants and whales, despite being many times larger than humans, suffer from tumors less frequently than humans (Peto et al., 1975; Abegglen et al., 2015). Less elegant in this sense remains the mystery of how it protects the giant whale body from developing tumors, in which the increased number of the p53 gene copies has not been found (Gaughran et al., 2016). However, even the fact that whales likely use another tumor suppressor mechanism in addition to p53 (Keane et al., 2015) cannot call into question the key role of the p53 protein in protecting against tumor development in humans.

The origin of the p53 tumor suppressor: Tubeworms

Although the function of p53 is clearly linked to protection against tumorigenesis, its origin undoubtedly predates the evolution of multicellular organisms. The evolutionarily oldest known organisms containing p53 are tubeworms (Nedelcu, Tan, 2007; King et al., 2008). Tubeworms (*Choanoflagellata*) are the closest relatives of multicellular animals (*Metazoa*). They are unicellular or colony-forming organisms in marine and freshwater biotypes. Their cells typically form collars. The collar consists of a ring of 15–50 projections of cytoplasm that form a funnel around the long flagellum. In colonies, the cells containing the collar face both outward and inward, and the interior of the colony is filled with amoeboid cells. Life in colonies allows initial specialization to non-dividing, flagellated, motile, and feeding cells on the surface of the colony and dividing, undifferentiated "stem" cells inside the colony. Such a colony resembles a simple structured animal. Molecular phylogenetic analyses have shown that tubeworms are not descended from metazoans, but represent an independent lineage that evolved prior to their differentiation. Because of their unique position in the phylogenetic tree, the study of tubeworms offers unique insights into the origin and evolution of multicellularity (Zrzavý, 2006; Srivastava et al., 2008; de Mendoza, Ruiz-Trillo, 2014). The ancestor of metazoans was a multicellular organism with differentiated cell types, epithelia, a body plan, and the capacity for regulated development including gastrulation. On the other hand, the last common ancestor of tubeworms and metazoans was a unicellular organism or an organism that formed simple colonies (Srivastava et al., 2008).

Evolution and tinkering

When did the p53 gene emerge in the course of evolution? Certainly, at the latest with the tubeworms. Thus, as already stated, it undoubtedly also arose in unicellular organisms, which does not fit with its function as a tumor suppressor. After all, tumors cannot arise in unicellular organisms! This suggests that the p53 protein must have originally had a different function and only later became the important guardian against tumor formation (see chapter 12A). But this is nothing unique or unusual in evolution. A similar mechanism of evolutionary tinkering is used repeatedly.

The mechanisms of evolution (mutation and natural selection) do not work like engineers, who work with a clear task and vision, which they try to fulfill as accurately as possible to the best of their knowledge and within the limits of the available technical possibilities. Rather, they work like a do-it-yourselfer, who often does not even know exactly what he is actually creating or what he will use it for. The do-it-yourselfer simply uses whatever he finds around him. A piece of wire, string, wood scraps or an old box and just makes something out of it. Something that can serve a variety of purposes. For what? Well, that depends on the circumstances and the immediate need. Similarly, the do-it-yourselfer may take an existing tool designed for a well-defined, clear and certain function and use it for a completely different, unexpected purpose. The results are far from perfect, but somehow they work: the toy car wobbles but drives, the cargo is spilled but part of it is transported to a new location… Everything can be further repaired and improved (Jacob 1977).

And somehow it was the same with the p53 protein. Originally, it mediated the cellular response to DNA damage (Srivastava et al., 2010), but gradually it accumulated more and more capabilities and skills until it became a versatile and indispensable guardian of multicellular organisms against tumorigenesis. A wise, responsible, and dutiful commander of the rescue system. But every gene had to originate somewhere. It had to emerge somewhere and appear for the first time, as an absolute evolutionary novelty. When did the tumor suppressor p53 appear? And what about other tumor suppressors and oncogenes? Does the timing of their appearance say anything about their function, purpose and importance?

Evolutionary origin of oncogenes and tumor suppressors

Genomic phylostratigraphy is a method with a somewhat cryptic name that deals with determining the evolutionary origin of founder genes. Founder genes represent evolutionary innovations – completely new genes that arose suddenly, for example, from non-coding regions of the genome. They do not arise by gene duplication and subsequent divergence of one of the genes or by rearrangement of existing genes or their parts into a new structure and adaptation to a new function (Levy, 2019). All of these mechanisms are considered evolutionary tinkering. Phylostratigraphy simply looks for the first appearance of genes during evolution. In general, the emergence of new founder genes is thought to be associated with significant evolutionary changes or turning points. The emergence and development of multicellularity, with which the risk of tumor development is closely associated, represents such a fundamental evolutionary turning point. Did the genes associated with tumorigenesis originate with the emergence of multicellularity?

Mutations involved in tumorigenesis can occur in several hundred genes, which are commonly divided into oncogenes and tumor suppressors (see chapter 2A; Sondka et al., 2018; Tate et al., 2019). Tumor suppressors are sometimes further subdivided into "caretakers," i.e. something like "administrator" genes, and "gatekeepers," i.e. "guardian" genes. Mutations in caretaker genes indirectly contribute to tumor development by increasing genetic instability and thus the rate of mutations occurring in gatekeeper genes. Mutations of gatekeepers directly participate in tumorigenesis, for example, by deregulating cell division, differentiation, growth, and cell death. Interestingly, some tumor suppressors, including the p53 protein, are both caretakers and gatekeepers.

The use of phylostratigraphy to determine when oncogenes, tumor suppressor genes, caretaker and gatekeeper genes appeared during evolution provided surprisingly clear and robust conclusions. Genes associated with tumor development appeared in two phases. One group of genes appeared quite as expected with the emergence of multicellularity. Surprisingly, the second group of genes already appeared with the formation of the first cell. The appearance of genes involved in the regulations related to tumors and thus multicellularity was hardly expected in this phase. This might be explained by the fact that the genes that already appeared in the formation of the first cell fall into the category of caretakers. Even the oldest cells, including prokaryotic cells, certainly needed to take care of the genome. The gatekeepers, on the other hand, appeared only with the advent of multicellularity (Domazet-Lošo, Tautz, 2010).

For a deeper understanding of the meaning of all this, how and if at all the development of tumors is related to the emergence of multicellularity, it is necessary to understand more precisely the origin of multicellularity. And this is what the next, the last chapter, will be about.

14A. Multicellularity

Unicellular organisms

For a long time after the origin of life on Earth, unicellular organisms were the only living creatures. They appeared on this planet about 3.45 billion years ago. They usually reproduce very quickly. For example, *Escherichia coli*, one of the best-studied bacteria, can divide within 20 minutes, or 72 times a day, under optimal conditions. The speed of division is limited only by the time needed to replicate the genome and, of course, by the availability of nutrients. When sufficient nutrients are available, they divide virtually continuously. If *Escherichia coli* could continually multiply exponentially, the mass of the resulting cells, although weighing only 10^{-12} grams, would exceed the mass of the Earth by 2,664 times in two days. Fortunately, bacteria spend most of their lives waiting for food (Lane, 2012). Unicellular organisms have demonstrated their enormous capacity for life and adaptation, colonizing every corner of our planet. Multicellularity evolved much later.

Simple and complex multicellular organisms

Multicellular organisms are either simple or complex. In simple organisms, all cells have contact with the external environment, while in complex organisms, some cells are internalized and have no contact with the external environment. This affects the size of the organism, results in the need to transport nutrients, and consequently limits any possible return to unicellularity. Complex multicellular organisms are characterized by the formation of cell-cell and cell-extracellular matrix adhesions, prominent intercellular communication, and complex three-dimensional arrangement. Cell differentiation, tissue development and morphogenesis are highly pronounced, which is associated with the development of intercellular signaling and regulation of gene expression. Integral to these processes is the regulation of cell division, growth, and programmed cell death. A complex body plan and precise organization of tissues are required for the transport of nutrients, oxygen, and signaling molecules, as well as for the general functioning of the organism (Srivastava et al., 2010; Knoll, 2011; Niklas, Newman, 2013).

Evolution of multicellular animals

Multicellular organisms can arise either by simple aggregation of originally independent, genetically distinct cells or clonally, i.e. by gradual mitotic division from a single initial cell resulting in cells that do not segregate but remain close together. The evolution of multicellular animals from unicellular ancestors required the development of numerous mechanisms that coordinate cell division, growth, differentiation, adhesion, death, intercellular communication, and much more. Where did the molecular tools of multicellularity come from? As mentioned in chapter 13B, some of them are the result of evolutionary tinkering, and others were an evolutionary novelty. Both processes are associated with the general increase in genetic material. Many molecular components of the tools of multicellularity are already present in unicellular relatives and therefore had to be present in their common ancestors. These include, for example, the proteins involved in adhesion. They were already present in choanoflagellates, where they probably served to "hunt a prey." Some of these pre-existing components have changed dramatically in the lineages of multicellular organisms, differing in number and function.

Some components are completely absent from unicellular relatives, likely representing evolutionary innovations associated with the development of multicellularity. These include, for example, key molecules in the signaling pathways that provide cellular communication. The number and complexity of transcription factors responsible for cell development and differentiation also increased greatly during the evolution of multicellularity (Rokas, 2008a, 2008b; Niklas, Newman, 2013; Alegado, King, 2014; Levin et al., 2014; Brunet, King, 2017). The current genomes of multicellular animals reflect the history of their origin and evolution. In the genomes of mammals, for example, about a quarter of the genes (23%) are identical to those of all other living organisms. About another quarter of the genes (29%) are common to all eukaryote cells. About 27% of the genes are related to the origin of multicellular animals, and only another fifth of the genes arose during the evolution of the animals themselves (Zrzavý, 2006).

Complex multicellularity has evolved only in eukaryotic cells. The oldest fossil evidence for the existence of eukaryotes is 2.1 billion years old, and multicellular eukaryotes evolved about 1.2 billion years ago. Complex multicellularity has evolved independently several times during evolution, reportedly at least seven times and in all kingdoms of organisms, including fungi, plants, and animals. Simple multicellularity has even evolved repeatedly more than twenty times (Knoll, 2011). Multicellular animals appeared on Earth about 800 to 635 million years ago. They have undeniable advantages over unicellular organisms. Their size allows them to escape predators more easily, they are more efficient at obtaining and utilizing food, and they penetrate new ecological niches more readily. Multicellularity generally reduces dependence on the external environment. One of the most important consequences of the emergence and development of multicellularity was an enormous increase in the diversity of species and forms. The emergence of multicellularity represents a fundamental milestone in the evolution of life and is one of the most important evolutionary transitions (Stanley, 1973; Rokas, 2008a, 2008b; Knoll, 2011; Niklas, Newman, 2013; Alegado, King, 2014; Brunet, King, 2017).

Evolution and major evolutionary transitions

Biological evolution is a long-term, spontaneous process by which terrestrial life originated, developed, and diverged. In general, evolution means "gradual development," "continuous development from lower to higher," "gradual development of the system." We also usually associate the process of biological evolution with the idea of gradual, "step-by-step" development. The driving forces of this process are, on the one hand, mutations as a source of differentiation and generation of new variants and, on the other hand, natural selection, which, according to various criteria, selects from a diverse group of individuals those that it suppresses or favors, depending on their ability to adapt to changing environmental conditions. In this sense, there is a constant competition, a constant rivalry, between the different variants. The ability of a particular individual to reproduce and pass on its genes to the next generation is an expression of its biological fitness. This constant play of mutations, natural selection and changing fitness is well compatible with the idea of a continuous, gradual evolution step by step, perhaps "trait by trait" or "mutation by mutation."

The evolution of life on Earth, which is accompanied by increasing complexity and hierarchy, does not, of course, proceed completely smoothly and uniformly. In the course of evolution there have been several (according to some authors seven or eight) major evolutionary transitions. These are key turning points that involved a significant increase in biological complexity and fundamentally affected the development of life on Earth and the evolutionary process itself. These major evolutionary transitions include, for example, the emergence of chromosomes from genes that originally replicated separately, the emergence of the first cell that is the common ancestor of all cells, the emergence of a eukaryotic cell with a nucleus, cell organelles, membranes, and cytoskeleton, the emergence of multicellular organisms mentioned earlier, and finally, the emergence of the diversely organized communities of multicellular organisms. At each of these transitions, the previously autonomously replicating units become part of a larger whole and can only be replicated as parts of it. In the process, a division labor occurs between the originally independent units, allowing for specialization and a change in the way genetic information is stored and transmitted. The evolution of multicellular organisms from previously independent single cells is considered by some authors to be the most spectacular of all major evolutionary transitions (Szathmáry, Smith, 1995; Gros-Berg, Strathmann, 2007; Szathmáry, 2015).

Major evolutionary transitions of the individual

Major evolutionary transitions are usually associated with major innovations. Individuals (for example, cells) that were originally capable of reproducing independently begin to cooperate, creating a new, more complex form of life, a new kind of individual of a higher hierarchy (multicellular organism). Within this newly created individual, there is no mutual conflict between the original individuals. This is not trivial, considering that until this transition, each cell dreamed its own dream of becoming two cells, and in this dream it competed with every other cell for resources and space. Therefore, the same question arises in every major transition: how are the selfish interests of individuals overcome so that a cooperative community based on reciprocity can emerge? And are there any commonalities between the various major evolutionary transitions?

The origin of multicellular organisms exhibits some general tendencies, and the paths leading to them follow similar steps. In the beginning, there is usually a simple aggregation that supports the biological fitness of the individual cells. In the next stage, cells become interdependent and the level affected by natural selection shifts. This is a major evolutionary transition in individuality. Natural selection ceases to operate at the level of individual cells and begins to operate at the level of the entire multicellular unit, which thus becomes a new evolutionary individual. Evolutionary individuals (sometimes called Darwinian individuals) are defined as integrated, indivisible units that differ by heritable variations in fitness and can evolve over time. An evolutionary transition in individuality produces a newly formed community of individuals that is sufficiently integrated to produce a new individual of a higher level. In the course of evolution, this can be illustrated by the transition from gene to genome, the emergence of the first cell, the transition from prokaryotic to eukaryotic cells, from a single cell to a multicellular organism, and further from independently living organisms to social organisms.

The first part of the road to multicellularity: Aggregation of undifferentiated cells

It is obvious that multicellular organisms have advantages over independently living single cells. But which of them were successful at the beginning of the evolution of multicellularity, when poorly organized clusters of undifferentiated cells formed? What "drove" evolution at that stage? And how can we study it retrospectively? We do not know a comprehensive and definitive answer, of course, and, moreover, it is difficult to seek it experimentally with existing unicellular or multicellular organisms. However, computer modeling and the use of digital organisms have yielded many interesting insights (Pfeifer, Bonhoeffer, 2003; Ofria, Wilke, 2004). Some of these findings are worth mentioning because they contribute to the understanding of the relationship between tumors and healthy cells. In addition, they are also very stimulating in terms of "overlaps."

Some of these take us back to chapter 8, which deals with cellular energy metabolism. For heterotrophic organisms that rely on external food sources, it is important to find a balance between speed and efficiency of obtaining energy (ATP production) because this affects their fitness. In general, they gain energy either quickly but inefficiently through glycolysis, or efficiently but slowly through oxidative phosphorylation. A series of experiments and computer simulations were conducted to investigate how the way cells gain energy affects their fitness depending on the availability of resources, cell motility, and their interactions. These experiments showed that under conditions of competition for shared resources, the winning cells can rapidly gain ATP, even at the expense of rapidly depleting resources and producing only a small cell population. Accelerated resource influx and greater cell motility also contribute to the preference for "fast" cells. The cells with lower mobility tend to be surrounded by cells of the same type, favoring "slow" cells that use resources slowly but effectively, resulting in higher density and more successful clustering. The ability of individuals within a cluster to exchange resources is another advantage. Clustering can also be a selective advantage for efficient obtaining of energy because it excludes "fast," selfish, non-cooperating individuals from the cluster community. Slow and efficient resource utilization by cells within communities (clusters) thus reflects their cooperation, whereas tumor cells with their switch to "fast" aerobic glycolysis may be perceived as cells leaving the realm of cooperation (Pfeiffer et al., 2001; Pfeifer, Bonhoeffer, 2003).

Probably the greatest benefits of multicellularity result from the division of labor among the various differentiated cells within the organism. Computer modeling has shown that the more complex the (digital) task to be solved, the greater the benefit of dividing the substeps among the cells. The main benefit of cell specialization is not to speed up these steps, but to save energy that would otherwise be needed to switch from one task to another. In addition, the cells gradually become interdependent and after some time even completely lose the ability to function independently. This type of modeling also contributed to the understanding of the basis of intercellular cooperation and the major evolutionary transition to multicellular organisms (Goldsby et al., 2012, 2014a).

The second part of the road to multicellularity: The transformation into a solid, integrated entity

Why and how are the basic traits of an evolutionary individual, such as heritability of fitness, indivisibility, and evolvability (i.e. the capacity for natural selection and evolution), transferred from the old to the new level? A typical sign of an evolutionary individual is the high degree of integration mentioned earlier. This includes the loss of individual autonomy of cells and the appearance of a substantial division of labor that accompanies cell differentiation. But why do cells within a multicellular organism cooperate so effectively? How do they manage and control their initial "competitiveness" and stop fighting each other? A multicellular animal, i.e. an already mature evolutionary individual, normally goes through a bottleneck phase during its individual (ontogenetic) development in which the number of its components, i.e. cells, sharply decreases. It can be reduced to a single cell, a fertilized egg, from which a whole new organism then develops clonally. The consequence of such a narrowing is a high degree of relatedness, i.e. a decrease in the genetic diversity of the cells and thus a reduction in natural selection within the organism.

A key factor in the transition of an evolutionary individual from a cell to a multicellular organism appears to be a specialization of reproductive and vegetative functions, i.e. the separation of germ and somatic cell lineages, termed germ-soma differentiation. Somatic cells proliferate, differentiate, and shape the body of an individual throughout its life and ensure its function and survival. Germ cells, on the other hand, divide much less frequently, develop into gametes, and provide sexual reproduction for the entire organism. The first stages of separation of reproductive and somatic functions were also observed and described in digital organisms using computer modeling. This separation appeared as a solution to a following task in the model: some cells need to perform activities for the host organism that can damage DNA. In this situation, a new division of labor occurred. Some cells were relieved of this "dirty" work and thus kept their DNA intact. Their task was now to take care of the reproduction of the entire community and give birth to offspring. In contrast, the cells that did the "dirty work" lost their reproductive capacity and even showed signs of aging (Goldsby et al., 2014b).

Germline segregation limits the reproductive possibilities of somatic cells. The "fitness interest" is thus linked to the germ line and further reduces competition between the cells of the organism. The germ-soma differentiation is the most important factor that transforms a group of cells into a true multicellular organism. After this differentiation, there is no way back. None of the cells in the community can survive and reproduce alone. The cell community is indivisible and becomes an individual. A return to unicellularity is no longer possible (Michod, 2007; Knoll, 2011; Niklas, Newman, 2013; Kennedy et al., 2014; West et al., 2015).

Cooperation and the Dawkins selfish gene

We often associate Darwin's theory of evolution primarily with the "struggle for survival," with competition and survival of the fittest individuals. We often do this in the context of selfish behavior. Every gene, cell, and organism is focused on its own evolutionary success, at the expense of its opponents. And yet, various forms of intraspecific and interspecific cooperation and altruism are undeniable and universal components of life. How to understand this paradox? Are these contradictions irreconcilable? Richard

Dawkins escalated the contradiction in his theory of the selfish gene. According to his theory, the objects of natural selection and thus of biological evolution are neither individuals nor families, populations nor species, but only different alleles of individual genes, which are the only materials passed on in unchanged form from generation to generation in sexually reproducing organisms. According to Dawkins, biological evolution is to be understood as a race between different alleles of a certain locus for the greatest possible occurrence in the gene pool (i.e. the set of all alleles of all individuals) of a given population. Organisms (and other hierarchically higher entities) are understood only as carriers, transporters, tools that genes have evolved to use to reproduce most efficiently. And for this purpose, individual alleles of different genes can cooperate in different ways, form coalitions, and develop other complicated strategies for successful survival and reproduction. According to some authors, the theory of the selfish gene enables unification of views on natural selection at all levels and may even explain altruistic behavior (Zrzavý, 1998; Zrzavý et al., 2004; Flegr, 2005).

It seems undeniable, however, that although mutations and natural selection are the basic principles of evolution, cooperation makes it "constructive." Cooperation is essential to the creation of new levels of organization. When competing individuals at lower levels begin to cooperate, new levels of organization emerge. Cooperation enables specialization and thus promotes biodiversity. Cooperation is the secret of the unlimited possibilities of the evolutionary process. Perhaps the ability to cooperate in a competitive world is the most remarkable aspect of evolution. Perhaps "natural cooperation" is the third fundamental principle of evolution, along with natural selection and mutation (Nowak, 2006). Cooperation is necessary for evolution and is also an inherent part of the evolution of multicellularity (Axelrod, Hamilton, 1981; Nowak, 2006).

Evolutionary innovation

The concept of major evolutionary transitions reflects the fact that in the otherwise continuous flow of life's development there are interruptions, fundamental changes, and transitions that significantly alter the form of life and the dynamics of its further development. The transition of an evolutionary (Darwinian) individual, i.e. the entity that is the "object" of natural selection, is considered a central component of these transitions.

In addition to the concept of major evolutionary transitions, there are other tools for understanding evolutionary mechanisms that reflect the variability of evolutionary dynamics and aim at discovering general principles that apply to the different phases of evolution. The concept of evolutionary innovations is one of them. It provides insights into the nature of multicellularity and the links between multicellularity and tumors. Innovations are understood as events in the evolutionary process that represent unpredictable qualitative changes in structure or process that increase efficiency and performance or contribute to the creation or expansion of a new niche. Innovations can be biological, such as the multicellularity mentioned earlier that enables acquisition of new resources and protection from predators and parasites, wings that enable flight, eyes and complex vision, or, for example, photosynthesis as a new way of generating energy. Innovations, however, are also phenomena of cultural evolution or technological improvements, such as the use of tools and fire, the domestication of plants, microprocessors that enable the development of computers and telephones, and more. In this case, similar to the concept of major evolutionary transitions, we are looking for answers to the questions of how evolutionary innovations arise, how they spread, how they affect their environment, and whether they can lead to subsequent adaptive processes. The question of whether there are unifying biological principles in this way is also interesting. Even more interesting and provocative (certainly in the context of "overlaps") is the question of whether such principles are also applicable to innovations in culture and technology (Hochberg et al., 2017; Aktipis, Maley, 2017).

Innovation, multicellularity, and cooperation

Multicellularity itself is one of the most important innovations. Furthermore, it fostered the emergence of other innovations, for example, in the organization, specialization, and regulation of cells forming a multicellular organism. The five most important innovations were gradually recognized as the essence of cooperation between cells in multicellular organisms. They were termed the five foundations of multicellularity (Fig. 49).

The first foundation of multicellularity is inhibition of proliferation. Regulation of proliferation is essential for the homeostasis of the tissues and organs of a multicellular organism and applies to almost all cells (some specific cells such as stem cells and progenitor cells are excluded). To control cell proliferation in multicellular organisms, several cell cycle checkpoints and mechanisms automatically trigger apoptosis or senescence when cells begin to divide in an uncontrolled manner. The second foundation is regulated cell death, which is critical for the development, organization, and maintenance of the correct structure of the organism. The third foundation of multicellularity is division of labor and the associated differentiation and specialization of cells into certain activities. The division of labor between cells in clusters occurs quite spontaneously. However, to maintain tissue homeostasis, the process of cell differentiation within a multicellular organism must be tightly regulated. The fourth foundation of multicellularity is the redistribution and transport of resources to provide adequate oxygen and nutrients to all cells in the organism, including those that have no contact with the external environment. The final fifth foundation of multicellularity is the formation and maintenance of an extracellular environment shared by cells within an organism. Waste produced by individual cells must be removed, dead cells must be recognized and recycled, and the extracellular matrix, which consists of a network of

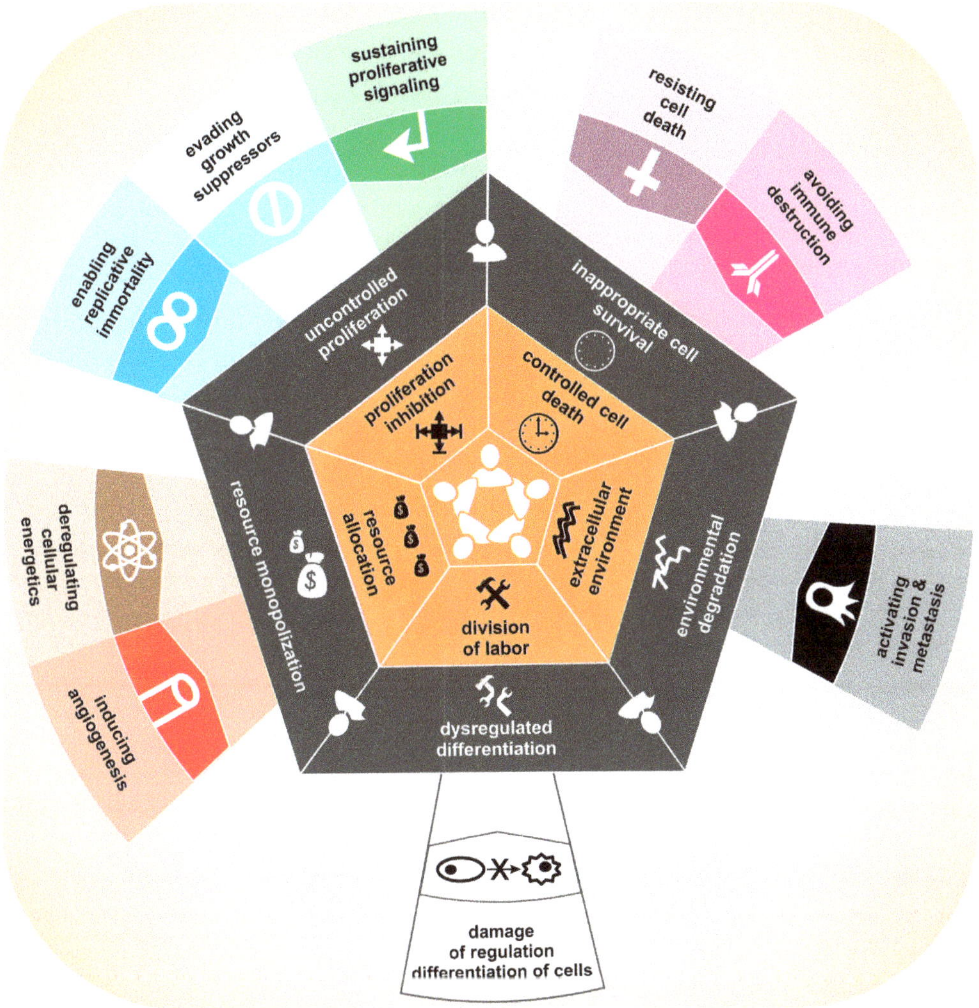

Fig. 49 The five foundations of cooperation, cheating, and typical features of tumors
The major innovations that are the essence of cooperation between cells within a multicellular organism, referred to as the foundations of cooperation: (1) inhibition of proliferation, (2) regulated cell death, (3) division of labor and associated cell differentiation, (4) redistribution and transport of resources, and (5) creation and maintenance of the extracellular environment shared by cells in the organism (central orange part of the diagram). Any form of cooperation can be disrupted by cheating (dark gray part of the diagram). Forms of cheating significantly overlap with the hallmarks of cancer (outer part of the diagram) described by Hanahan and Weinberg (2011), except that damage to differentiation regulation is not mentioned in their paper (adapted from Aktipis et al., 2015; Aktipis, Maley, 2017, modified).

support proteins such as the basement membrane, must be maintained (Aktipis et al., 2015; Aktipis, Maley, 2017).

Multicellularity and cheating

Cooperation is defined as a joint effort aimed at achieving a benefit for all parties involved. Cheating means behavior that results in obtaining benefits at the expense of others, i.e. the opposite – a violation or rejection of cooperation efforts. A cheater is a person who is uncooperative or less cooperative than would be fair, but benefits from the cooperation of others. A cheater is also someone who selfishly exploits shared resources for his or her own benefit. A cheater violates common rules and does not follow them. Tumors are typical cheaters in the world of multicellular organisms. Just as the five foundations of multicellularity have been described as five key rules for cooperation between cells in multicellular organisms, the potential for cheating – the violation of those same rules by cells – has also been recognized. And it is probably not surprising that these forms of cheating clearly overlap with the signs typical of tumors (Fig. 49).

The first foundation of multicellularity is inhibition of proliferation. Violation of this rule leads to uncontrolled, unregulated cell division. In tumors, this form of cheating is associated with three of its typical features: (1) insensitivity to cell cycle arrest signals, (2) self-sufficiency in their own proliferation signals, and (3) acquisition of unlimited replication potential. The second foundation of multicellularity is regulated cell death. Resistance to programmed cell death (4) has been recognized as a typical and almost universal feature of all tumors. The ability to evade immune system surveillance (5) also contributes significantly to tumor cell viability. The third foundation is division of labor and associated cell differentiation. These are the key features that typically accompany the emergence of multicellularity and provide the greatest advantages to the multicellular organism. In addition, gradual cell differentiation in the tissues of complex multicellular organisms reduces natural selection and competition among cells within the organism (Pepper et al., 2007). Tumor cells are usually insufficiently differentiated. Grading, one of the basic parameters of tumor classification, is based on the assessment of the degree of cell differentiation. The less differentiated the tumor, the higher it is graded, and the higher the grading, the more aggressive the tumor and the worse the prognosis. Surprisingly, violation of proper differentiation is absent from the list of typical signs of tumors. The fourth foundation of multicellularity is the redistribution of resources and transport. Cheating in this area of cooperation and mutual solidarity leads to (6) induction of angiogenesis and (7) deregulation of energy metabolism, i.e. misuse of resources for one's own benefit. The final, fifth foundation of multicellularity is the formation and maintenance of the extracellular environment shared by the cells of the organism. Tumor cells degrade and alter the extracellular matrix, facilitating (8) the formation of metastases (Aktipis et al., 2015; Aktipis, Maley, 2017).

Multicellularity and tumors

As mentioned earlier, multicellularity has evolved several times independently during evolution and in all kingdoms of organisms, i.e. fungi, plants, and animals. This shows that the order and key rules of effective intercellular cooperation, which are a necessary condition for successful multicellularity, have been found repeatedly in the history of life. And what about the occurrence of tumors? In what types of multicellular organisms do tumors form? We usually deal with tumors primarily in the context of serious human disease. Cancer has been recognized and defined anthropocentrically, based on its occurrence and progression in humans (and to a lesser extent, for example, in laboratory mice). Thus, we primarily associate a tumor with its typical and most severe manifestations in humans, such as metastases. However, apart from humans and other vertebrates such a view of cancer is unusual.

To recognize, understand, and define tumors in all forms of multicellular organisms, we must start from a more general framework of this phenomenon. One such starting point for defining tumors is the five foundations of cooperation necessary to maintain the integrity of the multicellular body, and in particular any form of cheating that disrupts them and directly contributes to the development of tumors. If we understand tumors in this broader and more general sense as all manifestations of neoplastic growth characterized by abnormal cell proliferation and differentiation, we surprisingly find them in almost all types of complex and even some types of simple multicellular organisms, including, for example, bacterial biofilms. Nevertheless, animals seem to be much more prone to tumor development than other types of multicellular organisms. (Although we cannot exclude the effect of much lower knowledge of other types of multicellular organisms than of animals.) The greater susceptibility of animals to tumor development may be related to a greater division of labor between cells (animal bodies are composed of a greater number of cell types compared to fungi, plants, and algae), a higher metabolic rate, a higher number of proliferating cells in some tissues (e.g. epithelial and immune cells), the fact that the transport systems of animals carry both resources and cells, which may facilitate metastasis, and, finally, the fact that animal cells do not have a cell wall (Aktipis et al., 2015).

Multicellularity and tumors – cooperation and cheating

The discovery that tumors occur in all types of multicellular organisms has supported the notion that the risk of developing a tumor is an inseparable, unavoidable part of the life of multicellular organisms. Just as evolution has repeatedly succeeded in establishing the rules of intercellular cooperation and creating effective, integrated multicellularity, this unity and integration has always been and continues to be repeatedly and inevitably threatened by cheating, i.e. the violation of these rules. Whenever some form of cooperation was successfully initiated and developed, sooner or later a way emerged though which this cooperation was violated and circumvented

(Aktipis et al., 2015; Aktipis, Maley, 2017). Jan Zrzavý, a Czech evolutionary biologist, writes: "The existence of cheaters of various kinds is not a coincidence, but a necessary consequence of the existence of cooperative behavior. Only in a society where people are willing to help each other can cheaters succeed." And he adds: "But if there are too many cheaters, the whole system of mutual help collapses into pure egoism" (Zrzavý et al., 2017, p. 115).

The evolution of multicellularity requires a limitation of evolution and innovation at the cellular level, which, if left unchecked, could lead to the development of tumors and death of the organism. Multicellular organisms have therefore evolved a variety of mechanisms to suppress somatic cheating at different levels. (1) At the level of autonomous cells, for example, this is the initiation of programmed cell death in response to DNA damage or insufficient cell proliferation. (2) Another form of cheating control is in tissue architecture. It helps reduce cheating, for example, through the rare occurrence of stem cells, which are reluctant to divide and are exempt from "dirty work," guaranteeing them a low level of damage to DNA and other cellular structures. The existence of stem cells also enables the separation of the abilities of self-renewal, which they possess, and intensive proliferation, which they lack. Combining the capabilities of self-renewal and effective proliferation would greatly facilitate somatic cheating. (3) Systemic measures to control somatic cheating include monitoring the function of the immune system, which can effectively eliminate potential of neoplastic cells (Pepper et al., 2007; Aktipis et al., 2015). Thus, a prerequisite for the successful development of a multicellular organism is not only effective cooperation between cells, but also effective prevention and elimination of cheating and conflict. Rules that are a prerequisite for successful multicellularity must therefore include both guidelines for cooperation and constraints on cell behavior that would lead to cheating and abuse of the organism. Successful multicellularity requires a set of innovations for cooperation and control of cell cheating.

The evolution of multicellularity and tumors, cooperation and cheating

Evolution can operate simultaneously at the level of cells within multicellular organisms and at the level of the organisms themselves. The tension between cooperation that favors the organism, and cheating that favors the individual cells, is crucial for the evolution of multicellularity on the one hand, and for the collapse of multicellularity by tumor evolution on the other. This tension is central both to the development of tumors in the body and to the evolution of mechanisms to suppress tumors at the level of the organism. This tension drives innovations that promote cooperation within the multicellular community and suppress cheating. At the same time, it drives innovations that encourage tumor cells to find new ways of cheating to evade the body's defenses (Aktipis, Maley, 2017).

Thus, this is a particular form of continuous, never-ending co-evolution of multicellular organisms and tumors. For every form of cooperation, sooner or later new form of cheating is found. For every form of cheating, sooner or later the organism finds an adequate response in the form of a newly created protective barrier, the newly established rule of cooperation. A constant, never-ending chain of innovations, a constant, never-ending evolution of cooperation and cheating. The never-ending "arms race," as some evolutionary biologists see co-evolution: each innovation (evolutionary novelty) that favors one side creates a selection pressure to which the opposing side must respond by developing an effective strategy, i.e. an adequate "counter-innovation" (Flegr, 2005). This explains why, as multicellular organisms become more complex, the complexity of the tumors found in them also increases. And this also implies that the occurrence of tumors is not only natural and cannot in principle be completely prevented, but that the cheating strategies of tumor cells and tumors will continue to deepen and improve. We can only hope that the strategies of cooperation and protective barriers of multicellular organisms will also deepen and improve quickly and efficiently enough.

14B. Social arrangement

Where is evolution heading?

Are we evolving from somewhere to somewhere? Does evolution have a direction? Is there an ideal target point we are headed toward? And is it even necessary to know the answer? Yes and no. Some need it, others avoid it. Some cannot live without an answer or without seeking it meaningfully and with sufficient motivation; others, on the other hand, do not want to bother with "such unnecessary considerations." If you have read this far, you probably belong to the first category. What is the direction of evolution?

Life in general seems to tend to create increasingly complex and multi-layered structures. Life evolves from simpler forms to more and more complex ones. The tendency toward greater complexity seems undeniable, and evolution, and the ability to evolve, is one of the fundamental characteristics of life. Is life capable of self-organization, of spontaneously assembling itself into ever more complex structures and hierarchically higher systems, without control from above? Life creates hierarchical structures in a manner reminiscent of fractals, self-similar figures and phenomena created by the repeated application of the same rules, principles, and schemes, each time at a higher level of hierarchy. This capacity for self-organization and tendency to create more complex structures occurs in physical, chemical, biological, and eventually also social systems (Veverka, 2013). Israeli historian Yuval Harari, who studies macro-historical processes, in his world-bestseller *From Animals into Gods: A Brief History of Humankind*, asks whether human history has a direction. He answers this question in the affirmative, noting a tendency toward unification. Larger and more complex civilizations are emerging and gradually absorbing the smaller cultures. Empires do break apart, but according to Harari, "these break-ups are temporary reversals in an inexorable trend towards unity" (Harari, 2014, p. 172).

The increase in complexity, the creation of hierarchically higher systems, the major evolutionary transitions, and even the major evolutionary transitions of the individual represent "a unifying narrative of the history of life" (Kennedy et al., 2014). The strength of this tendency is evidenced by the aforementioned fact that multicellularity has evolved repeatedly in the history of life, at least seven times in the case of complex multicellularity and even more than twenty times in the case of simple multicellularity (Knoll, 2011). It is noteworthy that a higher degree of integration, eusociality, has evolved independently at least eleven or twelve times in some animals, including social insects, according to some evolutionary biologists (Flegr, 2005; Žďárek, 2015).

Eusociality

The prefix "eu" in the word "eusociality" means "good, gladly." Eusociality, then, is something like a good, genuine, or the right social arrangement. An arrangement with such a sophisticated functional internal organization that it provides significant benefits to the community. The best-known eusocial animals are ants, honey bees, bumblebees, wasps, and termites, but they also include shrimp and, among mammals, some naked mole-rats. The eusociality of such animals is defined by three key features that distinguish them from other, less organized animal communities.

The first feature is the coexistence of several generations in a group. In bees, ants, and termites, at least two adult generations coexist in a nest, for example, when the queen is the mother of the workers. The mother cannot live permanently without the support of her daughters, and the workers would not be able to ensure the continuity of the colony without their mother. The second feature is joint custody of offspring. In eusocial communities, individuals who are not the parents also participate in the care of offspring. In bees and ants, the workers who feed, clean, and generally care for the larvae are their sisters. The third feature of eusociality is the specialization of some members of the community in reproduction. In a eusocial community, reproduction is the privilege of only a select group and not of all adult individuals. The nest society is divided into a reproductive, fertile "royal" caste and a nonreproductive, sterile "worker" caste. Workers have many different occupations, including nest builder, observer, feeder, explorer, carrier, nanny, caretaker, feeder, storekeeper, herder, fighter, mortician, milkmaid, and others. The queen has only one privilege and one task: to lay eggs. She is specially equipped for this task. Thus, after some time, the termite queen becomes a kind of "factory" for eggs, and can lay up to several million per year. Her abdomen grows larger and larger as the ovaries and fat tissue fill it. The queen is virtually immobile and many times larger than the workers. The task of the workers and other community members is to create suitable conditions for the queen and the entire colony and thus, together with the queen, ensure a new generation and the continuity of the nest (Hoelldobler, Wilson, 1997; Žďárek, 2015).

Formation and types of communities

There are two known ways of forming communities of organisms. The first is the communal path, in which unrelated individuals of the same species join together. This group is more resistant to environmental adversities and more suited to build and protect a common nest. Initially, females care for their own offspring. Later, they cooperate in this care, and no longer distinguish between their own offspring and those of another female. They gradually specialize; some females lose interest in caring for offspring and thus develop into the queen caste. The second path is that of the family. This is a social cohabitation of relatives based on the deepening of the relationship between daughters and their mother. The founding mother survives and cooperates with her daughters into adulthood. The formation of such insect families is only the first step toward establishing a community of a higher degree of complexity. In general, several forms of differently developed cohabitation are distinguished, from subsocial (parents care for immature offspring) to quasisocial (members of one generation live together and care for offspring together), semisocial (additional division of functions in reproduction), to true eusocial communities in which the fertile and sterile castes are strictly separated.

The evolutionary paths of social societies are analogous to the evolutionary paths of multicellular organisms. These may form by simple clustering, which corresponds to the communal path, or clonally, by mitotic division of the initial cell, which corresponds to the familial path. In this context, neither the communal nor the familial pathway need be strictly in one direction. There are individuals or populations of a given species that live at different levels of the social order and can return to a solitary or more socially independent way of life. A return is impossible only in a true eusocial species, where a sterile worker caste differing in morphology and behavior has evolved alongside a fertile royal caste (Žďárek, 2015). This strict and irreversible division into a fertile and a sterile caste is reminiscent of germ-soma differentiation, i.e. the separation of germ and somatic cell lineages in a multicellular organism. Germ-soma differentiation was a crucial step in the major evolutionary transition from unicellular to multicellular organisms, when the separation of the germ line terminated the reproductive capacity of somatic cells, tied their "fitness interest" to the germ line, and thus eliminated the possibility of a return to unicellularity. Thus, the emergence of a eusocial community with a separation into sterile and sexual castes and with a close linkage of their biological fitness, similar to the emergence of a multicellular organism, represents another major evolutionary transition (Szathmáry, Smith, 1995; Szathmáry, 2015).

The success of the social organisms

The emergence of eusociality is considered by some researchers to be one of the greatest miracles of nature and the highest social stage (Žďárek, 2015). In this, too, one can see some analogy with the claims of other scientists who consider the emergence of multicellularity as the most spectacular major evolutionary transition (chapter 14A). The eusociality of ants and termites is the main cause of their spectacular ecological dominance. In the Brazilian Amazon rainforest, the dry weight of all ants is about four times the dry weight of all terrestrial vertebrates, i.e. mammals, birds, reptiles, and amphibians combined. Although the proportion of insect species that live in well-organized colonies is only two percent of the nearly one million insect species described, it accounts for more than half of the biomass of all insects on Earth (Hoelldobler, Wilson, 1997). Why are ants so amazingly successful in the animal kingdom?

The reason is the quickly usable and enormous power resulting from the cooperation in the colony. The advantage of social insects, for example, lies in the efficiency that results from an effective division of labor and specialization. Specialized workers perform a variety of different tasks more reliably and quickly than solitary individuals that move from one task to the next. The magnificent strategy of social life is most evident in territorial disputes and competition for food. Ant workers take greater risks in these battles and act more decisively; they can behave almost like kamikaze ants. A protected mother queen, whose gene pool is largely identical to that of the workers, ensures the survival and reproduction of the colony. In contrast, a solitary ant has no such choice; its reproductive capacity ends with its eventual death. Moreover, the sophisticated construction of the nest allows a certain degree of independence and resistance to climatic and environmental conditions, which again is not at all comparable to what a solitary worker ant can achieve alone.

On the other hand, colony functioning requires more natural resources and is slower to reproduce and expand. Solitaries also have an advantage under conditions of limited food sources and nesting sites. Solitary insects are therefore better pioneers, and this is also the reason why they haven't been completely eliminated by eusocial species. Colonies take a long time to grow, they move slowly, but once they're underway they're difficult to stop. They occupy a central position in their habitat, and displace solitary insects to the periphery. Effective cooperation mediated by sophisticated chemical communication makes social organization one of the most successful strategies in the history of life evolution (Hoelldobler, Wilson, 1997; Žďárek, 2015).

Superorganism – a new evolutionary individual

Because the mature insect community functions like a large body, many researchers refer to it as a superorganism. They compare the activities of different caste members to the functions of different organs. Some search for food, others transport, process, or store it, others are engaged only in the reproduction of the community, others in its defense (Žďárek, 2015). The ability of a beehive to precisely maintain temperature, described in chapter 1B, is a collective thermoregulation reminiscent of and comparable to the thermoregulation of warm-blooded animals. The ants' pheromones, chemical substances that they use as important means of communication, including signaling to distant parts of the nest, can be compared to animal hormones. The air conditioner used to ventilate the termite nest, which acts not only as an air cooler but also as an exchanger of carbon dioxide and oxygen, represents a kind of lung of the termite nest. And there are more similar examples. The term superorganism is therefore used by some authors to evoke precisely this similarity and analogy to the human or another multicellular body (Tautz, 2009, p. 11; Žďárek, 2015).

The term "superorganism" was first used at the end of the 19th century by sociologist Herbert Spencer and referred to the similarity of human society to a biological organism (Spencer, 1885). The term was then used by myrmecologist William M. Wheeler in 1911 to describe ant colonies with morphologically distinct fertile and sterile castes. Wheeler noted that an animal colony is a true organism and not just an analogy: it behaves as a unit, has its own characteristics such as size, behavior, and organization, which are passed from one generation of colonies to the next; the queen is the reproductive organ, and the workers form the brain, heart, intestines, and other tissues (Wheeler, 1911; Hoelldobler, Wilson, 1997). However, as the 20th century progressed and the social communities of insects and other animals were studied in more detail, the definitions of the terms superorganism and eusociality became fuzzy, and different authors used them in slightly different meanings. Therefore, there was also an opinion that the concept of superorganism has become exhausted and obsolete, and has disappeared from the vocabulary of modern sociobiology, which deals with the study of the biological nature of social behavior (Hoelldobler, Wilson, 1997). Currently, however, this term seems to be experiencing something of a renaissance and is being used repeatedly in connection to questions about the future development of mankind. Therefore, it is imperative to define the term precisely in order to clarify when it is used as a mere literary, romantic, or, on the contrary, dystopian metaphor, and when it is used in serious thinking leading to firmly established, defensible conclusions.

From the point of view of evolutionary biology and in accordance with the clearly defined and delimited concept of major evolutionary transitions and major evolutionary transitions of individuals, this is possible. Thus, we can use the term superorganism to describe a community in which all individual members have undergone a lifelong irreversible morphological division into a sexual royal and a sterile worker caste. Such a community cannot be created by clustering, i.e. by the communal path. The only possibility is the clonal, familial path, which requires a 100% monogamous origin. The monogamous founding pair of a superorganism is analogous to a zygote, which is the proverbial bottleneck at the beginning of the clonal evolution of a multicellular organism. It also produces communities with similarly close familial ties, which helps to reduce conflicts inside them. The interdependence of the organisms that make up the superorganism is so strong that its emergence represents a path of no return, and separate, independent life of individual members is no longer possible. The superorganism becomes the new evolutionary (Darwinian) entity (Boomsma, Gawne, 2018).

Even higher units?

If a biological hierarchy of genes, cells, organisms, and communities gradually formed through major evolutionary transitions, and is the fundamental reality of the living world, it is quite legitimate to ask what is the next step in the evolution of life, what is the next level of integration? According to some authors, the possible candidates for the next individuation (the emergence of a new evolutionary individual) are ant supercolonies, large networks of functionally connected, integrated ant nests. In this case, the nests have multiple queens and are therefore formed by multiple lineages. Workers, broods, and resources are exchanged between nests and queens. Within the supercolony, ants regularly cooperate with unrelated individuals. Depending on the degree of genetic relatedness within supercolonies and the degree of competition between neighboring supercolonies, supercolonies can potentially represent socially stable structures and candidates for other individuations that represent the next possible transition of the evolutionary entity.

If systems in general can be considered either true evolutionary entities subject to Darwinian natural selection or only metaphorical individuals, i.e. as mere analogs of homeostatic integrated organisms, then ant supercolonies, after evaluating all adequate parameters characterizing the evolutionary entity, represent a gray area between these two possibilities. They are partially integrated, but lack some reproductive parameters typical of a Darwinian individual. To become a true evolutionary unit, they would need to increase the degree of integration (primarily through loss of individual nest autonomy), increase reproductive specialization, increase the frequency of passage through the bottleneck, increase the heritability of supercolony traits, and establish (or demonstrate) a clear link between heritable variation and fitness, i.e. reproductive success (Kennedy et al., 2014).

Mankind as a superorganism?

The organization of insect communities, superorganisms and supercolonies is repeatedly compared to human society and for some authors they are a prototype for the future. This comparison is often used in the context of the rapid development of information technologies, which are compared to the neural network and brain, the mastermind or global brain of the future human organization. In other cases, the human superorganism is invoked as a possible solution to the stresses that humans have caused by their way of life on planet Earth, both among themselves and in relation to the environment and other organisms (Havel, 1999; Kasl, 2002; Capra, 2002, 2004; Lipton, 2011). The human superorganism is apparently also a dream of many dictators, autocrats, and leaders who yearn for "order and peace at work." However, from the definition of a superorganism as an evolutionary individual, it is not only clear that mankind certainly does not meet the criteria of a superorganism at all, but that the emergence of a human superorganism in the precisely defined sense is not even an ideal aim to strive for (at least for most people) (Boomsma, Gawne, 2018). Myrmecologist Edward O. Wilson, who founded the field of sociobiology, considers the view of human community as a superorganism to be far-fetched. Wilson acknowledges that we live in communities based on cooperation, specialization, and abundant altruism. However, our division of labor is based on cultural relationships, not instinct, as in ants. We are also too selfish to act as cells within a single organism (Wilson, 2016, p. 83). An individual that is ejected from the higher organism has no chance of survival. Its individuality is completely suppressed in favor of the whole (Veverka, 2013; Žďárek, 2015).

Proponents of the concept of major evolutionary transitions also consider human society as a specific category distinct from eusocial insects. Whether the evolution of human society already represents a separate, unique, major evolutionary transition remains an open question. However, they list four typical features for human society. The first and most important is language, which is qualitatively a completely different level of communication from the incomparably simpler chemical communication of insects and other animals. The emergence and use of language became the basis of the human cognitive revolution and the very specific development of human society (Harari, 2014). The evolution of language then contributes to a high degree of human cooperation – which is the second feature – associated with the ability to negotiate and agree on division of labor. Another feature is human eusociality, which is incomparably milder than the eusociality of insects with their lifelong separate fertile and sterile castes. In humans eusociality is represented, for example, by grandmothers, who function as a non-reproductive but functionally useful "caste" (see grandmother theory – chapter 5B), although this is only a temporary role, not a lifelong role for women. The final, fourth, typical feature of human society is cultural group selection. Cultural groupings represent nonrelational, variable, dynamic entities that can cooperate, compete, develop, reproduce, and thus be subject to selection and evolution. Their contribution to the evolution of a complex human society is more pronounced and substantial than the contribution of direct genetic kinship and familial grouping (Szathmáry, Maynard Smith, 1995; Szathmáry, 2015).

Although human society is specific and differs in many ways from other animal communities, a good knowledge of all forms of social life can help clarify basic or general principles that govern the evolution of communities, including human society. In addition, knowledge of evolution and the rules of social life also contributes to understanding other general biological principles – the basic principles of life. These include, for example, the aforementioned rules of social evolution, the nature and importance of diversity (heterogeneity), plasticity, immunity, and also general principles of biological organization (Kennedy et al., 2017). Just as morphogenesis, with cell differentiation and tissue/organ formation, is the key process in the development of organisms, sociobiogenesis, which describes stages, transformations, and behaviors of individuals and castes, is the key process in the evolution of society.

Analogies or similarities of rules and algorithms between morphogenesis and sociobiogenesis are thus another example of developmental fractals (Hoelldobler, Wilson, 1997).

Same or different? The civil society

First, a long quote from experienced Czech entomologist Jan Žďárek, author of the book *Insect Families and States*, who is thoroughly familiar with the various forms of insect social life: "Insect states are organized and controlled in a way unknown to human society. In humans, every group, every club, every party, every association has a leader, whether we call him president, general, chairman, and it does not matter whether he has appointed himself dictatorially or has been elected democratically by majority vote. From him emanate orders, instructions, decrees and directives that descend through the ladder of subordination to the rank and file members. These then carry out the decisions made by the leader. Examples of such centrally controlled groups are the army, industrial enterprises or government agencies. It is unthinkable to let any peasant or worker do whatever he wants at the machine. And that makes a big difference. In insect societies, this freedom of choice is really guaranteed to the ranks and files – the workers – lo and behold, societies have thrived for millions of years despite or because of this freedom" (Žďárek, 2015, pp. 20–21).

Already in chapter 1 we asked the question how a complex, multicellular human organism is controlled. We came to the conclusion that the human body has no central control unit and is controlled according to the principle "from below." In the following chapters, we gradually deciphered the complex relationships between cells and cellular (tissue, organ) systems, recognized several coordinating, integrating systems of the body, including the immune, nervous, and endocrine systems, and understood the central importance of rich and accurate communication between cells for the healthy functioning of the whole organism.

All of this is consistent with general systems theory. Systems are not centrally controlled; their behavior and functioning result from interactions among the components that make them up. Is it possible that with the emergence of human societies, the "wheels of history" or "wheels of evolution" begin to turn in the opposite direction and form a central control in these extremely complex systems – unlike any other systems? Is not such a view just the result of "historical experience," insufficient distance and a kind of "operational blindness"?

We have encountered something similar before. Biologists have long believed (and many still do) that cells in a multicellular organism are in a state of quiescence and must be stimulated to proliferate by external stimuli. The idea that cells which are part of a multicellular system, much like unicellular organisms, are also constantly ready to multiply, and that their multiplication must be controlled by signals from the external environment, sought and seeks its way into the general consciousness of biologists only slowly and laboriously because it largely contradicts their "ordinary experience" (chapter 12).

Similarly, our "common experience" tells us that mankind has always had and continues to have its kings, presidents, prime ministers, chiefs, and directors, and that they are probably the ones who rule and decide and also determine and shape human history. But is that really the case? Maybe there is a good reason to create order and unifying/coordinating structures. Is it possible, however, that society is created and determined in fact from below? Would not this largely correspond to the concept of civil society? Civil society is considered a kind of transitional area between the state and the family. It includes organized groups and associations that are separate from the state, have a certain autonomy, and are created from below and voluntarily by members of society. Their goal is to protect and expand the interests and ideas of these groups and their individual members. They organize the activities of citizens without links to state structures or commercial organizations. If we trust the general systems theory and believe in its universality, we should recognize the importance of the concept of civil society. Its greatest benefit lies in recognizing the value of each member of human society and providing a platform to assert that value in a creative, free, and responsible manner for the well-being and interest of the community as a whole, as well as in its own.

What are the similarities and differences? Altruism

The willingness of individuals to subordinate their private interests to the broader interests of the community is a necessary concomitant of social organization and social life. How is this achieved? After all, before the emergence of a social order, "Darwinian" rules prevailed between individuals who rivaled each other. So where does the willingness to give up one's own advantage for the good of the whole come from? Where does the altruistic behavior come from that benefits the group as a whole but harms the individual member?

One of the possible answers was provided by the theory of relational selection. According to this theory, individuals help or prefer their relatives because they share the same genes. By helping their relatives (such as the queen mother) to effectively pass on their genes to the next generation, they simultaneously ensure the transmission and success of their own genes. In this context, inclusive fitness is mentioned. This includes not only the fitness of the organism itself (exclusive fitness) but also the fitness contribution of its relatives, where the contribution of the relatives is proportional to the relational coefficient.

The second theory explains the evolution of eusociality and altruism by multilevel selection. This acts on two levels. Individual selection is based on competition, i.e. the ability of an individual to prevail among other members of the same group; group selection occurs at the level of competing groups. Whereas in individual selection selfish individuals win over altruistic ones within a group, in group selection communities with a high degree of cooperation and altruism are more successful than groups of individualists. According to evolutionary biologists, however, neither group selection nor altruistic behavior is the case in social insects. The reproductive success of an individual is determined by the success of the entire community, such as a hive of bees, an anthill, or a termite nest. Workers and soldiers – all individuals that from sterile castes – are not true individuals from an evolutionary perspective, i.e. they are not evolutionary individuals because they do not pass on their genes to their offspring (Flegr, 2005).

Nor should we think of insect behavior as altruistic or selfish, because insect social behavior is largely innate and based on a genetically anchored program. Social supergenes, groups of functionally and structurally tightly linked genes that segregate as a whole into gametes, largely determine insect social organization and behavior (Linksvayer et al., 2013; Schwander et al., 2014; Weitekamp et al., 2017). Ants, bees, termites, and other individuals within superorganisms thus do not have much choice. But other animals, including humans, do have a choice!

Evolutionary biologists, when talking about altruism and cooperation, point out that while a community based on altruism is more successful than a community based on mutual competition among its individual members, in evolution the winners are not the best, but the most stable entities. And "pure altruism" is unstable because it is very fragile and vulnerable. One egoist, one cheater, appearing among altruists is sufficient to destroy everything, especially all the altruists. It is possible that opportunism – behavior based on adaptation to external circumstances, opportunities and immediate benefits rather than on fixed principles – is more successful than pure altruism (Zrzavý et al., 2004). This is somewhat consistent with the conclusions reached by Edward Wilson, one of the authors of the theory of multilevel selection, when he considered it in the context of altruism and selfishness in humans. He believes that both tendencies meet in every human being. Both the need and desire to succeed and excel individually, and a passionate interest in others and the urge to belong to a group, to a community. Wilson asserts that "there is an inherent conflict in each of us; we are all genetic hybrids, half saints and half sinners, fighters for truth and hypocrites. Individual selection promotes sin, while group selection promotes virtue" (Wilson, 2014, pp. 33).

The evolution of altruism

So where does altruistic behavior come from? What is its cause and origin? And most importantly, how has it been sustained? Why has collaboration become established and why have cheaters failed in destroying it? Altruism toward related individuals seems undeniable and can be very efficient. However, the effectiveness of relational selection decreases rapidly as the degree of relatedness decreases. For the altruist, in addition to relatedness and shared genes, the future utility of the recipient of altruism may be important. Group selection could certainly come into play in the evolution of altruism, although some biologists point to its low efficiency and certain limitations resulting, for example, from the strong influence of within-group and between-group competition. Altruistic individuals within the group are constantly threatened by selfish individuals and should be quickly eliminated in any competition with them.

An effective solution has been proposed with the concept of reciprocal altruism. According to this, cooperation based on the principle of exchange can be evolutionarily successful. The individual behaves altruistically toward another individual because he expects that this individual – or another individual in the population who witnesses the altruistic act – will reciprocate in the future: I'll help you today, and you'll help me when I need you. Many types of altruistic behavior can be better explained by reciprocity than by relatedness, for example. Human society is said to use a high degree of cooperation among unrelated individuals that is unusual in animals. About two-thirds of humans cooperate. The remaining third do not cooperate. Non-cooperative individuals either look out solely for themselves or even abuse the benefits of the community by parasitizing on it. Some individuals cooperate, but only on the condition that the others also cooperate. Therefore, strong reciprocity is very important for human cooperation (Flegr, 2005; Eldakar, Wilson, 2011; Žďárek, 2015; Zrzavý et al., 2017; Koukolík, 2019; Koubska, Koukolík, 2019).

Even so, altruism presents a certain evolutionary paradox. Although altruists are useful to the overall success of the group, they are weakened by being vulnerable to abuse by egoists. In other words, those who cooperate can be exploited by cheaters. So why has altruism been preserved? Why have altruists not been completely supplanted by egoists? Punishment is an important mechanism that can effectively reduce egoism within the group. The possibility of egoism being punished keeps it in check. But punishment is not free; it often comes at a cost to the punisher. Punishing the egoist for the good of the whole group, even at the cost of his own losses, is actually a form of altruism and is therefore called altruistic punishment. It has been experimentally confirmed that the rate of cooperation within the group increases when the members of the group know that any dirty trick will be punished immediately and appropriately. "People who are able to make sacrifices and who are willing to punish violators of social norms altruistically are the inner glue of all human groups" (Koukolík, 2019).

Somewhat surprisingly, it turns out that in addition to altruistic punishment, there is also selfish punishment. Selfish punishment involves egoists who themselves primarily cheat and help maintain altruism by punishing other egoists. Even cheaters seem to have good reasons to punish other cheaters. In a group with too many cheaters, there would apparently not be enough altruists to abuse. Punishing cheaters by other cheaters is quite hypocritical, but the community can tolerate it if ensures success and stability. If the punishment is costly, selfish punishment can be seen as compensation paid by the cheater to the group for his own cheating. Selfish punishment can be viewed as a self-limiting strategy that allows for balanced coexistence of egoists and altruists (Eldakar, Wilson, 2008). The risk of selfish punishment lies in its potential abuse by selfish punishers, especially if they occupy positions of power. Consequently, on the one hand, selfish punishment can promote altruism and community stability, but when it is associated with significant power, it can also significantly compromise it (Eldekar et al., 2018).

Computational models have shown that altruism can be maintained in a group without selfish punishment, or the group can switch from selfish punishment to altruistic punishment. When the cost of punishing cheaters decreases, the number of selfish punishers also decreases. Because when punishment is cheap, the number of those who punish increases, and so does the number of altruists who punish. And there is a whole range of cheap ways to keep people in the right frame. From outright ostracism to coalition closures, coordinated actions, mass lawsuits, and good old gossip (El-Dakar et al., 2013). This brings us back to chapter 3B and the theory of gossip. According to this theory, gossip plays an important role in maintaining social order by helping to identify who is who within the group. Gossip, which also plays a role as a form of altruistic punishment, is actually a behavior that essentially conditions mutual altruism within the group. Gossip creates and maintains knowledge of who is altruistic and who is cheating, who is willing to help and share and who is selfish and thinks only of himself. Loss of reputation and prestige can serve as a really effective tool of punishment and efficiently supports altruists in the community. At the same time, it reaffirms the importance of truthfulness and truthful communication, the importance of not "bearing false witness against your neighbor," as the Ninth Commandment prescribes.

Altruism and game theory

The evolution of altruistic behavior is based on competition between alternative patterns of behavior. This process can be studied using the mathematical apparatus of game theory. This is a discipline of applied mathematics that studies a variety of conflict situations. Game theory describes the decision-making process of the players involved and determines the best possible solutions in mutual competition. Evolutionary game theory is the application of game theory to estimate the evolution of populations with different decision-making methods. The evolutionary version of game theory does not require players to play rationally, only to adopt a particular strategy. Success is measured in the competition against other strategies. The best or most successful strategy is that which results in a larger number of descendants in the next run of the game; less successful strategies go extinct. Thus, it is a simulation of survival in nature. Evolutionary game theory can also help explain the possible success or failure of cooperation, cheating, altruistic behavior, selfishness, punishment, as well as the conditions under which they apply. In addition, it helps to find or propose explanations for other phenomena that we observe in various kinds of social animals and humans. A number of different player strategies have been formulated and modeled in differently designed games used to test their success (Flegr, 2005; Zrzavý et al., 2017; Kruml, 2018).

What conclusions have the games produced? It turns out that evolutionary development can be quite unpredictable. It is not necessarily directed toward a goal in which perfect individuals living in mutual cooperation prevail. Evolution can go in circles as phases alternate with different strategies. It is as if nature is not evolving toward an optimal, stable distribution of strategies and gains among the actors involved, but is simply seeks to settle into one or other of the many possible states of equilibrium. Evolutionarily stable strategies do not seem to exist. The key to finding and maintaining equilibrium lies in diversity, i.e. population heterogeneity and a mix of several different strategies. Evolutionary games have shown that the presence of selfish strategies also contributes to population stability. They prevent dominant cooperative strategies from becoming dangerously weak. In other words, "Good wins, but it takes a pinch of evil to survive" (Kruml, 2018, p. 106). Cheaters can also help keep the peace in the form of a Nash equilibrium. This is a situation in game theory where neither player can unilaterally change the chosen strategy to improve their situation and reverse the outcome of the game (Kruml, 2018).

Selfishness, cheating, fraud forever?

Why has the human species not succeeded in eradicating selfishness and envy? Why is there always cheating and stealing? The results of evolutionary game theory suggest that we have to reconcile with a fair dose of cheating. It is precisely the presence of cheating that, in a social system in which cooperation and altruism are predominant, maintains a reasonable caution in us, increases our resilience and ability to distinguish when to be nice and when to be strict. It thus prevents dramatic swings and keeps us from deviating too far from the desired balance (Kruml, 2018).

This is virtually the same conclusion reached by biologists who have studied the cooperation and cheating of cells within a multicellular organism. They also find that cell cooperation is a basic requirement for the development of multicellularity and described five innovations, five key rules of cooperation, called the foundations of multicellularity. At the same time, they describe forms of cheating that violate the rules of cooperation, and recognize tumors as typical cheaters that selfishly abuse the benefits of cooperation between cells in multicellular organisms. And they also conclude that the risk of tumor formation is an inevitable, inherent part of multicellularity (Aktipis et al., 2015; Aktipis, Maley, 2017).

Research on altruism in social systems has shown that to maintain it, there must be mechanisms that effectively reduce selfishness, for example, through altruistic or selfish punishment (Eldekar, Wilson, 2008; Eldekar et al., 2013, 2018). Similarly, in multicellular organisms, a variety of mechanisms have evolved to effectively control somatic cell cheating at multiple levels.

And the efforts are not over yet! A never-ending "arms race" is underway – the co-evolution of forms of cell cheating on the one hand and mechanisms to protect cell cooperation on the other is unceasing (Aktipis et al., 2015; Aktipis, Maley, 2017). Authors primarily concerned with multicellularity and tumors have generalized their knowledge in the concept of evolutionary novelties, which has allowed them to recognize many parallels between multicellular and human societies and to identify a number of "overlaps." According to them, innovations in new forms of cooperation create opportunities for new forms of cheating. In multicellular organisms, the changing, evolving, constantly emerging tumors are these new cheating forms. In human society, innovations in the form of new ideas and technologies are changing the way we work, how and what kinds of relationships we form, and how we reproduce. These innovations bring new opportunities, offer many benefits, but also expose us to new risks. The innovation of market capitalism, for example, brings benefits to human society based on the division of labor, but at the same time poses risks of environmental damage and new forms of cheating, such as white-collar crime. On the other hand, cheating sometimes brings unexpected social benefits. It is often the trigger for further innovation, both in human societies and in technical systems. Damage to the system sometimes facilitates its reorganization and restructuring, which then improves the functioning of the whole system (Aktipis, Maley, 2017).

Dan Ariely reached very similar conclusion in his study of creativity in human cheating. Creativity as a source of original ideas is one of the virtues and an important engine of social progress that helps us find new solutions to complex problems. On the other hand, it also allows us to find original ways to circumvent the rules and cheat, using new, original and better arguments to justify ourselves. This can not only give us advantages, but also makes us feel good. This is exactly as Ariely puts it in his theory of the correction factor (chapter 12B), in which he reminds us that whenever the economy breaks through a technological barrier – whether it is the invention of the postal service, the telephone, the radio, the computer, or mortgage-backed commercial instruments – the progress allows people to cross existing boundaries, both in technology and in dishonesty. For example, one of the first ways the U.S. mail was used was to improperly sell non-existent goods. It took a while to figure that out, and the problem of mail fraud eventually led to the introduction of a whole range of various specific measures. And this is generally true. Only after the possibilities, effects, and limitations of technology have been demonstrated in practice can the tools for its use and protection against abuse be defined (Ariely, 2012). Co-evolution, in short. A never-ending, relentless "arms race" between collaboration and cheating, between altruism and selfishness.

15. What do tumors teach us?

"This is not a world, but rather the materials for a world. God has given us not so much the colors of a picture as the colors of a palette. But He has also given us a subject, a model, a fixed vision. We must be clear about what we want to paint. This adds a further principle to our previous list of principles. We have said we must be fond of this world, even in order to change it. We now add that we must be fond of another world (real or imaginary) in order to have something to change it to."

(Gilbert Keith Chesterton, 2018, p. 93)

Man is neither a cell nor an ant

Where is evolution heading? What is its goal? We do not know. The direction of evolution is unpredictable, indeterminate, open. We can only look back and try to draw lessons from what has been and what is now. We look for general principles and algorithms, for what has worked and what has not. Looking back, we can identify stages of development and layers of evolution (Szathmáry, Smith, 1995; Veverka, 2013; Szathmáry, 2015). Each higher layer is characterized by qualitatively new principles and properties that cannot be derived or reduced to the properties of the lower layers. The development of new properties is called emergence. But each layer not only has its own specific emergent properties, it also repeats elements from lower layers. Evolution works with schemes that have been verified at lower levels (Veverka, 2013).

Thus, man carries within himself the history of all living beings, but at the same time cannot be reduced to a cell, an ant, or any other organism that evolved before him or in parallel with him. And human society cannot be reduced to a superorganism or to other communities of other organisms. In the course of his evolution, man underwent several fundamental changes, literally evolutions – cognitive, agricultural, scientific (Harari, 2014). Although they did not exclude him from the realm of living organisms, they dramatically changed his possibilities and capabilities.

The development of mankind is determined not only by biological evolution, which is based on the transmission of genetic information from parents to offspring and is relatively slow, but also by cultural evolution, which is much faster. Similarly as genes are considered the units of genetic information, information that shapes a particular culturally transmitted character is usually referred to as memes. The spread of memes is much more efficient than the spread of genes. The ability to transmit information non-genetically is not only a human privilege, but it was humans who developed the ability to transmit what they learned very effectively, through spoken language and later through written script, not only vertically to their descendants, but also horizontally to other individuals living in human society. Human evolution provides more and more room for choice and decision making by the individual. The potential of individuals to consciously influence their own future and the future of the community to which they belong, the global human civilization, is growing. With this, the responsibility of each individual also grows.

What do tumors teach us?

At the beginning of chapter 1A, we started from the knowledge that the incidence of tumors in humans is not rare. It seemed that the potential for tumor development lay at the core of the human body, in the way it is constructed and functions. We wondered if this was really the case, and gradually stumbled upon the answer. From an evolutionary point of view, the development of tumors as a result of cheating in a multicellular organism is really inevitable. It is a direct consequence of the cooperation that has developed between cells, which was the starting point and prerequisite for the emergence and development of multicellularity. Cancer cannot be eradicated. Just as there is no way to stop cheating in humans and their quest for self-enrichment at the expense of others. And just as we fight cancer, try to prevent it and learn to treat it, we must never give up trying to prevent the emergence of "social cancer," human cheating and selfishness. Just as cancer can kill the body in which it develops, human selfishness can destroy entire societies. Vigilance and, yes, responsibility are called for! And why not get a lesson and inspiration from tumors?

We have given an overview of eleven recognized typical signs of tumors, which are: (1) self-sufficiency in growth signal production, (2) insensitivity to cell cycle arrest signals, (3) impaired apoptosis, (4) unlimited replication potential, (5) induction of angiogenesis, (6) metastatic potential, (7) increased genetic instability, (8) reprogramming of energy metabolism, (9) evasion of immune surveillance, (10) presence of chronic inflammation, and (11) tumor microenvironment. Much like a single renegade cell is at the origin of a biological tumor, a single selfish individual, a single "rogue" person can be the beginning of a catastrophic process that can engulf society. A single selfish individual may be strong enough to gradually undermine and destroy a community of cooperating, altruistic individuals. How can you recognize this renegade person? How can one recognize the circumstances in which the apostate thrives? This is exactly where our knowledge of tumors and their typical characteristics could help us.

I offer an overview of eleven non-deadly human sins inspired by tumor characteristics that may contribute to or create conditions for serious problems in our communities. These are sins that are not deadly in themselves, but they are a warning. What do we commit?

(1) We compete with ourselves.
(2) We lie and do not value the truth.
(3) Death is a taboo for us.
(4) We are in denial about aging.
(5) We do not respect the rules.
(6) We selfishly abuse resources.
(7) We waste.
(8) We underestimate the value of relationships and home.
(9) We circumvent the law.
(10) We mobilize constantly.
(11) We are proud, introspective individualists.

Four levels of responsibility

The study of cell cooperation and cheating within multicellular organisms has shown that innovation and cooperation are not enough. At the same time, it is essential to effectively punish cheating, combat it, and prevent cheating cells from jeopardizing the health of the organism. Similarly, it has been shown that it is necessary to punish selfishness to keep it under control and preserve altruism in the community. Without the looming possibility of punishment, the egoist will quickly destroy the cooperation of altruists and disturb the balance of the whole community.

At the end of chapter 12B, referring to the eleven typical signs of tumors, we summarized the three levels of personal responsibility of community members to defend against the possible development of social cancer. The first level of responsibility commands: do not be a cancer person yourself! The second level of responsibility commands: do not cooperate with cancer persons! The urgent commandment of the third level is: participate in maintaining a good, healthy environment! The fourth level of our personal responsibility was mentioned in chapter 14B and deals with maintaining the healthy condition of our community to defend effectively against the cancer that would grow through and eventually destroy it. The clear lesson is to punish the cheaters! This is not a call for vigilantism, lynching, or an eye for an eye, tooth for a tooth revenge. It is a call for vigilance and awareness of the importance of not underestimating the destructiveness of cheating, which must not be tolerated at any level. It is a call to be concerned, to watch closely for cheating, to name it appropriately and to do everything possible to ensure that it does not go unpunished.

Just a cell?

Where does cancer come from? By now at the latest, after reading this book, it should be clear that cancer does not attack the body from the outside. Cancer is not a foreign invader, not a mysterious, unrecognizable demon. It is a "home-made" problem, an internal fault that has occurred in one's own gears. At first the error is inconspicuous, seemingly small, but it gradually gains strength and can aggressively permeate the whole system, abusing it and forcing its "wheels" to work for its own interests. In this context, how can one even speak of a "personal" responsibility of each individual cell? The adult human body consists of $3.72 \pm 0.81 \times 10^{13}$ of them (chapter 1A). What can a single cell do? There are almost eight billion people living on the planet Earth today (7.86×10^9). Each of us can see how selfish, cheating people try to get more from the system than they give to it. Is the individual strong enough to do something about it?

This whole book is about cancer. How it starts, how it develops and gains more and more characteristics, how it gets stronger until it is so powerful that it wins: it kills its host and dies with it. Fascinating? It may be. But at the same time, and perhaps above all, this book is about the truly fascinating world of healthy, physiological cells that form a multicellular organism. About the magical, and in many ways still mysterious, world of $3.72 \pm 0.81 \times 10^{13}$ cells, all of which are endowed with nearly identical genetic information and potential, and yet are able to form the incredibly diverse, complex, and perfectly functioning system of the body. How does each individual cell live in this system? How does the cell cope with its position, its task and its commitment? Sure, it's just a rhetorical, anthropocentric question, but why not ask it? Why give up the magic of metaphor, which can offer insight into a complex tangle of elusive questions and puzzles? Probably none of the body's cells can imagine and realize what it is a part of, how enormously the "project" transcends it, in the gears of which it is a tiny cog. And yet its good functioning helps to create the whole and keep it moving. How can it succeed? A healthy cell in the body accepts its role. The place and environment it is located in determine its identity – what kind of cell it is. A healthy cell accepts its given identity, including its limits. And it becomes sovereign, specific and entire by itself. This is its individual responsibility, its personal contribution, its personal 10^{-13} share in the success of the whole project, the multicellular organism, the human body. In this way, the "heroism of small work" can also be understood (Petrusek, 2000). The heroism of humble acceptance of a task, a position, a vocation. How different are the tumor cells! They are the titans, the most successful, fastest, most efficient, resistant and ultimately "victorious" cells of the body… but really victorious?

We are human beings, not cells! We are endowed with the ability to perceive, to think, to feel, to imagine, to desire, to dream… Unlike cells, we have the ability to want more and to go further. We have the ability, the possibility and the duty to make free choices. Unlike cells, we also have the ability to survey, or at least somewhat imagine, what we are involved in, what our "project" is about. And we can decide how to deal –

also in relation to this project – with our given circumstances: our potential, our position and our dreams. No one should tell us that we are "just" cells, "just" obedient cogs in the wheel of a complex system that we do not understand. Let us pursue our dreams with humility, with respect for the greatness and interest of the common project, and let us defend it with courage. Let us also pursue our dreams with the will to live a good, free life, and with the confidence that our personal and common, transcending interests, have a point of intersection. Let us not miss it.

Glossary

Allele – one of several alternative forms of a gene; a gene determines a trait (e.g. eye color), while an allele determines an expression of the trait (e.g. blue or brown eye color)

Aneuploid – a condition in which particular chromosomes or chromosomal regions are present in extra or fewer copies

Anoikis – form of apoptosis that is triggered by the failure of a cell to establish anchorage to a solid substrate, such as the extracellular matrix or by loss of such anchorage

Antagonistic pleiotropy – a hypothesis proposing that genes or processes that have evolved to promote the health or fitness of a young organism may have harmful effects on older organisms and further contribute to their aging

Antigen – a molecule, that can be specifically recognized and bound by an antibody or a T-cell receptor or that provokes the production of an antibody

Antigen-presenting cell – cell that displays foreign antigen complexed with an major histocompatibility complex on its surface for presentation to T lymphocytes

Apoptosis – the most common form of physiological, genetically programmed death of eukaryotic cells

Astrocytes – branched star-shaped feeding cells with long protrusions belonging to the group of neuroglial cells; usually, they attach to the capillary wall with one projection touch the surface of a neuron with the other to supply nutrients

Autophagy – the intracellular uptake of cytoplasm (proteins, nucleic acids, small molecules, whole organelles, etc.) into the lysosome and its subsequent degradation to simple building blocks (e.g. amino acids, nucleotides) that can be recycled

Basement membrane – thin mat of extracellular matrix that separates epithelial cells, and many other types of cells from connective tissue

Caspases – protein degrading enzymes that act as mediators of programmed cell death (apoptosis); they are activated by proteolytic cleavage

Cell cycle checkpoint – a control mechanism in the eukaryotic cell cycle which ensure its proper progression

Cell proliferation – fast and repeated cell division

Chromatin – the aggregate of DNA and histone proteins that make up an eukaryotic chromosome

Chromosomal translocation – a genetic alteration in which segments of two non-homologous chromosomes are reciprocally exchanged

Chromosome – a discrete unit of the genome composed of DNA proteins present in nucleus of eukaryotic cell

Clone – a collection of organisms or cells derived from a single parent that is genetically identical with that parent

Codon – a sequence of three nucleotides in a DNA or mRNA molecule that codes for a particular amino acid to be incorporated into a growing polypeptide chain, or a signal for the beginning or end of its synthesis

Commensal microorganisms – microorganisms that are commonly present in a human body

Cyclic AMP – nucleotide that is generated from ATP and used in multiple cell signaling pathways as secondary messenger

Cytokines – signal proteins that mediate communication between the cells of immune system and participate in immune response

Dalton (Da) – an atomic mass unit that is equal to one twelfth of the mass of a free carbon-12 atom at rest

Differentiation – process of structural and functional diversification of a cell into a uniquely specialized cell type

DNA recombination – the exchange of genetic material either between multiple chromosomes or between different regions of the same chromosome

Dominant mutation – a mutation that determines the phenotype in a heterozygous state independently of the presence of the other – non-mutated – allele

Endothelial cell – flattened cell type that forms a sheet lining of all blood and lymphatic vessels

Epigenetics – a discipline of genetics that deals with inherited changes in gene expression that are not due to mutations

Epigenome – a collection of all epigenetic marks on DNA in a single cell

Evolvability – an organism´s capacity to undergo adaptive evolution, e.g. the ability to engage in natural selection

Exosomes – small endocytic membrane vesicles released by various cells (healthy, infected, or tumor cells) to the environment; they are capable of removing unwanted proteins from cells, mediating cell-cell communication, signal transduction, and transmission of pathogens

Extracellular matrix – a mesh of secreted proteins, largely glycoproteins and proteoglycans, that surrounds most cells within tissues and creates structure in the intercellular space

Fibroblast – common cell type of fusiform shape found in various parts of the human body; secretes an extracellular matrix rich in collagen; the least specialized cell type in connective tissues

Fibrotization – conversion of a tissue into a fibrotic state, which leads to a restriction of tissue function

Fitness – the cost of the individual in terms of evolution; a measure of the average ability of organisms with a given genotype to survive and reproduce

Gene – the hereditary unit that contains the genetic information about a specific function and is located at a specific site of a chromosome

Gene amplification – increase in the number of copies of a gene or a set of genes

Gene pool – collection of alleles present among the genomes of all members of a species

Genome – total genetic information possessed by an individual organism

Glial cells – cells involved in the formation of the neural system with neurons, where they account for about 90% of all cells; they support neuronal networks, nourish neurons, produce myelin and are capable of phagocytosis

Gliomas – tumors of the glial cells in brain or spinal cord, such as astrocytomas, oligodendrogliomas, and ependymomas

Glutamine – one of the twenty-three amino acids incorporated into newly formed proteins during translation; it is also important for detoxification of ammonia and maintenance of nitrogen balance

Grading – the degree of tumor differentiation determined by microscopy with prognostic and predictive value; as a rule, a less differentiated tumor indicates greater aggressiveness, but also more susceptibility to treatment

Heterotroph – an organism that, unlike autotrophic organisms, cannot synthesize its components from inorganic compounds and must obtain nutrients and energy by consuming other plants or animals

Homeostasis – a self-regulating process by which organisms maintain their internal stability; it is necessary for their functioning and survival in a variable external environment

Hypoxia – state of lower-than-normal oxygen tension in tissues

Immunodeficiency – a condition, in which an organism´s ability to respond to antigens and other stimuli is impaired

Immunogenicity – ability of a compound to stimulate an immune response

Immunosuppressive drug – an agent that inhibits or reduces the function of the immune system

Ischemia – a condition of local anemia in a particular tissue or organ that can lead to its damage or necrosis; it is caused mainly by lack of oxygen (hypoxia)

Kinetochore – nucleoprotein complex that is associated with the centromeric DNA of eukaryotic chromosome and is responsible during mitosis (or meiosis) for forming a physical connection between the chromosome and the microtubules of the spindle fibers

Macrophages – phagocytic cells derived from blood monocytes; they have both scavenger and antigen-presenting functions in immune responses

Matrix metalloproteinases – extracellular or membrane proteases that degrade matrix proteins

Mesenchyme – the meshwork of loosely organized embryonic animal connective tissue of undifferentiated cells from which most tissues arise; provides support, nutrition, and filling functions

Microtubules – hollow protein fibers in the cytoplasm of eukaryotic cells; as part of the cytoskeleton, they are involved in motoric functions and form the mitotic spindle

Mitochondria – organelles found in most eukaryotic cells where the Krebs cycle and aerobic phosphorylation, respiration, take place for energy production

Mitogen – an agent that provokes cell proliferation

Mitotic catastrophe – a form of cell death resulting from aberrant mitosis

Mutation – a permanent, heritable change in genetic information; depending on the type of cell affected, a distinction is made between gamete mutations (in the germ line), which affect the gametes and are transmissible to the offspring, and somatic mutations, which affect the cells of other organs and tissues and are not transmissible to the offspring

Myrmecology – the branch of entomology that deals with ants

Natural selection – process in which, on the basis of various criteria, those are selected from a group of different individuals who are preferred and those who are suppressed

Necrosis – a process of cell death that occurs inside the organism in response to irreversible injury

Normoxia – an oxygen level equal to that normally experienced by cells in tissues

Nucleotide – basic unit of nucleic acids; phosphorylated nucleoside consisting of a purine or pyrimidine base linked to a sugar moiety (ribose or deoxyribose) via an ester bond

Oncogene – an altered protooncogene whose product can act in a dominant fashion to help make a cell cancerous

Organicism – the explanation of life and living processes in terms of the levels of organization of living systems rather than in terms of the properties of their smallest components

Oxidative stress – a phenomenon caused by an imbalance between the production of reactive oxygen species (free radicals), which are a byproduct of oxidization and metabolism, and the ability of biological system to readily detoxify the reactive intermediates

Palliative care – complex care that focuses on the quality of life of people living with an advanced or incurable life-threatening illness; it aims to optimize quality of life and relieve both physical and psychological suffering, preserve the patient´s dignity, and support the family

Pericytes – cells closely related to smooth muscle cells that surround capillaries and provide the capillary walls formed by endothelial cells with tensile strength and contractility

Phenomenology – the study of "phenomena," but instead of examining their nature, it is concerned with the appearances of things, or with things as they appear in our experience

Phosphorylation – reaction in which a phosphate group is covalently coupled to another molecule

Post-translational modifications – alterations of a protein that occur after the initial polymerization of the polypeptide backbone on the ribosome; they change the properties, stabilize the conformation, and contribute to the regulation of the protein´s function

Progenitor cell – a cell that can differentiate into a specific cell type and renew itself; an intermediate step in the formation of a mature, specialized cell from a stem cell

Proteases – enzymes, that can cleave proteins

Protooncogene – normal gene, usually concerned with the regulation of cell proliferation, that can be converted into an oncogene

Ras oncogenes – a family of genes encoding small monomeric proteins that can bind GTP and act as GTPases

Recessive mutation – a mutation that is obscured in the phenotype of a heterozygote by the other – dominant allele

Reductionism – the belief that complex phenomena and problems can be explained by breaking them down into smaller components and then analysing them; it is based on the assumption that the whole is "no more" than the sum of its parts

Retinoblastoma – a malignant tumor that originates from the developing retina in the first three years of life

Sequence-specific transcription factor – a protein that controls the rate of transcription of genetic information from DNA into RNA by binding to a specific DNA sequence

Sociobiology – a field of biology that aims to explain social behavior in terms of biology and to study the role of the social environment in the evolution of behavior

Sporadic tumors – tumors occurring randomly in a large population without any apparent predisposition, such as one caused by a heritable genetic susceptibility

Stem cell – undifferentiated cell that can continue to divide indefinitely; the resulting daughter cells can either differentiate into specialized cell types or remain a stem cell in the process of self-renewal

Stroma – a connective tissue that supports an organ

Symbiosis – a relationship of dissimilar organisms living together in a mutual beneficial interaction

Teratocarcinoma – embryonal carcinoma – a form of malignant tumor derived from undifferentiated embryonal carcinoma cells and other tissues that grow in wrong positions

Teratogenesis – a prenatal toxicity characterized by a structural or functional defects in the developing embryo or fetus

"Thinking outside the box" – an idiom associated with the "nine-dot puzzle" – a puzzle whose task is to connect nine squarely arranged dots with a pen by four (or fewer) straight lines without lifting the pen; the task can only be solved by ignoring conventional boundaries and drawing the lines outside the virtual "box" specified by the dots

Tissue culture – procedure of propagating cells outside of living tissues *in vitro*

Tumor suppressor gene – a gene whose partial or complete inactivation leads to an increased likelihood of cancer development

Zygote – a fertilized egg, diploid cell formed by fusion of a male and a female gamete

References

ABEGGLEN, L. M., CAULIN, A. F., CHAN, A., LEE, K., ROBINSON, R., CAMBELL, M. S., KISO, W. K., SCHMITT, D. L., WADDELL, P. J., BHASKARA, S., JENSEN, S. T., MALEY, C. C., SCHIFFMAN, J. D. Potential mechanisms for cancer resistance in elephants and comparative cellular response to DNA damage in humans. *JAMA*, Nov 2015, vol. 314, no. 17, p. 1850–1860. DOI: https://doi.org/10.1001/jama.2015.13134.

AGUIRRE-GHISO, J. A. Models, mechanisms and clinical evidence for cancer dormancy. *Nature Reviews. Cancer*, Nov 2007, vol. 7, no. 11, p. 834–846. DOI: https://doi.org/10.1038/nrc2256.

AKTIPIS, C. A., BODDY, A. M., JANSEN, G., HIBNER, U., HOCHBERG, M. E., MALEY, C. C., WILKINSON, G. S. Cancer across the tree of life: cooperation and cheating in multicellularity. *Philosophical Transactions of the Royal Society of London. Series B, Biological Sciences*, Jul 2015, vol. 370, no. 1673, 201402219. DOI: https://doi.org/10.1098/rstb.2014.0219.

AKTIPIS, C. A., MALEY, C. C. Cooperation and cheating as innovation: insight from cellular societies. *Philosophical Transactions of the Royal Society of London. Series B, Biological Sciences*, Oct 2017, vol. 372, no. 1735, 20160421. DOI: https://doi.org/10.1098/rstb.2016.0421.

ALBERTINI, R. J., NICKLAS, J. A., O'NEILL, J. P., ROBINSON, S. H. In vivo somatic mutations in humans: measurement and analysis. *Annual Review of Genetics*, Dec 1990, vol. 24, p. 305–336. DOI: https://doi.org/10.1146/annurev.ge.24.120190.001513.

ALBERTS, B. Model organisms and human health. *Science*, Dec 2010, vol. 330, no. 6012, p. 1724. DOI: https://doi.org/10.1126/science.1201826.

ALBERTS, Bruce, BRAY, Dennis, JOHNSON, Alexander, LEWIS, Julian, RAFF, Martin, ROBERTS, Keith, WALTER, Peter. *Essential Cell Biology*. New York : Garland Publishing, 1998. 630 p. ISBN 0-8153-2045-0.

ALBUQUERQUE, C., BALTAZAR, C., FILIPE, B., PENHA, F., PEREIRA, T., SMITS, R., CRAVO, M., LAGE, P., FIDALGO, P., CLARO, I., RODRIGUES, P., VEIGA, I., RAMOS, J. S., FONSECA, I., LEITAO, C. N., FODDE, R. Colorectal cancers show distinct mutation spectra in members of the canonical WNT signaling pathway according to their anatomical location and type of genetic instability. *Genes, Chromosomes and Cancer*, Aug 2010, vol. 49, no. 8, p. 746–759. DOI: https://doi.org/10.1002/gcc.20786.

ALEGADO, R. A., KING, N. Bacterial influences on animal origins. *Cold Spring Harbor Perspectives in Biology*, Oct 2014, vol. 6, no. 11, 6:a016162. DOI: https://doi.org/10.1101/cshperspect.a016162.

AMARAL, J. D., XAVIER, J. M., STEER, C. J., RODRIGUES, C. M. P. The role of p53 in apoptosis. *Discovery Medicine*, Feb 2010, vol. 9, no. 45, p. 145–152.

AMÉRY, Jean. O stárnutí. *Revolta a rezignace*. 1. vyd. Praha : Prostor, 2008. 178 p. ISBN 978-80-7260-207-0.

AMÉRY, Jean. Vztáhnout na sebe ruku. *Rozprava o dobrovolné smrti*. 1. vyd. Praha : Prostor, 2010. 192 p. ISBN 978-80-7260-230-8.

ANDRYSIK, Z., GALBRAITH, M. D., GUARNIERI, A. L., ZACCARA, S., SULLIVAN, K. D., PANDEY, A., MACBETH, M., INGA, A., ESPINOSA, J. M. Identification of a core TP53 transcriptional program with highly distributed tumor suppressive activity. *Genome Research*, Oct 2017a, vol. 27, no. 10, p. 1645–1657. DOI: https://doi.org/10.1101/gr.220533.117.

ANDRYSÍK, Zdeněk. Hlavního strážce před rakovinou známe už 40 let. Ale neumíme ho využít. *Technet.cz* [online], 2017b. <https://technet.idnes.cz/pricina-vznik-rakovina-gen-tp53-david-lane-fpv-/veda.aspx?-c=-A171110_135316_veda_mla>.

ANTONIOU, A., HÉBRANT, A., DOM, G., DUMONT, J. E., MAENHAUT, C. Cancer stem cells, a fuzy evolving concept: A cell population or a cell property? *Cell Cycle*, Dec 2013, vol. 12, no. 24, p. 3743–3748. DOI: https://doi.org/10.4161/cc.27305.

ARANDA-ANZALDO, A., DENT, M. A. R. Reassessing the role of p53 in cancer and ageing from an evolutionary perspective. *Mechanism of Ageing and Development*, Apr 2007, vol. 128, no. 4, p. 293–302. DOI: https://doi.org/10.1016/j.mad.2007.01.001.

ARIELY, Dan. *Jak drahá je nepoctivost?* 1. vyd. Praha : Práh, 2012. 214 p. ISBN 978-80-7252-395-5.

ASSAILY, W. p53: guardian of multicellularity? A short essay on multicellular evolution and p53. *Hypothesis Journal*, 2005, vol. 1, no. 1, p. 14–15. DOI: https://doi.org/10.5779/hypothesis.v1i1.7.

AXELROD, R., HAMILTON, W. D. The evolution of cooperation. *Science*, Mar 1981, vol. 211, no. 4489, p. 1390–1396. DOI: https://doi.org/10.1126/science.7466396.

BAJGAR, M. Návrat do zlatého věku rentiérů. *Lidové noviny*, 14. června 2014, p. V25.

BAKER, S. J., MARKOWITZ, S., FEARON, E. R., WILLSON, J. K., VOGELSTEIN, B. Suppression of human colorectal carcinoma cell growth by wild-type p53. *Science*, Aug 1990, vol. 249, no. 4971, p. 912–915. DOI: https://doi.org/10.1126/science.2144057.

BARBOSA, I. A., MACHADO, N. G., SKILDUM, A. J., SCOTT, P. M., OLIVEIRA, P. J. Mitochondrial remodeling in cancer metabolism and survival: Potential for new therapies. *Biochimica et Biophysica Acta*, Aug 2012, vol. 1826, no. 1, p. 238–254. DOI: https://doi.org/10.1016/j.bbcan.2012.04.005.

BÁRTEK, J. Dilema: Stáří či rakovina? *MF Dnes*, únor 2007.

BARTEK, J., LUKAS, J. DNA damage checkpoints: from initiation to recovery or adaptation. *Current Opinion in Cell Biology*, Apr 2007, vol. 19, no. 2, p. 238–245. DOI: https://doi.org/10.1016/j.ceb.2007.02.009.

BATLLE, E., CLEVERS, H. Cancer stem cells revisited. *Nature Medicine*, Oct 2017, vol. 23, no. 10, p. 1124–1134. DOI: https://doi.org/10.1038/nm.4409.

BAYLEY, J.-P., DEVILEE, P. The Warburg effect in 2012. *Current Opinion in Oncology*, Jan 2012, vol. 24, no. 1, p. 62–67. DOI: https://doi.org/10.1097/CCO.0b013e32834deb9e.

BECKMAN, R. A., LOEB, L. A. Efficiency of carcinogenesis with and without a mutator mutation. *Proceedings of the National Academy of Sciences of the USA*, Sep 2006, vol. 103, no. 38, p. 14140–14145. DOI: https://doi.org/10.1073/pnas.0606271103.

BELYI, V. A., AK, P., MARKERT, E., WANG, H., HU, W., PUZIO-KUTER, A., LEVINE, A. J. The origins and evolution of the p53 family of genes. *Cold Spring Harbor Perspectives in Biology*, Jun 2010, vol. 2, no. 6, a001198. DOI: https://doi.org/10.1101/cshperspect.a001198.

BERKERS, C. R., MADDOCKS, O. D., CHEUNG, E. C., MOR, I., VOUSDEN, K.H. Metabolic regulation by p53 family members. *Cell Metabolism*, Nov 2013, vol. 18, no. 5, p. 617–633. DOI: https://doi.org/10.1016/j.cmet.2013.06.019.

BEUTLER, B., MAHONEY, J., LE TRANG, N., PEKALA, P., CERAMI, A. Purification of cachectin, a lipoprotein lipase-suppressing hormone secreted by endotoxin-induced RAW 264.7 cells. *Journal of Experimental Medicine*, May 1985, vol. 161, no. 5, p. 984–995. DOI: https://doi.org/10.1084/jem.161.5.984.

BHAT, R., BISSELL, M. J. Of plasticity and specificity: dialectics of the microenvironment and macroenvironment and the organ phenotype. *Wiley interdisciplinary reviews. Developmental Biology*, 2014, vol. 3, no. 2, p. 147–163. DOI: https://doi.org/10.1002/wdev.130.

BIANCONI, E., PIOVESAN, A., FACCHIN, F., BERAUDI, A., CASADEI, R., FRABETTI, F., VITALE, L., PELLERI, M. C., TASSANI, S., PIVA, F., PEREZ-AMODIO, S., STRIPPOLI, P., CANAIDER, S. An estimation of the number of cells in the human body. *Annals of Human Biology*, Nov–Dec 2013, vol. 40, no. 6, p. 463–471. DOI: https://doi.org/10.3109/03014460.2013.807878.

BINNEWIES, M., ROBERTS, E. W., KERSTEN, K., CHAN, V., FEARON, D. F., MERAD, M., COUSSENS, L. M., GABRILOVICH, D. I., OSTRAND-ROSEN BERG, S., HEDRICK, C. C., VONDERHEIDE, R. H., PITTET, M. J., JAIN, R. K., ZOU, W., HOWROFT, T. K., WOODHOUSE, E. C., WEINBERG, R. A., KRUMMEL, M. F. Understanding the tumor immune microenvironment (TIME) for effective therapy. *Nature Medicine*, May 2018, vol. 24, no. 5, p. 541–550. DOI: https://doi.org/10.1038/s41591-018-0014-x.

BISSELL, Mina J. Experiments that point to a new understanding of cancer. *TED talks* [online]. 2012. https://www.ted.com/talks/mina_bissell_experiments_that_point_to_a_new_understanding_of_cancer#t-6472>.

BISSELL, M. J. Goodbye flat biology – time for the 3rd and the 4th dimensions. *Journal of Cell Science*, Jan 2017, vol. 130, no. 1, p. 3–5. DOI: https://doi.org/10.1242/jcs.200550.

BISSELL, M. J., HINES, W. C. Why don't we get more cancer? A proposed role of the microenvironment in restraining cancer progression. *Nature Medicine*, Mar 2011, vol. 17, p. 320–329. DOI: https://doi.org/10.1038/nm.2328.

BISSELL, M. J., INMAN J. Reprogramming stem cell is a microenvironmental task. *Proceedings of the National Academy of Sciences of the USA*, Oct 2008, vol. 105, no. 41, p. 15637–15638. DOI: https://doi.org/10.1073/pnas.0808457105.

BLAGOSKLONNY, M. V. Geroconversion: irreversible step to cellular senescence. *Cell Cycle*, Dec 2014, vol. 13, no. 23, p. 3628–3635. DOI: https://doi.org/10.4161/15384101.2014.98 5507.

BLAGOSKLONNY, M. V. Tumor suppression by p53 without apoptosis and senescence: conundrum or rapalog-like gerosuppression? *Aging*, Jul 2012, vol. 4, no. 7, p. 450–455. DOI: https://doi.org/10.18632/aging.100475.

BONIZZI, G., CICALESE, A., INSINGA, A., PELICCI, P. G. The emerging role of p53 in stem cells. *Trends in Molecular Medicine*, Jan 2012, vol. 18, no. 1, p. 6–12. DOI: https://doi. org/10.1016/j.molmed.2011.08.002.

BOOMSMA, J. J., GAWNE, R. Superorganismality and caste differentiation as points of no return: how the major evolutionary transitions were lost in translation. *Biological Reviews of the Cambridge Philosophical Society*, Feb 2018, vol. 93, no. 1, p. 28–57. DOI: https://doi. org/10.1111/brv.12330.

BREIVIK, J. Commentary: Cancer – evolution within. *International Journal of Epidemiology*, Oct 2006, vol. 35, no. 5, p. 1161–1162. DOI: https://doi.org/10.1093/ije/dyl187.

BREIVIK, J. Don't stop for repairs in a war zone: Darwinian evolution units genes and environment in cancer development. *Proceeding in the National Academy of Sciences of the USA*, May 2001, vol. 98, no. 10, p. 5379–5381. DOI: https://doi.org/10.1073/pnas.101137698.

BREIVIK, J. The evolutionary origin of genetic instability in cancer development. *Seminars in Cancer Biology*, Feb 2005, vol. 15, no. 1, p. 51–60. DOI: https://doi.org/10.1016/ j.semcancer.2004.09.008.

BROSH, R., ROTTER, V. When mutants gain new powers: news from the mutant p53 field. *Nature Reviews. Cancer*, Oct 2009, vol. 9, no. 10, p. 701–713. DOI: https://doi.org/10.1038/ nrc2693.

BRUNET, T., KING, N. The origin of animal multicellularity and cell differentiation. *Developmental Cell*, Oct 2017, vol. 43, no. 2, p. 124–140. DOI: https://doi.org/10.1016/ j.devcel.2017.09.016.

BUCK, C. A. Adhesion mechanisms controlling cell-cell and cell-matrix interactions during the metastatic process. p. 172–205. In: MENDELSOHN, John, HOWLEY, Peter M., ISRAEL, Mark A., LIOTTA, Lance A. *The molecular basis of cancer*. Philadelphia: W. B. Saunders Co Ltd, 1995. 574 p. ISBN 978-0721-664-835.

BUNTING, S. F., NUSSENZWEIG, A. End-joining, translocations and cancer. *Nature Reviews. Cancer*, Jul 2013, vol. 13, p. 443–454. DOI: https://doi.org/10.1038/nrc3537.

BURTON, D. G. A., KRIZHANOVSKY, V. Physiological and pathological consequences of cellular senescence. *Cellular and Molecular Life Sciences*, Nov 2014, vol. 71, no. 22, p. 4373–4386. DOI: https://doi.org/10.1007/s00018-014-1691-3.

BYOCK, Ira. *Dobré umírání. Možnosti pokojného konce života.* 2. vyd. Praha : Vyšehrad, 2013, 328 p. ISBN 978-80-7429-134-0.

CAMPISI, J. Cancer and aging: rival demons? *Nature Reviews Cancer*, May 2003, vol. 3, p. 339–347. DOI: https://doi.org/10.1038/nrc1073.

CAPRA, Fritjof. *Bod obratu. Věda, společnost a nová kultura.* 1. vyd. Praha : DharmaGaia, 2002. 520 p. ISBN 80-85905-42-6 and ISBN 80-7287-024-6.

CAPRA, Fritjof. *Tkáň života. Nová syntéza mysli a hmoty.* 1. vyd. Praha : Academia, 2004. 296 p. ISBN 80-200-1169-2.

CARMELIET, P., JAIN, R. K. Angiogenesis in cancer and other diseases. *Nature*, Sep 2000, vol. 407, no. 6801, p. 249–257. DOI: https://doi.org/10.1038/35025220.

CARMELIET, P., JAIN, R. K. Principles and mechanisms of vessel normalization for cancer and other angiogenic diseases. *Nature Reviews. Drug Discovery*, Jun 2011, vol. 10, no. 6, p. 417–427. DOI: https://doi.org/10.1038/nrd3455.

CARROLL, Sean B. *Nekonečné, nesmírně obdivuhodné a překrásné.* 1. vyd. Praha : Academia, 2010. 368 p. ISBN 978-80-200-1800-7.

CHAFFER, C. L., WEINBERG, R. A. How does multistep tumorigenesis really proceed? *Cancer Discovery*, Jan 2015, vol. 5, no. 1, p. 22–24. DOI: https://doi.org/10.1158/2159-8290. CD-14-0788.

CHAMBERS, A. F., GROOM, A. C., MACDONALD, I. C. Dissemination and growth of cancer cells in metastatic sites. *Nature Reviews. Cancer*, Aug 2002, vol. 2, no. 8, p. 563–572. DOI: https://doi.org/10.1038/nrc865.

CHANDHOK, N. S., PELLMAN, D. A little CIN may cost a lot: revisiting aneuploidy and cancer. *Current Opinion in Genetics and Development*, Feb 2009, vol. 19, no. 1, p. 74–81. DOI: https://doi.org/10.1016/j.gde.2008.12.004.

CHATTERJEE, A., RODGER, E. J., ECCLES, M. R. Epigenetic drivers of tumourigenesis and cancer metastasis. *Seminars in Cancer Biology*, Aug 2018, vol. 51, p. 149–159. DOI: https://doi.org/10.1016/j.semcancer.2017.08.004.

CHESTERTON, Gilbert Keith. *Orthodoxy.* London, United Kingdom : Global Grey Ebooks, 2018. 145 p.

CHIPUK, J. E., KUWANA, T., BOUCHIER-HAYES, L., DROIN, N. M., NEWMEYER, D. D., SCHULLER, M., GREEN, D. R. Direct activation of Bax by p53 mediates mitochondrial membrane permeabilization and apoptosis. *Science*, Feb 2004, vol. 303, no. 5660, p. 1010–1014. DOI: https://doi.org/10.1126/science.1092734.

CHVÁLA, Vladislav. Nevydáš proti svému bližnímu falešné svědectví. *Náš Liberec* [online]. 2014, [referred February 6, 2014]. https://www.nasliberec.cz/komentare/komentare-zastupitelu/54-vladislav-chvala/1295-nevydáš-proti-svému-bližnímu-falešné-svědectví>.

CÍLEK, V. V kaňonu vran. *Respekt*, leden 2015, no. 4, p. 35.

CLEAVER, J. E. Defective repair replication of DNA in xeroderma pigmentosum. *Nature*, May 1968, vol. 218, no. 5142, p. 652–656. DOI: https://doi.org/ 10.1038/218652a0.

COBB, J. A., BJERKBAEK, L., GASSER, S. M. RecQ helicases: at the heart of genetic stability. *FEBS Letters*, Oct 2002, vol. 529, no. 1, p. 43–48. DOI: https://doi.org/10.1016/ S0014-5793(02)03269-6.

COBURN, C., ALLMAN, E., MAHANTI, P., BENEDETTO, A., CABREIRO, F., PINCUS, Z., MATTHIJSSENS, F., ARAIZ, C., MANDEL, A., VLACHOS, M., EDWARDS, S. A., FISCHER, G., DAVIDSON, A., PRYOR, R. E., STEVENS, A., SLACK, F. J., TAVERNARAKIS, N., BRAECKMAN, B. P., SCHROEDER, F. C., NEHRKE, K., GEMS, D. Anthranilate fluorescence marks a calcium-propagated necrotic wave that promotes organismal death in C. elegans. *PLoS Biology*, Jul 2013, vol. 11, no. 7, e1001613. DOI: https://doi.org/10.1371/journal.pbio.1001613.

COMEN, E., NORTON, L. Self-seeding in cancer. *Recent Results in Cancer Research* 2012, vol. 195, p. 13–23. DOI: https://doi.org/10.1007/978-3-642-28160-0_2.

COOKS, T., HARRIS, C. C., OREN, M. Caught in the cross fire: p53 in inflammation. *Carcinogenesis*, Aug 2014, vol. 35, no. 8, p. 1680–1690. DOI: https://doi.org/10.1093/carcin/ bgu134.

CORTEZ, D., ELLEDGE, S. J. Conducting the mitotic symphony. *Nature*, Jul 2000, vol. 406, no. 6794, p. 354–356. DOI: https://doi.org/10.1038/35019227.

CORTHAY, A. Does the immune system naturally protect against cancer? *Frontiers in Immunology*, May 2014, vol. 5, article 197. DOI: https://doi.org/10.3389/fimmu.2014.00197.

CORY, S., ADAMS, J. M. The Bcl2 family: regulators of the cellular life-or-death switch. *Nature Reviews. Cancer*, Sep 2002, vol. 2, no. 9, p. 647–654. DOI: https://doi.org/10.1038/ nrc883.

COTRAN, Ramzi S., KUMAR, Vinay, COLLINS, Tucker, ROBBINS, Stanley L. *Robbins Pathologic Basis of Disease*. 6th ed., 1999. Philadelphia : W. B. Saunders, Co. 1425 p. ISBN 978-0721673356.

COX, T. R., GARTLAND, A., ERLER, J. T. The pre-metastatic niche: is metastasis random? *BoneKEy Reports*, May 2012, vol. 1, article 80. DOI: https://doi.org/10.1038/ bonekey.2012.80.

CURTIN, N. J. DNA repair dysregulation from cancer driver to therapeutic target. *Nature Reviews. Cancer*, Dec 2012, vol. 12, no. 12, p. 801–817. DOI: https://doi.org/10.1038/ nrc3399.

CZARNECKA, A. M., GAMMAZZA, A. M., DI FELICE, V., ZUMMO, G., CAPPELLO, F. Cancer as a „mitochondriopathy". *Journal of Cancer Molecules*, 2007, vol. 3, no. 3, p. 71–79.

ČAPEK, Karel. *Hovory s T. G. Masarykem.* 1. souborné vyd. Praha : Československý spisovatel, 1990. 592 p. ISBN 80-202-0170-X.

ČAPEK, Karel. *Hry. Loupežník, R.U.R., Věc Makropulos, Bílá nemoc, Matka.* 1. vyd. Praha : Československý spisovatel, 1956. 446 p.

DAMERON, K. M., VOLPERT, O. V., TAINSKY, M. A., BOUCK, N. Control of angiogenesis in fibroblasts by p53 regulation of thrombospondin-1. *Science,* Sep 1994, vol. 265, no. 5178, p. 1582–1584. DOI: https://doi.org/10.1126/science.7521539.

DE LAURENZI, V., MELINO, G. Apoptosis. The little devil of death. *Nature,* Jul 2000, vol. 406, no. 6792, p. 135–136. DOI: https://doi.org/10.1038/35018190.

DE MAGALHAES, J. P. How ageing processes influence cancer. *Nature Reviews. Cancer,* May 2013, vol. 13, no. 5, p. 357–365. DOI: https://doi.org/10.1038/nrc3497.

DE MENDOZA, A., RUIZ-TRILLO, I. Forward genetics for back-in-time questions. *eLife,* Nov 2014, vol. 3, e05218. DOI: https://doi.org/10.7554/eLife.05218.

DENG, Y., CHANG, S. Role of telomeres and telomerases in genomic instability, senescence and cancer. *Laboratory Investigation,* Nov 2007, vol. 87, no. 11, p. 1071–1076. DOI: https://doi.org/10.1038/labinvest.3700673.

DE VISSER, K. E., EICHTEN, A., COUSSENS, L. M. Paradoxical roles of immune system during cancer development. *Nature Reviews. Cancer,* Jan 2006, vol. 6, no. 1, p. 24–36. DOI: https://doi.org/10.1038/nrc1782.

DOLBERG, D. S., BISSELL, M. J. Inability of Rous sarcoma virus to cause sarcomas in the avian embryo. *Nature,* Jun 1984, vol. 309, no. 5968, p. 552–556. DOI: https://doi.org/10.1038/309552a0.

DOMAZET-LOŠO, T., TAUTZ, D. Phylostratigraphic tracking of cancer genes suggests a link to the emergence of multicellularity in metazoa. *BMC Biology,* May 2010, vol. 8, no. 66, p. 1–10. DOI: https://doi.org/10.1186/1741-7007-8-66.

DONGRE, A., WEINBERG, R. A. New insights into the mechanisms of epithelial mesenchymal transition and implications for cancer. *Nature Reviews. Molecular and Cell Biology,* Feb 2019, vol. 20, no. 2, p. 69–84. DOI: https://doi.org/10.1038/s41580-018-0080-4.

DURAJ, Matouš. *Fašizující tendence v agrární straně ve třicátých letech dvacátého století : bakalářská práce.* Praha : Univerzita Karlova : Filozofická fakulta, 2013. 41 p.

EGEBLAD, M., WERB, Z. New functions for the matrix metalloproteinase in cancer progression. *Nature Reviews. Cancer,* Mar 2002, vol. 2, no. 3, p. 163–176. DOI: https://doi.org/10.1038/nrc745.

ELDAKAR, O. T., GALLUP, A. C., DRISCOLL, W. W. When hawks give rise to doves: The evolution and transition of enforcement strategies. *Evolution,* Jun 2013, vol. 67, no. 6, p. 1549–1560. DOI: https://doi.org/10.1111/evo.12031.

ELDAKAR, O. T., KAMMEYER, J. O., NAGABANDI, N., GALLUP, A. C. Hypocrisy and corruption: How disparities in power shape the evolution of social control. *Evolutionary Psychology,* Apr–Jun 2018, vol. 16, no. 2, p. 1–12. DOI: https://doi.org/10.1177/1474704918756993.

ELDAKAR, O. T., WILSON, D. S. Eight criticisms not to make about group selection. *Evolution*, Jun 2011, vol. 65, no. 6, p. 1523–1526. DOI: https://doi.org/10.1111/j.1558-5646.2011.01290.x.

ELDAKAR, O. T., WILSON, D. S. Selfishness as second-order altruism. *Proceedings of the National Academy of Sciences of the USA*, May 2008, vol. 105, no. 19, p. 6982–6986. DOI: https://doi.org/10.1073/pnas.0712173105.

ELINAV, E., NOWARSKI, R., THAISS, C. A., HU, B., JIN, C., FLAVELL, R. A. Inflammation-induced cancer: crosstalk between tumours, immune cells and microorganisms. *Nature Reviews. Cancer*, Nov 2013, vol. 13, no. 11, p. 759–771. DOI: https://doi.org/10.1038/nrc3611.

ELLIS, S., FRANKS, D. W., NATTRASS, S., CANT, M. A., BRADLEY, D. L., GILES, D., BALCOMB, K. C., CROFT, D. P. Postreproductive lifespan are rare in mammals. *Ecology and Evolution*, Jan 2018a, vol. 8, no. 5, p. 2482–2494. DOI: https://doi.org/10.1002/ece3.3856.

ELLIS, S., FRANKS, D. W., NATTRASS, S., CURRIEW, T. E., CANT, M. A., GILES, D., BALCOMB, K. C., CROFT, D. P. Analyses of ovarian activity reveal repeated evolution of post-reproductive lifespans in toothed whales. *Scientific Reports*, Aug 2018b, vol. 8, no. 1, p. 12833. DOI: https://doi.org/10.1038/s41598-018-31047-8.

FEARON, K. C. H., GLASS, D. J., GUTTRIDGE, D. C. Cancer cachexia: Mediators, signaling, and metabolic pathways. *Cell Metabolism*, Aug 2012, vol. 16, no. 2, p. 153–166. DOI: https://doi.org/10.1016/j.cmet.2012.06.011.

FEINBERG, A. P. Epigenetic stochasticity, nuclear structure and cancer: the implications for medicine. *Journal of Internal Medicine*, Jul 2014, vol. 276, no. 1, p. 5–11. DOI: https://doi.org/10.1111/joim.12224.

FEOLA, S., CHIARO, J., MARTINS, B., CERULLO, V. Uncovering the tumor antigen landscape: what to know about the discovery process. *Cancers*, Jun 2020, vol. 12, no. 6, p. 1660. DOI: https://doi.org/10.3390/cancers12061660.

FINN, O. J. Immuno-oncology: understanding the function and dysfuction of the immune system in cancer. *Annals of Oncology*, Sep 2012, vol. 23 (Suppl 8), p. viii6–viii9. DOI: https://doi.org/10.1093/annonc/mds256.

FISCHER, M. Census and evaluation of p53 target genes. *Oncogene*, Jul 2017, vol. 36, no. 28, p. 3943–3956. DOI: https://doi.org/10.1038/onc.2016.502.

FLEGR, Jaroslav. *Evoluční biologie*. 1. vyd. Praha : Academia, 2005. 559 p. ISBN 80-200-1270-2.

FODDE, R., SMITS, R., CLEVERS, H. APC, signal transduction and genetic instability in colorectal cancer. *Nature Reviews. Cancer*, Oct 2001, vol. 1, no. 1, p. 55–66. DOI: https://doi.org/10.1038/35094067.

FOLKMAN, J. Angiogenesis inhibitors, a new class of drugs. *Cancer Biology and Therapy*, Jul–Aug 2003, vol. 2, no. 4 (Suppl 1), p. S127–S133. DOI: https://doi.org/10.4161/cbt.212.

FOLKMAN, J. Tumor angiogenesis: Therapeutic implications. *The New England Journal of Medicine*, Nov 1971, vol. 285, no. 21, p. 1182–1186. DOI: https://doi.org/10.1056/NEJM197111182852108.

FOLKMAN, J., HAHNFELDT, P., HLATKY, L. Cancer: looking outside the genome. *Nature Reviews. Molecular Cell Biology*, Oct 2000, vol. 1, no. 1, p. 76–79. DOI: https://doi.org/10.1038/35036100.

FRANKISH, A., DIEKHANS, M., FERREIR, A.-M., JOHNSON, R., JUNGREIS, I., LOVELAND, J., MUDGE, J. M., SISU, C., WRIGHT, J., ARMSTRONG, J., BARNES, I., BERRY, A., BIGNELL, A., SALA, S. C., CHRAST, J., CUNNINGHAM, F., DI DOMENICO, T., DONALDSON, S., FIDDES, I. T., GIRÓN, C. G., GONZALES, J. M., GREGO, T., HARDY, M., HOURLIER, T., HUNT, T., IZUOGU, O. G., LAGARDE, J., MARTIN, F. J., MARTÍNEZ, L., MOHANAN, S., MUIR, P., NAVARRO, F. C. P., PARKER, A., PEI, B., POZO, F., RUFFIER, M., SCHMITT, B. M., STAPLETON, E., SUNER, M.-M., SYCHEVA, I., USZCZYNSKA-RATAJCZAK, B., XU, J., YATES, A., ZERBINO, D., ZHANG, Y., AKEN, B., CHOUDHARY, J. S., GERSTEIN, M., GUIGÓ, R., HUNNARD, T. J. P., KELLIS, M., PATEN, B., REYMOND, A., TRESS, M. L., FLICEK, P. Gencode reference annotation for the human and mouse genomes. *Nucleic Acids Research*, Jan 2019, vol. 47, no. D1, p. D766–D773. DOI: https://doi.org/10.1093/nar/gky955.

FREED-PASTOR, W. A., PRIVES, C. Mutant p53: one name, many proteins. *Genes and Development*, Jun 2012, vol. 26, no. 12, p. 1268–1286. DOI: https://doi.org/10.1101/gad.190678.112.

FRIAS, C., PAMPALONA, J., GENESCA, A., TUSELL, L. Telomere dysfunction and genome instability. *Frontiers in Bioscience*, Jun 2012, vol. 17, p. 2181–2196. DOI: https://doi.org/10.2741/4044.

FRIEDBERG, E. C. How nucleotide excision repair protects against cancer. *Nature Reviews. Cancer*, Oct 2001, vol. 1, no. 1, p. 22–33. DOI: https://doi.org/10.1038/35094000.

FRIEDMANN-MORVINSKI, D., VERMA, I. M. Dedifferentiation and reprogramming: origins of cancer stem cells. *EMBO Reports*, Mar 2014, vol. 15, no. 3, p. 244–253. DOI: https://doi.org/10.1002/embr.201338254.

FU, C., JIANG, A. Dendritic cells and CD8 cell immunity in tumor microenvironment. *Frontiers in Immunology*, Dec 2018, vol. 9, p. 3059. DOI: https://doi.org/10.3389/fimmu.2018.03059.

FUKASAWA, K. Introduction. Centrosome. *Oncogene*, Sep 2002, vol. 21, no. 40, p. 6140–6145. DOI: https://doi.org/10.1038/sj.onc.1205771.

GASPAR, C., FRANKEN, P., MOLENAAR, L., BREUKEL, C., VAN DER VALK, M., SMITS, R., FODDE, R. A targeted constitutive mutation in the Apc tumor suppressor gene underlies mammary but not intestinal tumorigenesis. *PLoS Genetics*, Jul 2009, vol. 5, no. 7, e1000547. DOI: https://doi.org/10.1371/journal.pgen.1000547.

GAUGHRAN, S. J., PLESS, E., STEARNS, S. C. How elephants beat cancer. *eLife*, Oct 2016, vol. 5, e21864. DOI: https://doi.org/10.7554/eLife.21864.

GEMS, D. Evolution of sexually dimorphic longevity in humans. *Aging*, Feb 2014b, vol. 6, no. 2, p. 84–91. DOI: https://doi.org/10.18632/aging.100640.

GEMS, D. The aging-disease false dichotomy: understanding senescence as pathology. *Frontiers in Genetics*, Jun 2015, vol. 6, p. 212. DOI: https://doi.org/10.3389/fgene.2015.00212.

GEMS, D. What is anti-aging treatment? *Experimental Gerontology*, Oct 2014a, vol. 58, p. 14–18. DOI: https://doi.org/10.1016/j.exger.2014.07.003.

GEORGES, Bertrand. *Umění rozlišovat*. 1. vyd. Kostelní Vydří : Karmelitánské nakladatelství, 2011. 192 p. ISBN 978-80-7195-511-5.

GERLINGER, M., SWANTON, C. How Darwinian models inform therapeutic failure initiated by clonal heterogeneity in cancer medicine. *British Journal of Cancer*, Oct 2010, vol. 103, no. 8, p. 1139–1143. DOI: https://doi.org/10.1038/sj.bjc.6605912.

GHAJAR, C. M., BISSELL, M. J. Extracellular matrix control of mammary gland morphogenesis and tumorigenesis: insights from imaging. *Histochemistry and Cell Biology*, Dec 2008, vol. 130, no. 6, p. 1105–1118. DOI: https://doi.org/10.1007/s00418-008-0537-1.

GILBERT, S. F., SARKAR, S. Embracing complexity: Organicism for the 21st century. *Developmental Dynamics*, Sep 2000, vol. 219, no. 1, p. 1–9. DOI: https://doi.org/10.1002/1097-0177(2000)9999:9999<::AID-DVDY1036>3.0.CO;2-A.

GIORGI, C., BONORA, M., MISSIROLI, S., MORGANTI, C., MORCIANO, G., WIECKOWSKI, M. R., PINTON, P. Alterations in mitochondrial and endoplasmatic reticulum signaling by p53 mutants. *Frontiers in Oncology*, Feb 2016, vol. 6, p. 42. DOI: https://doi.org/10.3389/fonc.2016.00042.

GOLDSBY, H. J., DORNHAUS, A., KERR, B., OFRIA, C. Task-switching costs promote the evolution of division of labor and shifts in individuality. *Proceedings of the National Academy of Sciences of the USA*, Aug 2012, vol. 109, no. 34, p. 13686–13691. DOI: https://doi.org/10.1073/pnas.1202233109.

GOLDSBY, H. J., KNOESTER, D. B., KERR, B., OFRIA, C. The effect of conflicting pressures on the evolution of division of labor. *PLoS One*, Aug 2014a, vol. 9, no. 8, p. e102713. DOI: https://doi.org/10.1371/journal.pone.0102713.

GOLDSBY, H. J., KNOESTER, D. B., OFRIA, C., KERR, B. The evolutionary origin of somatic cells under the dirty work hypothesis. *PLoS Biology*, May 2014b, vol. 12, no. 5, p. e1001858. DOI: https://doi.org/10.1371/journal.pbio.1001858.

GOYAL, L. Cell death inhibition: keeping caspases in check. *Cell*, Mar 2001, vol. 104, no. 6, p. 805–808. DOI: https://doi.org/10.1016/S0092-8674(01)00276-8.

GRANGER, M. P., WRIGHT, W. E., SHAY, J. W. Telomerase in cancer and aging. *Critical Reviews in Oncology/Hematology*, Jan 2002, vol. 41, no. 1, p. 29–40. DOI: https://doi.org/10.1016/S1040-8428(01)00188-3.

GREGOR, Miloš, VEJVODOVÁ, Petra. *Nejlepší kniha o fake news*. 1. vyd. Brno : Computer Press, 2018. 144 p. ISBN 978-80-264-1805-4.

GROSBERG, R. K., STRATHMANN, R. R. The evolution of multicellularity: A minor major transition? *Annual Reviews of Ecology, Evolution, and Systematics*, Dec 2007, vol. 38, p. 621–654. DOI: https://doi.org/10.1146/annurev.ecolsys.36.102403.114735.

GRÜN, Anselm. *Umění stárnout*. 1. vyd. Kostelní Vydří : Karmelitánské nakladatelství, 2009. 143 p. ISBN 978-80-7195-316-6.

GRÜN, Anselm. *V půli cesty*. 1. vyd. Kostelní Vydří : Karmelitánské nakladatelství, 2001. 78 p. ISBN 978-80-7192-569-9.

GRÜN, Anselm. *Život je teď: Umění stárnout*. 1. vyd. Praha : Portál, 2010. 136 p. ISBN 978-80-7367-749-7.

GUDKOV, A. V., GUROVA, K. V., KOMAROVA, E. A. Inflammation and p53: A tale of two stresses. *Genes and Cancer*, Apr 2011, vol. 2, no. 4, p. 503–516. DOI: https://doi.org/10.1177/1947601911409747.

HACKETT, J. A., GREIDER, C. W. Balancing instability: dual roles for telomerase dysfunction in tumorigenesis. *Oncogene*, Jan 2002, vol. 21, no. 4, p. 619–626. DOI: https://doi.org/10.1038/sj.onc.1205061.

HAHNFELDT, P. Significance of tumor self-seeding as an augmentation to the classic metastasis paradigm. *Future Oncology*, May 2010, vol. 6, no. 5, p. 681–685. DOI: https://doi.org/10.2217/fon.10.43.

HALAZONETIS, T. D., GORGOULIS, V. G., BARTEK, J. An oncogene-induced DNA damage model for cancer development. *Science*. Mar 2008, vol. 319, no. 5868, p. 1352–1355. DOI: https://doi.org/10.1126/science.1140735.

HANAHAN, D. Rethinking the war on cancer. *The Lancet Oncology*, Feb 2014, vol. 383, no. 9916, p. 558–563. DOI: https://doi.org/10.1016/S0140-6736(13)62226-6.

HANAHAN, D., COUSSENS, L. M. Accessories to the crime: Functions of cells recruited to the tumor microenvironment. *Cancer Cell*, Mar 2012, vol. 21, no. 3, p. 309–322. DOI: https://doi.org/10.1016/j.ccr.2012.02.022.

HANAHAN, D., FOLKMAN, J. Patterns and emerging mechanisms of the angiogenic switch during tumorigenesis. *Cell*, Aug 1996, vol. 86, no. 3, p. 353–364. DOI: https://doi.org/10.1016/S0092-8674(00)80108-7.

HANAHAN, D., WEINBERG, R. A. Hallmarks of cancer: The next generation. *Cell*, Mar 2011, vol. 144, no. 5, p. 646–674. DOI: https://doi.org/10.1016/j.cell.2011.02.013.

HANAHAN, D., WEINBERG, R. A. The hallmarks of cancer. *Cell*, Jan 2000, vol. 100, no. 1, p. 57–70. DOI: https://doi.org/10.1016/S0092-8674(00)81683-9.

HARARI, Yuval Noah. *Sapiens: A Brief History of Humankind*. London, United Kingdom : Signal, McClelland and Stewart Publ 2014. 439 p. ISBN 978-0-7710-3850-1.

HARLEY, C. B. Telomerase is not an oncogene. *Oncogene*, Jan 2002, vol. 21, no. 4, p. 494–502. DOI: https://doi.org/10.1038/sj.onc.1205076.

HAVEL, Ivan M. *Jitro kyberkultury*. 1999. <http://www.cts.cuni.cz/~havel/work/jitro.html>.

HAWKES, K. Grandmothers and the evolution of human longevity. *American Journal of Human Biology*, May–Jun 2003, vol. 15, no. 3, p. 380–400. DOI: https://doi.org/10.1002/ajhb.10156.

HAYFLICK, Leonard. *Jak a proč stárneme*. 1. vyd. Praha : Columbus, 1997. 426 p. ISBN 80-7176-536-8.

HENSON, J. D., NEUMANN, A. A., YEAGER, T. R., REDDEL, R. R. Alternative lengthening of telomeres in mammalian cells. *Oncogene*, Jan 2002, vol. 21, no. 4, p. 598–610. DOI: https://doi.org/10.1038/sj.onc.1205058.

HESKETH, Robin. *Introduction to Cancer Biology*. 1st printing, New York : Cambridge University Press, 2013. 630 p. ISBN 97-8110-760-1482.

HICKSON, I. D. RecQ helicases: caretakers of the genome. *Nature Reviews. Cancer*, Mar 2003, vol. 3, no. 3, p. 169–178. DOI: https://doi.org/10.1038/nrc1012.

HOCHBERG, M. E., MARQUET, P. A., BOYD, R., WAGNER, A. Process and pattern in innovations from cells to societies. *Philosophical Transactions of The Royal Society B Biological Sciences*, Dec 2017, vol. 372, 20160414. DOI: https://doi.org/10.1098/rstb.2016.0414.

HÖLLDOBLER, Bert, WILSON, Edward O. *Cesta k mravencům*. 1. vyd. Praha : Academia, 1997. 198 p. ISBN 80-200-0612-5.

HOLT-LUNSTAD, J., SMITH, T. B., BAKER, M., HARRIS, T., STEPHENSON, D. Loneliness and social isolation as risk factors for mortality: A meta-analytic review. *Perspectives on Psychological Science*, Mar 2015, vol. 10, no. 2, p. 227–237. DOI: https://doi.org/10.1177/1745691614568352.

HOLT-LUNSTAD, J., SMITH, T. B., LAYTON, J. B. Social relationships and mortality risk: A meta-analytic review. *PLoS Medicine*, Jul 2010, vol. 7, no. 7, e1000316. DOI: https://doi.org/10.1371/journal.pmed.1000316.

HONORÉ, Carl. *Chvála pomalosti*. 1. vyd. Praha : Nakladatelství 65. pole, 2012. 293 p. ISBN 978-80-87506-22-6.

HORAK, C. E., LEE, J. H., MARSHALL, J.-C., SHREEVE, S. M., STEEG, P. S. The role of metastasis suppressor genes in metastatic dormancy. *APMIS: Acta Pathologica, Microbiologica, et Immunologica Scandinavica*, Jul–Aug 2008, vol. 116, no. 7–8, p. 586–601. DOI: https://doi.org/10.1111/j.1600-0463.2008.01027.x.

HORKÝ, P. Slyšet jen svoji ozvěnu. *Respekt*, 28. listopadu – 4. prosince 2016, no. 48, p. 22–23.

HOŘEJŠÍ, V. Imunitní systém a nádory – komplikovaný vztah. *Onkologická Revue*, vol. 6, suppl. 01 Imunoterapie, 2019, p. 1–4.

HOŘEJŠÍ, Václav, BARTŮŇKOVÁ, Jiřina, BRDIČKA, Tomáš, ŠPÍŠEK, Radek. *Základy imunologie*. 6. vyd. Praha : Stanislav Juhaňák – Triton, 2017. 297 p. ISBN 978-80-7553-250-3.

HOUDEK, P. Svědomí snadno letí přes palubu. *MF Dnes*, 24. dubna 2010, p. 38–40.

HUNTINGTON, N. D., CURSONS, J., RAUTELA, J. The cancer-natural killer cell immunity cycle. *Nature Reviews. Cancer*, Aug 2020, vol. 20, no. 8, p. 437–454. DOI: https://doi.org/10.1038/s41568-020-0272-z.

HURST, D. R., WELCH, D. R. Metastasis suppressor genes at the interface between the environment and tumor cell growth. *International Review of Cell and Molecular Biology*, Feb 2011, vol. 286, p. 107–180. DOI: https://doi.org/10.1016/B978-0-12-385859-7.00003-3.

IGNEY, F. H., KRAMMER, P. H. Death and anti-death: tumour resistance to apoptosis. *Nature Reviews. Cancer*, Apr 2002, vol. 2, no. 4, p. 277–288. DOI: https://doi.org/10.1038/nrc776.

ISAACSON, Walter. *Steve Jobs*. New York, NY, USA : Simon and Schuster, 2011. 670 p. ISBN 978-1-4516-4853-9.

JACKSON, S. P., BARTEK, J. The DNA-damage response in human biology and disease. *Nature*, Oct 2009, vol. 461, no. 7267, p. 1071–1078. DOI: https://doi.org/10.1038/nature08467.

JACOB, F. Evolution and tinkering. *Science*, Jun 1977, vol. 196, no. 4295, p. 1161–1166. DOI: https://doi.org/10.1126/science.860134.

JAKOBY, Bernard. *Tajemství umírání*. 1. vyd. Liberec : Dialog, 2005. 127 p. ISBN 80-86271-42-8.

JALLEPALLI, P. V., LENGAUER, C. Chromosome segregation and cancer: cutting through the mystery. *Nature Reviews. Cancer*, Nov 2001, vol. 1, no. 2, p. 109–117. DOI: https://doi.org/10.1038/35101065.

JÍLEK, Petr. *Imunologie stručně, jasně, přehledně*. 4. vyd. Praha : Grada, 2014. 96 p. ISBN 978-80-247-4822-1.

JOHMURA, Y., NAKANISHI, M. Multiple facets of p53 in senescence induction and maintenance. *Cancer Science*, Nov 2016, vol. 107, no. 11, p. 1550–1555. DOI: https://doi.org/10.1111/cas.13060.

KALLURI, R., WEINBERG, R. A. The basics of epithelial-mesenchymal transition. *The Journal of Clinical Investigation*, Jun 2009, vol. 119, no. 6, p. 1420–1428. DOI: https://doi.org/10.1172/JCI39104.

KAMP, W. M., WANG, P.-Y., HWANG, P. M. TP53 mutation, mitochondria and cancer. *Current Opinion in Genetics and Development*, Jun 2016, vol. 38, p. 16–22. DOI: https://doi.org/10.1016/j.gde.2016.02.007.

KAPLAN, R. N., RAFII, S., LYDEN, D. Preparing the "soil": The premetastatic niche. *Cancer Research*, Dec 2006, vol. 66, no. 23, p. 11089–11093. DOI: https://doi.org/10.1158/0008-5472.CAN-06-2407.

KARRAN, P., BIGNAMI, M. DNA damage tolerance, mismatch repair and genome instability. *Bioessays*, Nov 1994, vol. 16, no. 11, p. 833–839. DOI: https://doi.org/10.1002/bies.950161110.

KASL, T. Sociální superorganismus a jeho globální mozek. *Vesmír*, červen 2002, vol. 81, no. 6, p. 314–316.

KASTAN, M. B., BARTEK, J. Cell-cycle checkpoints and cancer. *Nature*, Nov 2004, vol. 432, no. 7015, p. 316–323. DOI: https://doi.org/10.1038/nature03097.

KEANE, M., SEMEIKS, J., WEBB, A. E., LI, Y. I., QUESADA, V., CRAIG, T., MADSEN, L. B., VAN DAM, S., BRAWAND, D., MARQUES, P. I., MICHALAK, P., KANG, L., BHAK, J., YIM, H.-S., GRISHIN, N. V., NIELSEN, N. H., HEIDE--JORGENSEN, M. P., OZIOLOR, E. M., MATSON, C. W., CHURCH, G. M., STUART, G. W., PATTON, J. C., GEORGE, J. C., SUYDAM, R., LARSEN, K., LÓPEZ-OTÍN, C., O'CONNELL, M. J., BICKHAM, J. W., THOMSEN, B., DE MAGALHAES, J. P. Insights into the evolution of longevity from the bowhead whale genome. *Cell Reports*, Jan 2015, vol. 10, no. 1, p. 112–122. DOI: https://doi.org/10.1016/j.celrep.2014.12.008.

KEIBEL, A., SINGH, V., SHARMA, M. C. Inflammation, microenvironment, and the immune system in cancer progression. *Current Pharmaceutical Design*, Dec 2009, vol. 15, no. 17, p. 1949–1955. DOI: https://doi.org/10.2174/138161209788453167.

KEIM, C., KAZADI, D., ROTSCHILD, G., BASU, U. Regulation of AID, the B-cell genome mutator. *Genes and Development*, Jan 2013, vol. 27, no. 1, p. 1–17. DOI: https://doi.org/10.1101/gad.200014.112.

KEIZER, K., LINDENBERG, S., STEG, L. The spreading of disorder. *Science*, Dec 2008, vol. 322, no. 5908, p. 1681–1685. DOI: https://doi.org/10.1126/science.1161405.

KELLER, Jan. *Posvícení bezdomovců*. 1. vyd. Praha : Sociologické nakladatelství (SLON), 2013. 286 p. ISBN 978-80-7419-155-8.

KELLER, Jan. *Tři sociální světy : Sociální struktura postindustriální společnosti*. 2. vyd. Praha : Sociologické nakladatelství (SLON), 2011. 211 p. ISBN 978-80-7419-044-5.

KENNEDY, P., BARON, G., QIU, B., FREITAK, D., HELENTERÄ, H., HUNT, E. R., MANFREDINI, F., O'SHEA-WHELLER, T., PATALANO, S., PULL, C. D., SASAKI, T., TAYLOR, D., WYATT, C. D. R., SUMNER, S. Deconstructing superorganisms and societies to address big questions in biology. *Trends in Ecology and Evolution*, Nov 2017, vol. 32, no. 11, p. 861–872. DOI: https://doi.org/10.1016/j.tree.2017.08.004.

KENNEDY, P., ULLER, T., HELENTERÄ, H. Are ant supercolonies crucibles of a new major transition in evolution? *Journal of Evolutionary Biology*, Sep 2014, vol. 27, no. 9, p. 1784–1796. DOI: https://doi.org/10.1111/jeb.12434.

KENNY, P. A., LEE, G. Y., BISSELL, M. J. Targeting the tumor microenvironment. *Frontiers in Bioscience*, May 2007, vol. 12, p. 3468–3474. DOI: https://doi.org/10.2741/2327.

KERBEL, R., FOLKMAN, J. Clinical translation of angiogenesis inhibitors. *Nature Reviews. Cancer*, Oct 2002, vol. 2, no. 10, p. 727–739. DOI: https://doi.org/10.1038/nrc905.

KIECHLE, Stefan. *Umět se rozhodnout*. 1. vyd. Kostelní Vydří : Karmelitánské nakladatelství, 2016. 72 p. ISBN 978-80-7195-869-7.

KIM, K.-W., LEE, S.-J., KIM, W.-Y., SEO, J. H., LEE, H. Y. How can we treat cancer disease not cancer cells. *Cancer Research and Treatment*, Jan 2017, vol. 49, no. 1, p. 1–9. DOI: https://doi.org/10.4143/crt.2016.606.

KIM, M.-Y., OSKARSSON, T., ACHARYYA, S., NGUYEN, D. X., ZHANG, X. H., NORTON, L., MASSAQUÉ, J. Tumor self-seeding by circulating cancer cells. *Cell*, Dec 2009, vol. 139, no. 7, p. 1315–1326. DOI: https://doi.org/10.1016/j.cell.2009.11.025.

KING, N., WESTBROOK, M. J., YOUNG, S. L., KUO, A., ABEDIN, M., CHAPMAN, J., FAIRCLOUGH, S., HELLSTEN, U., ISOGAI, Y., LETUNIC, I., MARR, M., PINCUS, D., PUTNAM, N., ROKAS, A., WRIGHT, K. J., ZUZOW, R., DIRKS, W., GOOD, M., GOODSTEIN, D., LEMONS, D., LI, W. Q., LYONS, J. B., MORRIS, A., NICHOLS, S., RICHTER, D. J., SALAMOV, A., BORK, P., LIM, W. A., MANNING, G., MILLER, W. T., MCGINNIS, W., SHAPIRO, H., TJIAN, R., GRIGORIEV, I. V., ROKHSAR, D. The genome of the choanoflagellate Monisoga brevicolis and the origin of metazoans. *Nature*, Feb 2008, vol. 451, no. 7180, p. 783–788. DOI: https://doi.org/10.1038/nature06617.

KLAUSNER, R. D. The fabric of cancer cell biology – Weaving together the strands. *Cancer Cell*, Feb 2002, vol. 1, no. 1, p. 3–10. DOI: https://doi.org/10.1016/S1535-6108(02)-00020-X.

KLENER, P. Angiogeneze jako součást nádorového „ekosystému" a možnosti jejího ovlivnění. *Klinická onkologie*, 2010, vol. 23, no. 1, p. 14–20.

KLIMEŠ, Jeroným. VIII. Nepromluvíš křivého svědectví proti bližnímu svému. *Christnet. eu* [online]. 2014, [referred November 10, 2014]. <http://www.christnet.eu/clanky/5397/nepromluvis_kriveho_svedectvi_proti_ bliznimu_svemu.url>.

KNOLL, A. H. The multiple origins of complex multicellularity. *Annual Reviews of Earth and Planetary Sciences*, May 2011, vol. 39, p. 217–239. DOI: https://doi.org/10.1146/annurev.earth.031208.100209.

KOLATA, G. Grant system leads cancer researchers to play it safe. *The New York Times*, June 27, 2009. <https://www.nytimes.com/2009/06/28/health/research/28cancer.html>.

KOMAROVA, N. L., WODARZ, D. The optimal rate of chromosome loss for the inactivation of tumor suppressor genes in cancer. *Proceedings of the National Academy of Sciences of the USA*, May 2004, vol. 101, no. 18, p. 7017–7021. DOI: https://doi.org/10.1073/pnas.0401943101.

KOUBSKÁ, Libuše, KOUKOLÍK, František. *Všechno dopadne jinak.* 2. vyd. Praha : Vyšehrad, 2019. 128 p. ISBN 978-80-7601-093-2.

KOUCKÁ, Pavla, PREKOPOVÁ, Jiřina, ŠTURMA, Jaroslav. *Výchova láskou.* 1. vyd. Praha : Portál, 2012. 179 p. ISBN 978-80-262-0077-2.

KOUKOLÍK, František. Koukej spolupracovat. *Medical Tribune* [online]. 2019, XV(5):D7 [referred March 19, 2019]. <https://www.tribune.cz/clanek/44454-koukej-spolupracovat>.

KOUKOLÍK, František. *Mocenská posedlost.* 1. vyd. Praha : Karolinum, 2010. 280 p. ISBN 978-80-246-1825-8.

KOUKOLÍK, František. *Sociální mozek.* 2. vyd. Praha : Karolinum, 2016. 308 p. ISBN 978-80-246-2850-9.

KOUKOLÍK, František, DRTILOVÁ, Jana. *Vzpoura deprivantů : Nestvůry, nástroje, obrana.* 1. přepracované vyd. Praha : Galén, 2006. 327 p. ISBN 80-7262-410-5.

KOUKOLÍK, F., RIEBAUEROVÁ, M. Politika přitahuje částečné psychopaty. *MF Dnes*, 5. ledna 2013, p. 12A.

KREJSEK, Jan, KOPECKÝ, Otakar. *Klinická imunologie.* 1. vyd. Hradec Králové : Nucleus HK, 2004. 941 p. ISBN 80-86225-50-X.

KREUZALER, P., WATSON, C. J. Killing a cancer: what are the alternatives? *Nature Reviews. Cancer*, May 2012, vol. 12, no. 6, p. 411–424. DOI: https://doi.org/10.1038/nrc3264.

KROEMER, G., GALLIZZI, L., VANDENABEELE, P., ABRAMS, J., ALNEMRI, E. S., BAEHRECKE, E. H., BLAGOSKLONNY, M. V., EL-DEIRY, W. S., GOLSTEIN, P., GREEN, D. R., HENGARTNER, M., KNIGHT, R. A., KUMAR, S., LIPTON, S. A., MALORNI, W., NUNEZ, G., PETER, M. E., TSCHOPP, J., YUAN, J., PIACENTINI, M., ZHIVOTOVSKY, B., MELINO, G. Classification of cell death: recommendations of the Nomenclature Committee on Cell Death 2009. *Cell Death and Differentiation*, Jan 2009, vol. 16, no. 1, p. 3–11. DOI: https://doi.org/10.1038/cdd.2008.150.

KRUML, David. *Vězeň to má spočítané : lekce z teorie her.* 1. vyd. Brno : Masarykova univerzita, 2018. 126 p. ISBN 978-80-210-8832-0.

KÜBLER-ROSSOVÁ, Elisabeth, KESSLER, David. *Lekce života : o tajemstvích lidského bytí.* 1. vyd. Brno : Jota, 2013. 252 p. ISBN 978-80-7462-327-1.

KUHN, Thomas Samuel. *Struktura vědeckých revolucí.* Dotisk 1. vyd. Praha : Oikoymenh, 2008. 206 p. ISBN 80-86005-54-2.

LABARGE, M. A., PETERSEN, O. W., BISSELL, M. J. Of microenvironments and mammary stem cells. *Stem Cell Reviews*, Jun 2007, vol. 3, no. 2, p. 137–146. DOI: https://doi.org/10.1007/s12015-007-0024-4.

LAMBERT, A. W., PATTABIRAMAN, D. R., WEINBERG, R. A. Emerging biological principles of metastasis. *Cell*, Feb 2017, vol. 168, no. 4, p. 670–691. DOI: https://doi.org/10.1016/j.cell.2016.11.037.

LANE, D. P. Cancer, p53, guardian of the genome. *Nature*, Jul 1992, vol. 358, no. 6381, p. 15–16. DOI: https://doi.org/10.1038/358015a0.

LANE, D. P., CHEOK, C. F., BROWN, C., MADHUMALAR, A., GHADESSY, F. J., VERMA, C. Mdm2 and p53 are highly conserved from placozoans to man. *Cell Cycle*, Feb 2010, vol. 9, no. 3, p. 540–547. DOI: https://doi.org/10.4161/cc.9.3.10516.

LANE, Nick. *Power, Sex, Suicide. Mitochondria and the Meaning of Life.* New York, NY, USA : Oxford University Press, 2005. 354 p. ISBN 978-0-19-280481-5.

LAPTENKO, O., PRIVES, C. Transcriptional regulation by p53: one protein, many possibilities. *Cell Death and Differentiation*, Jun 2006, vol. 13, no. 6, p. 951–961. DOI: https://doi.org/10.1038/sj.cdd.4401916.

LAUDER, S. Mateřství u ledu. *Respekt*, 3.–9. listopadu 2014, no. 45, p. 19–22.

LAZEBNIK, Y. Can a biologist fix a radio? – or, what I learned while studying apoptosis. *Biochemistry (Moscow)*, Dec 2004, vol. 69, no. 12, p. 1403–1406. DOI: https://doi.org/10.1007/s10541-005-0088-1.

LAZEBNIK, Y. What are the hallmarks of cancer? *Nature Reviews. Cancer*, Apr 2010, vol. 10, no. 4, p. 232–233. DOI: https://doi.org/10.1038/nrc2827.

LENGAUER, C., KINZLER, K. W., VOGELSTEIN, B. Genetic instabilities in human cancer. *Nature*, Dec 1998, vol. 396, no. 6712, p. 643–649. DOI: https://doi.org/10.1038/25292.

LETAI, A. G. Diagnosing and exploiting cancer's addiction to blocks in apoptosis. *Nature Reviews. Cancer*, Feb 2008, vol. 8, no. 2, p. 121–132. DOI: https://doi.org/10.1038/nrc2297.

LEVIN, T. C., GREANEY, A. J., WETZEL, L., KING, N. The Rosetteless gene controls development in the choanoflagellate S. rosetta. *eLife*, Oct 2014, vol. 3, p. 3e04070. DOI: https://doi.org/10.7554/eLife.04070.

LEVINE, A. J., OREN, M. The first 30 years of p53: growing ever more complex. *Nature Reviews. Cancer*, Oct 2009, vol. 9, no. 10, p. 749–758. DOI: https://doi.org/10.1038/nrc2723.

LEVY, A. How evolution builds genes from scratch. *Nature*, Oct 2019, vol. 574, no. 7778, p. 314–316. DOI: https://doi.org/10.1038/d41586-019-03061-x.

LIN, T., LIN, Y. p53 switches off pluripotency on differentiaton. *Stem Cell Research and Therapy*, Feb 2017, vol. 8, no. 1, p. 44. DOI: https://doi.org/10.1186/s13287-017-0498-1.

LINDNER, T. Nejde tady o závist. *Respekt*, 14.–20. září 2015, no. 38, p. 22–23.

LINKSVAYER, T. A., BUSCH, J. W., SMITH, C. R. Social supergenes of superorganisms: do supergenes play important roles in social evolution? *Bioessays*, Aug 2013, vol. 35, no. 8, p. 683–689. DOI: https://doi.org/10.1002/bies.201300038.

LIPOLD, Jan. S pravdou a láskou na věčné časy a někdy jinak. Rok 2019, nebo 1975? *Seznam Zprávy* [online]. 2019, [referred November 15, 2019]. <https://www.seznamzpravy. cz/clanek/komentar-s-pravdou-a-laskou-na-vecne-casy-a-nekdy-jinak-rok-2019-nebo-1975-82716?dop-ab-variant=100&seq-no=-5&source=hp&utm_medium=z-boxiku&utm_source=www.seznam.cz>.

LIPTON, Bruce H. *The Biology of Belief: Unleashing the Power of Consciousness, Matter and Miracles*. 1st Ed. Santa Rosa, CA, USA, 2005. 224 p. ISBN 0-9759914-7-7

LOBERG, R. D., BRADLEY, D. A., TOMLINS, S. A., CHINNAIYAN, A. M., PIENTA, K. J. The lethal phenotype of cancer: the molecular basis of death due to malignancy. *A Cancer Journal for Clinicians*, Jul–Aug 2007, vol. 57, no. 4, p. 225–241. DOI: https://doi. org/10.3322/canjclin.57.4.225.

LOEB, L. A. Mutator phenotype may be required for multistage carcinogenesis. *Cancer Research*, Jun 1991, vol. 51, no. 12, p. 3075–3079.

LÓPEZ-OTÍN, C., BLASCO, M. A., PARTRIDGE, L., SERRANO, M., KROEMER, G. The hallmarks of aging. *Cell*, Jun 2013, vol. 153, no. 6, p. 1194–1217. DOI: https://doi. org/10.1016/j.cell.2013.05.039.

LORENZ, Konrad. *Civilized Man's Eight Deadly Sins*. 1st Ed. Harcourt, New York, NY, USA, 1974. 107 p. ISBN 0-15-118061-X.

LOWE, J., SHATZ, M., RESNICK, M. A., MENENDEZ, D. Modulation of immune responses by the tumor suppressor p53. *BioDiscovery*, May 2013, vol. 8, no. 2, p. e8947. DOI: https://doi.org/10.7750/BioDiscovery.2013.8.2.

LU, W.-J., AMATRUDA, J. F., ABRAMS, J. M. p53 ancestry: gazing through an evolutionary lens. *Nature Reviews. Cancer*, Oct 2009, vol. 9, no. 10, p. 758–762. DOI: https://doi. org/10.1038/nrc2732.

LUNDBERG, A. S., WEINBERG, R. A. Control of the cell cycle and apoptosis. *European Journal of Cancer*, Apr 1999, vol. 35, no. 4, p. 531–539. DOI: https://doi.org/10.1016/S0959-8049(99)00046-5.

MACHÁČEK, J. Poučení z amerického vývoje. *Respekt*, 2. srpna 2011. <https://www.respekt.cz/audit-jana-machacka/pouceni-z-americkeho-vyvoje>.

MALATERRE, C. Organicism and reductionism in cancer research: Towards a systemic approach. *International Studies in the Philosophy of Science*, May 2007, vol. 21, no. 1, p. 57–73. DOI: https://doi.org/10.1080/02698590701305792.

MARLOWE, F. The patriarch hypothesis: an alternative explanation of menopause. *Human Nature*, Mar 2000, vol. 11, no. 1, p. 27–42. DOI: https://doi.org/10.1007/s12110-000-1001-7.

MARUSYK, A., ALMENDRO, V., POLYAK, K. Intra-tumour heterogeneity: a looking glass for cancer? *Nature Reviews. Cancer*, Apr 2012, vol. 12, no. 5, p. 323–334. DOI: https://doi.org/10.1038/nrc3261.

MARUSYK, A., POLYAK, K. Tumor heterogeneity: causes and consequences. *Biochimica et Biophysica Acta*, Jan 2010, vol. 1805, no. 1, p. 105–117. DOI: https://doi.org/10.1016/j.bbcan.2009.11.002.

MATĚJČEK, Zdeněk. *Psychologické eseje (z konce kariéry)*. 1. vyd. Praha : Karolinum, 2004. 209 p. ISBN 80-246-0892-8.

MATHON, N. F., LLOYD, A. C. Cell senescence and cancer. *Nature Reviews. Cancer*, Dec 2001, vol. 1, no. 3, p. 203–213. DOI: https://doi.org/10.1038/35106045.

MATSUURA, S., ITO E., TAUCHI, H., KOMATSU, K., IKEUCHI, T., KAJII, T. Chromosomal instability syndrome of total premature chromatid separation with mosaic variegated aneuploidy is defective in mitotic-spindle checkpoint. *American Journal of Human Genetics*, Aug 2000, vol. 67, no. 2, p. 483–486. DOI: https://doi.org/10.1086/303022.

MCALLISTER, S. S., WEINBERG, R. A. The tumor-induced systemic environment as a critical regulator of cancer progression and metastasis. *Nature Cell Biology*, Aug 2014, vol. 16, no. 8, p. 717–727. DOI: https://doi.org/10.1038/ncb3015.

MEEK, D. W. Regulation of the p53 response and its relationship to cancer. *The Biochemical Journal*, Aug 2015, vol. 469, no. 3, p. 325–346. DOI: https://doi.org/10.1042/BJ20150517.

MENENDEZ, D., INGA, A., RESNICK, M. A. The expanding universe of p53 targets. *Nature Reviews Cancer*, Oct 2009, vol. 9, no. 10, p. 724–737. DOI: https://doi.org/10.1038/nrc2730.

MENON, V., POVIRK, L. Involvement of p53 in the repair of DNA double strand breaks: Multifaceted roles of p53 in homologous recombination repair (HRR) and non-homologous end joining (NHEJ). *Sub-cellular Biochemistry*, Jul 2014, vol. 85, p. 321–336. DOI: https://doi.org/10.1007/978-94-017-9211-0_17.

MICHOD, R. E. Evolution of individuality during the transition from unicellular to multicellular life. *Proceedings of the National Academy of Sciences of the USA*, May 2007, vol. 104, Suppl. 1, p. 8613–8618. DOI: https://doi.org/10.1073/pnas.0701489104.

MILGRAM, Stanley. *Poslušnost vůči autoritě*. 1. vyd. Praha : Portál, 2017. 238 p. ISBN 978-80-262-1238-6.

MILLER, Gary W. *The exposome : a primer*. 1st printing, Oxford (UK) : Academic Press – Elsevier, 2014. 118 p. ISBN 978-0-12-417217-3.

MULLER, P. A. J., VOUSDEN, K. H., NORMAN, J. C. p53 and its mutants in tumor cell migration and invasion. *The Journal of Cell Biology*, Jan 2011, vol. 192, no. 2, p. 209–218. DOI: https://doi.org/10.1083/jcb.201009059.

MUNOS-ESPÍN, D., SERRANO, M. Cellular senescence: from physiology to pathology. *Nature Reviews. Molecular Cell Biology*, Jul 2014, vol. 15, no. 7, p. 482–496. DOI: https://doi.org/10.1038/nrm3823.

MUNZAROVÁ, Marta. Zamyšlení nad současnou problematikou euthanasie. *Dialog Evropa XXI.*, 1996, no. 2, p. 34–36. ISSN 1210-8332.

NARENDRA, B. L., REDDY, K. E., SHANTIKUMAR, S., RAMAKRISHNA, S. Immune system: a double-edged sword in cancer. *Inflammation Research*, Sep 2013, vol. 62, no. 9, p. 823–834. DOI: https://doi.org/10.1007/s00011-013-0645-9.

NAROD, S. A., FOULKES, W. D. BRCA1 and BRCA2: 1994 and beyond. *Nature Reviews. Cancer*, Oct 2004, vol. 4, no. 9, p. 665–677. DOI: https://doi.org/10.1038/nrc1431.

NEDELCU, A. M., TAN, C. Early diversification and complex evolutionary history of the p53 tumor suppressor gene family. *Development Genes and Evolution*, Dec 2007, vol. 217, no. 11–12, p. 801–806. DOI: https://doi.org/10.1007/s00427-007-0185-9.

NEGRINI, S., GORGOULIS, V. G., HALAZONETIS, T. D. Genomic instability – an evolving hallmark of cancer. *Nature Reviews. Cancer*, Mar 2010, vol. 11, no. 3, p. 220–228. DOI: https://doi.org/10.1038/nrm2858.

NELSON, C. M., BISSELL, M. J. Of extracellular matrix, scaffolds, and signaling: tissue architecture regulates development, homeostasis, and cancer. *Annual Review of Cell and Developmental Biology*, Nov 2006, vol. 22, p. 287–309. DOI: https://doi.org/10.1146/annurev.cellbio.22.010305.104315.

NEUMAJER, O. Jak vykouknout z informační bubliny. *Řízení školy*. Praha: Wolters Kluwer, 2017, vol. 14, no. 6, p. 47–50. ISSN 1214-8679.

NIKLAS, K. J., NEWMAN, S. A. The origins of multicellular organisms. *Evolution and Development*, Jan 2013, vol. 15, no. 1, p. 41–52. DOI: https://doi.org/10.1111/ede.12013.

NORDLING, C. O. A new theory on cancer – inducing mechanism. *British Journal of Cancer*, Mar 1953, vol. 7, no. 1, p. 68–72. DOI: https://doi.org/10.1038/bjc.1953.8.

NOWAK, M. A. Five rules for the evolution of cooperation. *Science*, Dec 2006, vol. 314, no. 5805, p. 1560–1563. DOI: https://doi.org/10.1126/science.1133755.

OFRIA, C., WILKE, C. O. Avida: a software platform for research in computational evolutionary biology. *Artificial Life*, Spring 2004, vol. 10, no. 2, p. 191–229. DOI: https://doi.org/10.1162/106454604773563612.

O'REILLY, M. S., BOEHM, T., SHING, Y., FUKAI, N., VASIOS, G., LANE, W. S., FLYNN, E., BIRKHEAD, J. R., OLSEN, B. R., FOLKMAN, J. Endostatin: an endogenous inhibitor of angiogenesis and tumor growth. *Cell*, Jan 1997, vol. 88, no. 2, p. 277–285. DOI: https://doi.org/10.1016/S0092-8674(00)81848-6.

O'REILLY, M. S., HOLMGREN, L., SHING, Y., CHEN, C., ROSENTHAL, R. A., MOSES, M., LANE, W. S., CAO, Y., SAGE, E. H., FOLKMAN, J. Angiostatin: a novel angiogenesis inhibitor that mediates the suppression of metastases by a Lewis lung carcinoma. *Cell*, Oct 1994, vol. 79, no. 2, p. 315–328. DOI: https://doi.org/10.1016/0092-8674(94)90200-3.

OREN, M., ROTTER, V. Mutant p53 gain-of-function in cancer. *Cold Spring Harbor Perspectives in Biology*, Feb 2010, vol. 2, no. 2, p. a001107. DOI: https://doi.org/10.1101/cshperspect.a001107.

ORR, B., COMPTON, D. A. A double-edged sword: how oncogenes and tumor suppressor genes can contribute to chromosomal instability. *Frontiers in Oncology*, Jun 2013, vol. 3, p. 164. DOI: https://doi.org/10.3389/fonc.2013.00164.

O'SULLIVAN, R. J., ALMOUZNI, G. Assembly of telomeric chromatin to create ALTernative endings. *Trends in Cell Biology*, Nov 2014, vol. 24, no. 11, p. 675–685. DOI: https://doi.org/10.1016/j.tcb.2014.07.007.

OVIDIUS Naso, Publius. *The Metamorphoses*. The second book. English translation by J. J. Howard. 1974. EBook 28621, p. 80, https://www.gutenberg.org/files/28621/28621-h/28621-h.htm#chapter3.

PARK, J.-H., ZHUANG, J., LI, J., HWANG, P. M. p53 as guardian of the mitochondrial genome. *FEBS Letters*, Apr 2016, vol. 590, no. 7, p. 924–934. DOI: https://doi.org/10.1002/1873-3468.12061.

PATTABIRAMAN, D. R., WEINBERG, R. A. Tackling the cancer stem cells – what challenges do they pose? *Nature Reviews. Drug Discovery*, Jul 2014, vol. 13, no. 7, p. 497–512. DOI: https://doi.org/10.1038/nrd4253.

PECORINO, Lauren. *Molecular biology of cancer*. 3rd printing, Oxford : University Press, 2012. 336 p. ISBN 978-0-19-957717-0.

PEHE, Jiří. Jak zadusit demokracii. *Novinky.cz* [online]. 2018, [referred May 9, 2018]. <https://www.novinky.cz/clanek/kultura-salon-jiri-pehe-jak-zadusit-demokracii-13814>.

PEINADO, H., ZHANG, H., MATEI, I. R., COSTA-SILVA, B., HOSHINO, A., RODRIGUES, G., PSAILA, B., KAPLAN, R. N., BROMBERG, J. F., KANG, Y., BISSELL, M. J., COX, T. R., GIACCIA, A. J., ERLER, J. T., HIRATSUKA, S., GHAJAR, C. M., LYDEN, D. Pre-metastatic niches: organ-specific homes for metastases. *Nature Reviews. Cancer*, May 2017, vol. 17, no. 5, p. 302–317. DOI: https://doi.org/10.1038/nrc.2017.6.

PELENGARIS, S., KHAN, M. Apoptosis, p. 251–278. In: PELENGARIS, Stella, KHAN, Michael (eds.) *The molecular biology of cancer*. 1st printing, 2006. Oxford (UK) : Blackwell Publishing. 531 p. ISBN 978-1-4051-1814-9.

PENET, M.-F., BHUJWALLA, Z. M. Cancer cachexia, recent advances, and future directions. *Cancer Journal*, Mar–Apr 2015, vol. 21, no. 2, p. 117–122. DOI: https://doi.org/10.1097/PPO.0000000000000100.

PEPPER, J. W., SPROUFFSKE, K., MALEY, C. C. Animal cell differentiation patterns suppress somatic evolution. *PLoS Computational Biology*, Dec 2007, vol. 3, no. 12, p. e250. DOI: https://doi.org/10.1371/journal.pcbi.0030250.

PEREIRA, B., FERREIRA, M. G. Sowing the seeds of cancer: telomeres and age-associated tumorigenesis. *Current Opinion in Oncology*, Jan 2013, vol. 25, no. 1, p. 93–98. DOI: https://doi.org/10.1097/CCO.0b013e32835b6358.

PERT, Candance B. *Molekuly emocí*. 2. dotisk. Olomouc : ANAG, 2016. 368 p. ISBN 978-80-7554-049-2.

PERT, C. B., DREHER, H. E., RUFF, M. R. The psychosomatic network: foundations of mind-body medicine. *Alternative Therapies in Health and Medicine*. Jul 1998, vol. 4, no. 4, p. 30–41.

PETO, R., ROE, F. J., LEE, P. N., LEVY, L., CLACK, J. Cancer and ageing in mice and men. *British Journal of Cancer*, Oct 1975, vol. 32, no. 4, p. 411–426. DOI: https://doi.org/10.1038/bjc.1975.242.

PETRUSEK, Miloslav. Masaryk na prahu nového tisíciletí. *Univerzita Karlova* [online]. 2000, [referred March 6, 2000]. <http://www.cuni.cz/UK-1245.html>.

PETRUZZELLI, M., WAGNER, E. F. Mechanisms of metabolic dysfunction in cancer-associated cachexia. *Genes and Development*, Mar 2016, vol. 30, no. 5, p. 489–501. DOI: https://doi.org/10.1101/gad.276733.115.

PETRŽELKA, Alexandr. NASA: Tahle civilizace spěje ke svému konci. *Právo* [online]. 2014, [referred April 12, 2014]. <https://www.novinky.cz/veda-skoly/clanek/nasa-tahle-civilizace-speje-ke-svemu-konci-225118>.

PFEIFFER, T., BONHOEFFER, S. An evolutionary scenario for the transition to undifferentiated multicellularity. *Proceedings of the National Academy of Sciences of the USA*, Feb 2003, vol. 100, no. 3, p. 1095–1098. DOI: https://doi.org/10.1073/pnas.0335420100.

PFEIFFER, T., SCHUSTER, S., BONHOEFFER, S. Cooperation and competition in the evolution of ATP-producing pathways. *Science*, Apr 2001, vol. 292, no. 5516, p. 504–507. DOI: https://doi.org/10.1126/science.1058079.

PIKETTY, Thomas. *Capital in the Twenty-First Century*. 1st printing, Belknap Press of Harvard University Press, 2014. 816 p. ISBN 978-0674430006.

PLATONI, Kara. Thinking outside the cell. *East Bay Express* [online]. 2007, [referred September 3, 2011]. <http://www.eastbayexpress.com/oakland/thinking-outside-the-cell/Content?oid=1087861/>.

POLYAK, K., WEINBERG, R. A. Transition between epithelial and mesenchymal states: acquisition of malignant and stem cell traits. *Nature Reviews. Cancer*, Apr 2009, vol. 9, no. 4, p. 265–273. DOI: https://doi.org/10.1038/nrc2620.

PORPORATO, P. E. Understanding cachexia as a cancer metabolism syndrome. *Oncogenesis*, Feb 2016, vol. 5, no. 2, p. e200. DOI: https://doi.org/10.1038/oncsis.2016.3.

POTTER, M., NEWPORT, E., MORTEN, K. J. The Warburg effect: 80 years on. *Biochemical Society Transactions*, Oct 2016, vol. 44, no. 5, p. 1499–1505. DOI: https://doi.org/10.1042/BST20160094.

POTTS, M. B., CAMERON, S. Cell lineage and cell death: Caenorhabditis elegans and cancer research. *Nature Reviews. Cancer*, Jan 2011, vol. 11, no. 1, p. 50–58. DOI: https://doi.org/10.1038/nrc2984.

PREKOPOVÁ, Jiřina. *Aby láska v rodinách proudila*. 1. vyd. Brno : Monika Vadasová-Elšíková, 2012. 199 p. ISBN 978-80-904991-0.

PREKOPOVÁ, Jiřina. *Nese mě řeka lásky*. 1. vyd. Brno : Cesta, 2007. 216 p. ISBN 978-80-7295-092-8.

PRUDIČOVÁ, Ivana. Jak přežít stárnutí. *Právo* [online]. 2015, [referred August 4, 2015]. <https://www.novinky.cz/zena/styl/374987-jak-prezit-starnuti.html>.

PŘIBÁŇ, Jiří. *Tyranizovaná spravedlnost*. 1. vyd. Praha : Portál, 2013. 280 p. ISBN 978-80-262-0491-6.

RADOGNA, F., DICATO, M., DIEDERICH, M. Cancer-type-specific crosstalk between autophagy, necroptosis and apoptosis as a pharmacological target. *Biochemical Pharmacology*, Mar 2015, vol. 94, no. 1, p. 1–11. DOI: https://doi.org/10.1016/j.bcp.2014.12.018.

REICHE, E. M. V., NUNES, S. O. V., MORIMOTO, H. K. Stress, depression, the immune system, and cancer. *The Lancet Oncology*, Oct 2004, vol. 5, no. 10, p. 617–625. DOI: https://doi.org/10.1016/S1470-2045(04)01597-9.

RELICHOVÁ, J. Genetická variabilita. Proč jsme každý jiný. *Vesmír*, červenec 1997, vol. 76, no. 7, p. 368.

RELICHOVÁ, Jiřina: *Genetika populací*. 1. vyd. Brno : Masarykova univerzita, 2009. 187 p. ISBN 978-80-210-4795-2.

RIBEIRO, A. L., OKAMOTO, O. K. Combined effects of pericytes in the tumor microenvironment. *Stem Cells International*, Apr 2015, vol. 2015, article ID 868475. DOI: https://doi.org/10.1155/2015/868475.

RIEDL, Jiří. *Sedm hlavních hříchů : diplomová práce*. Olomouc : Univerzita Palackého, Cyrilometodějská teologická fakulta, 2011. 70 p.

RINGEN, Stein. *Národ ďáblů*. 1. vyd. Brno : Masarykova univerzita, 2017. 232 p. ISBN 978-80-210-8506-0.

ROBBIANI, D. F., NUSSENZWEIG, M. C. Chromosome translocation, B cell lymphoma, and activation-induced cytidine deaminase. *Annual Review of Pathology*, Jan 2013, vol. 8, p. 79–103. DOI: https://doi.org/10.1146/annurev-pathol-020712-164004.

ROBERTS, Monty. *Muž, který naslouchá koním*. 2. vyd. Praha : Knižní klub, 2010a. 288 p. ISBN 978-80-242-2786-3.

ROBERTS, Monty. *O koních a lidech : muž, který naslouchá koním a mluví s lidmi.* 1. vyd. Praha : Ikar, 2010b. 280 p. ISBN 978-80-249-1425-1.

RODRÍGUEZ-PAREDES, M., ESTELLER, M. Cancer epigenetics reaches mainstream oncology. *Nature Medicine*, Mar 2011, vol. 17, no. 3, p. 330–339. DOI: https://doi.org/10.1038/nm.2305.

ROKAS, A. The molecular origins of multicellular transitions. *Current Opinion in Genetics and Development*, Dec 2008a, vol. 18, no. 6, p. 472–478. DOI: https://doi.org/10.1016/j.gde.2008.09.004.

ROKAS, A. The origins of multicellularity and the early history of the genetic toolkit for animal development. *Annual Review of Genetics*, Dec 2008b, vol. 42, p. 235–251. DOI: https://doi.org/10.1146/annurev.genet.42.110807.091513.

RYSKA, P., PRŮŠA, J. Co je příčinou korupce. *MF Dnes*, 16.–17. března 2013, p. 33.

ŘÍČAN, Pavel. *Agresivita a šikana mezi dětmi : jak dát dětem ve škole pocit bezpečí.* 1. vyd. Praha : Portál, 1995. 95 p. ISBN 80-7178-049-9.

SAETZLER, K., SONNENSCHEIN, C., SOTO, A. M. Systems biology beyond networks: generating order from disorder through self-organization. *Seminars in Cancer Biology*, Jun 2011, vol. 21, no. 3, p. 165–174. DOI: https://doi.org/10.1016/j.semcancer.2011.04.004.

SAWA-WEJKSZA, K., KANDEFER-SZERSZEŃ, M. Tumor-associated macrophages as target for antitumor therapy. *Archivum Immunologiae et Therapiae Experimentalis (Warsz)*, Apr 2018, vol. 66, no. 2, p. 97–111. DOI: https://doi.org/10.1007/s00005-017-0480-8.

SEDLÁČEK, Tomáš. *Ekonomie dobra a zla.* 1. vyd. Praha : 65. pole, 2009. 272 p. ISBN 978-80-903944-3-8.

SENGUPTA, S., HARRIS, C. C. p53: traffic cop at the crossroads of DNA repair and recombination. *Nature Reviews. Cancer*, Jan 2005, vol. 6, no. 1, p. 44–55. DOI: https://doi.org/10.1038/nrm1546.

SERRANO, D., D'AMOURS, D. When genome integrity and cell cycle decision collide: roles of polo kinases in cellular adaptation to DNA damage. *Systems and Synthetic Biology*, Sep 2014, vol. 8, no. 3, p. 195–203. DOI: https://doi.org/10.1007/s11693-014-9151-9.

SHALTIEL, I. A., KRENNING, L., BRUINSMA, W., MEDEMA, R. H. The same, only different – DNA damage checkpoints and their reversal throughout the cell cycle. *Journal of Cell Science*, Feb 2015, vol. 128, no. 4, p. 607–620. DOI: https://doi.org/10.1242/jcs.163766.

SHARMA, P., HU-LIESKOVAN, S., WARGO, J. A., RIBAS, A. Primary, adaptive and acquired resistance to cancer immunotherapy. *Cell*, Feb 2017, vol. 168, no. 4, p. 707–723. DOI: https://doi.org/10.1016/j.cell.2017.01.017.

SHAY, J. W., WRIGHT, W. E. Senescence and immortalization: role of telomeres and telomerase. *Carcinogenesis*, May 2005, vol. 26, no. 5, p. 867–874. DOI: https://doi.org/10.1093/carcin/bgh296.

SHEEN, Fulton John. *Pokoj v duši.* Řím : Křesťanská akademie, 1969. 292 p.

SHILOH, Y., ZIV, Y. The ATM protein kinase: regulating the cellular response to genotoxic stress, and more. *Nature Reviews. Cancer*, Apr 2013, vol. 14, no. 4, p. 197–210. DOI: https://doi.org/10.1038/nrm3546.

SHIMA, D. T., RUHRBERG, C. Angiogenesis. p. 411–423. In: PELENGARIS, Stella, KHAN, Michael (eds.) *The molecular biology of cancer*. 1st printing, Oxford (UK) : Blackwell Publishing, 2006. 531 p. ISBN 978-1-4051-1814-9.

SHUMAN MOSS, L. A., JENSEN-TAUBMAN, S., STETLER-STEVENSON, W. G. Matrix metalloproteinases: changing roles in tumor progression and metastasis. *The American Journal of Pathology*, Dec 2012, vol. 181, no. 6, p. 1895–1899. DOI: https://doi.org/10.1016/j.ajpath.2012.08.044.

SCHNEIDER, O. Kapitalismus a nerovnost. Proč je Pikettyho kniha tak populární. *Respekt*, 26. 5.–1. 6. 2014, no. 22, p. 37.

SCHREIBER, David Servan. *Jak čelit rakovině*. 1. vyd. Praha : Portál, 2010. 310 p. ISBN 978-80-7367-785-5.

SCHREIBER, R. D., OLD, L. J., SMYTH, M. J. Cancer immunoediting: integrating immunity's roles in cancer suppression and promotion. *Science*, Mar 2011, vol. 331, no. 6024, p. 1565–1570. DOI: https://doi.org/10.1126/science.1203486.

SCHWANDER, T., LIBBRECHT, R., KELLER, L. Supergenes and complex phenotypes. *Current Biology*, Mar 2014, vol. 24, no. 7, p. R288–R294. DOI: https://doi.org/10.1016/j.cub.2014.01.056.

SILBERGNAL, Stefan, DESPOPOULOS, Agamemnon. *Atlas fyziologie člověka*. Praha : Grada Publishing, 1993. 352 p. ISBN 80-85623-79-X.

SJÖBLOM, T., JONES, S., WOOD, L. D., PARSONS, D. W., LIN, J., BARBER, T. D., MANDELKER, D., LEARY, R. J., PTAK, J., SILLIMAN, N., SZABO, S., BUCKHAULTS, P., FARRELL, C., MEEH, P., MARKOWITZ, S. D., WILLIS, J., DAWSON, D., WILLSON, J. K., GAZDAR, A. F., HARTIGAN, J., WU, L., LIU, C., PARMIGIANI, G., PARK, B. H., BACHMAN, K. E., PAPADOPOULOS, N., VOGELSTEIN, B., KINZLER, K. W., VELCULESCU, V. E. The consensus coding sequences of human breast and colorectal cancers. *Science*, Oct 2006, vol. 314, no. 5797, p. 268–274. DOI: https://doi.org/10.1126/science.1133427.

SKLOOT, Rebecca. *Immortal Life of Henrietta Lacks*. New York : Random House Publishing, 2011. 381 p. ISBN 1400052181.

SOKOL, Jan. Rozhovor o lži. *KT* [online]. 2014, [referred March 10, 2014]. <http://www.jansokol.cz/2014/03/rozhovor-o-lzi/>.

SONDKA, Z., BAMFORD, S., COLE, C. G., WARD, S. A., DUNHAM, I., FORBES, S. A. The COSMIC cancer gene census: describing genetic dysfunction across all human cancers. *Nature Reviews. Cancer*, Nov 2018, vol. 18, no. 11, p. 696–705. DOI: https://doi.org/10.1038/s41568-018-0060-1.

SONNENSCHEIN, C., SOTO, A. M. An integrative approach toward biology, organisms, and cancer. *Methods in Molecular Biology*, Oct 2018, vol. 1702, p. 15–26. DOI: https://doi.org/10.1007/978-1-4939-7456-6_2.

SONNENSCHEIN, C., SOTO, A. M. Are times a "changing" in carcinogenesis? *Endocrinology*, Jan 2005, vol. 146, no. 1, p. 11–12. DOI: https://doi.org/10.1210/en.2004-1376.

SONNENSCHEIN, C., SOTO, A. M. Carcinogenesis explained within the context of a theory of organisms. *Progress in Biophysics and Molecular Biology*, Oct 2016, vol. 122, no. 1, p. 70–76. DOI: https://doi.org/10.1016/j.pbiomolbio.2016.07.004.

SONNENSCHEIN, C., SOTO, A. M. Somatic mutation theory of carcinogenesis: why it should be dropped and replaced. *Molecular Carcinogenesis*, Dec 2000, vol. 29, no. 4, p. 205–211. DOI: https://doi.org/10.1002/1098-2744(200012)29:4<205::AID-MC1002>3.0.CO;2-W.

SONNENSCHEIN, C., SOTO, A. M. The death of the cancer cell. *Cancer Research*, Jul 2011, vol. 71, no. 13, p. 4334–4337. DOI: https://doi.org/10.1158/0008-5472.CAN-11-0639.

SONNENSCHEIN, C., SOTO, A. M. *The society of cells: Cancer and control of cell proliferation.* New York : Springer-Verlag, 1999. 154 p. ISBN 0 387 91583 4.

SONNENSCHEIN, C., SOTO, A. M. Theories of Carcinogenesis: an emerging perspective. *Seminars in Cancer Biology*, Oct 2008, vol. 18, no. 5, p. 372–377. DOI: https://doi.org/10.1016/j.semcancer.2008.03.012.

SONTAG, Susan. *Illness as Metaphor*. New York, USA : Farrar, Straus and Giroux, 1977. 88 p. ISBN 0-374-17443-1.

SONTAG, Susan. *AIDS and its Metaphors*. New York, USA : Farrar, Straus and Giroux, 1989. 95 p. ISBN 0374102570. Referred from https://monoskop.org/images/d/d3/Susan_Sontag_AIDS_and_Its_Metaphors_1989.pdf, pp. 6–7.

SOTO, A. M., LONGO, G., MONTÉVIL, M., SONNENSCHEIN, C. The biological default state of cell proliferation with variation and motility, a fundamental principle for a theory of organisms. *Progress in Biophysics and Molecular Biology*, Oct 2016, vol. 122, no. 1, p. 16–23. DOI: https://doi.org/10.1016/j.pbiomolbio.2016.06.006.

SOTO, A. M., SONNENSCHEIN, C. Emergentism as a default: Cancer as a problem of tissue organization. *Journal of Biosciences*, Feb 2005, vol. 30, no. 1, p. 103–118. DOI: https://doi.org/10.1007/BF02705155.

SOTO, A. M., SONNENSCHEIN, C. One hundred years of somatic mutation theory of carcinogenesis: Is it time to switch? *Bioessays*, Jan 2014, vol. 36, no. 1, p. 118–120. DOI: https://doi.org/10.1002/bies.201300160.

SOTO, A. M., SONNENSCHEIN, C. Reductionism, organicism, and causality in the biomedical sciences: a critique. *Perspectives in Biology and Medicine*, 2018, vol. 61, no. 4, p. 489–502. DOI: https://doi.org/10.1353/pbm.2018.0059.

SOTO, A. M., SONNENSCHEIN, C. The somatic mutation theory of cancer: growing problems with the paradigm? *BioEssays*, Oct 2004, vol. 26, no. 10, p. 1097–1107. DOI: https://doi.org/10.1002/bies.20087.

SOUKALOVÁ, K. Kleveta jako prostředek sociální kontroly. *Antropowebzin*, 2015, no. 1–2, p. 19–29. ISSN 1801-8807.

SOUSSI, T. The history of p53. A perfect example of the drawbacks of scientific paradigms. *EMBO Reports*, Nov 2010, vol. 11, no. 11, p. 822–826. DOI: https://doi.org/10.1038/embor.2010.159.

SPENCER, H. *The principles of sociology*. 3rd edition. London/Edinburg : Williams and Norgate, 1885. 919 p.

SPENCER, V. A., XU, R., BISSELL, M. J. Gene expression in the third dimension: The ECM-nucleus connection. *Journal of Mammary Gland Biology and Neoplasia*, Mar 2010, vol. 15, no. 1, p. 65–71. DOI: https://doi.org/10.1007/s10911-010-9163-3.

SRIVASTAVA, M., BEGOVIC, E., CHAPMAN, J., PUTNAM, N. H., HELLSTEN, U., KAWASHIMA, T., KUO, A., MITROS, T., SALAMOV, A., CARPENTER, M. L., SIGNOROVITCH, A. Y., MORENO, M. A., KAMM, K., GRIMWOOD, J., SCHMUTZ, J., SHAPIRO, H., GRIGORIEV, I. V., BUSS, L. W., SCHIERWATER, B., DELLAPORTA, S. L., ROKHSAR, D. S. The Trichoplax Genome and the Nature of Placozoans. *Nature*, Aug 2008, vol. 454, no. 7207, p. 955–960. DOI: https://doi.org/10.1038/nature07191.

SRIVASTAVA, M., SIMAKOV, O., CHAPMAN, J., FAHEY, B., GAUTHIER, M. E. A., MITROS, T., RICHARDS, G. S., CONACO, C., DACRE, M., HELLSTEN, U., LARROUX, C., PUTNAM, N. H., STANKE, M., ADAMSKA, M., DARLING, A., DEGNAN, S. M., OAKLEY, T. H., PLACHETZKI, D. C., ZHAI, Y., ADAMSKI, M., CALCINO, A., CUMMINS, S. F., GOODSTEIN, D. M., HARRIS, C., JACKSON, D. J., LEYS, S. P., SHU, S., WOODCROFT, B. J., VERVOORT, M., KOSIK, K. S., MANNING, G., DEGNAN, B. M., ROKHSAR, D. S. The Amphimedon queenslandica genome and the evolution of animal complexity. *Nature*, Aug 2010, vol. 466, no. 7307, p. 720–726. DOI: https://doi.org/10.1038/nature09201.

STANLEY, S. M. An ecological theory for sudden origin of multicellular life in the late precambrian. *Proceedings of the National Academy of Sciences of the USA*, May 1973, vol. 70, no. 5, p. 1486–1489. DOI: https://doi.org/10.1073/pnas.70.5.1486.

STEFANO, G. B., KREAM, R. M. Cancer: Mitochondrial origins. *Medical Science Monitor*, Dec 2015, vol. 21, p. 3736–3739. DOI: https://doi.org/10.12659/MSM.895990.

STEWART, T. A., MINTZ, B. Successive generations of mice produced from an established culture line of euploid teratocarcinoma cells. *Proceedings of the National Academy of Sciences of the USA*, Oct 1981, vol. 78, no. 10, p. 6314–6318. DOI: https://doi.org/10.1073/pnas.78.10.6314.

STRÁNSKÝ, M. Definice lži. *Pro-Fil*, 2012, vol. 13, no. 1, p. 29–38. ISSN 1212-9097. DOI: https://doi.org/10.5817/pf13-1-287.

SYLJUASEN, R. G. Checkpoint adaptation in human cells. *Oncogene*, Aug 2007, vol. 26, no. 40, p. 5833–5839. DOI: https://doi.org/10.1038/sj.onc.1210402.

SZATHMÁRY, E. Toward major evolutionary transitions theory 2.0. *Proceedings of the National Academy of Sciences of the USA*, Aug 2015, vol. 112, no. 33, p. 10104–10111. DOI: https://doi.org/10.1073/pnas.1421398112.

SZATHMÁRY, E., SMITH, J. M. The major evolutionary transitions. *Nature*, May 1995, vol. 374, no. 6519, p. 227–232. DOI: https://doi.org/10.1038/374227a0.

ŠIKLOVÁ, Jiřina. Připravte se na šedivější svět. Sen o prodloužení lidského věku se naplňuje. *iDNES.cz* [online]. 2010, [referred Fabruary 2, 2010]. <http://zpravy.idnes.cz/jirina-siklova-pripravte-se-na-sedivejsi-svet-sen-o-prodlouzeni-lidskeho-veku-se-naplnuje-gwq-/zpr_archiv.aspx?c=A100128_095846_kavarna_chu>.

ŠIKLOVÁ, Jiřina. *Vyhoštěná smrt*. 1. vyd. Praha : Kalich, 2013. 128 p. ISBN 978-80-7017-197-4.

ŠIMEČKA, Martin M. Kapitál je zpátky. *Respekt*, 12.–18. květen 2014, no. 20, p. 38–40.

ŠKABRAHA, Martin. Nad knihou Thomase Pikettyho Kapitál v 21. století: Čas jsou peníze. *Salon, Právo* [online]. 2015, [referred September 16, 2015]. <https://www.novinky.cz/kultura/salon/379079-nad-knihou-thomase-pikettyho-kapital-v-21-stoleti-cas-jsou-penize.html>.

ŠŤASTNÝ, M., ŘÍHOVÁ, B. Únikové strategie nádorů pozornosti imunitního systému. *Klinická onkologie*, 2015, vol. 28, Suppl. 4, p. 4S28–4S37. DOI: https://doi.org/10.14735/amko20154S28.

TAM, W. L., WEINBERG, R. A. The epigenetics of epithelial-mesenchymal plasticity in cancer. *Nature Medicine*, Nov 2013, vol. 19, no. 11, p. 1438–1449. DOI: https://doi.org/10.1038/nm.3336.

TANG, S., NING, Q., YANG, L., MO, Z., TANG, S. Mechanisms of immune escape in the cancer immune cycle. *International Immunopharmacology*, Sep 2020, vol. 86, p. 106700. DOI: https://doi.org/10.1016/j.intimp.2020.106700.

TATE, J. G., BAMFORD, S., JUBB, H. C., SONDKA, Z., BEARE, D. M., BINDAL, N., BOUTSELAKIS, H., COLE, C. G., CREATORE, C., DAWSON, E., FISH, P., HARSHA, B., HATHAWAY, C., JUPE, S. C., KOK, C. Y., NOBLE, K., PONTING, L., RAMSHAW, C. C., RYE, C. E., SPEEDY, H. E., STEFANCSIK, R., THOMPSON, S. L., WANG, S., WARD, S., CAMBELL, P. J., FORBES, S. A. COSMIC: the catalogue of somatic mutations in cancer. *Nucleic Acids Research*, Jan 2019, vol. 47, no. D1, p. D941–D947. DOI: https://doi.org/10.1093/nar/gky1015.

TAUTZ, Jürgen. *Fenomenální včely : biologie včelstva jako superorganizmu*. 1. vyd. Praha : Brázda, 2009. 286 p. ISBN 978-80-209-0379-2.

TEODORO, J. G., EVANS, S. K., GREEN, M. R. Inhibition of tumor angiogenesis by p53: a new role for the guardian of the genome. *Journal of Molecular Medicine (Berlin, Germany)*, Nov 2007, vol. 85, no. 11, p. 1175–1186. DOI: https://doi.org/10.1007/s00109-007-0221-2.

THEOHARIDES, T. C., CONTI, P. Mast cells: the JEKYLL and HYDE of tumor growth. *Trends in Immunology*, May 2004, vol. 25, no. 5, p. 235–241. DOI: https://doi.org/10.1016/j.it.2004.02.013.

THOMMEN, D. S., SCHUMACHER, T. N. T cell dysfunction in cancer. *Cancer Cell*, Apr 2018, vol. 33, no. 4, p. 547–562. DOI: https://doi.org/10.1016/j.ccell.2018.03.012.

THORNBERRY, N. A., LAZEBNIK, Y. Caspases: enemies within. *Science*, Aug 1998, vol. 282, no. 5381, p. 1312–1316. DOI: https://doi.org/10.1126/science.281.5381.1312.

TIMP, W., FEINBERG, A. P. Cancer as a dysregulated epigenome allowing cellular growth advantage at the expense of the host. *Nature Reviews. Cancer*, Jul 2013, vol. 13, no. 7, p. 497–510. DOI: https://doi.org/10.1038/nrc3486.

TOMLINSON, I. P., NOVELLI, M. R., BODMER, W. F. The mutation rate and cancer. *Proceedings of the National Academy of Sciences of the USA*, Dec 1996, vol. 93, no. 25, p. 14800–14803. DOI: https://doi.org/10.1073/pnas.93.25.14800.

TOOR, S. M., NAIR, V. S., DECOCK, J., ELKORD, E. Immune checkpoints in the tumor microenvironment. *Seminars in Cancer Biology*, Oct 2020, vol. 65, p. 1–12. DOI: https://doi.org/10.1016/j.semcancer.2019.06.021.

TRACHTOVÁ, Zdeňka, JIŘIČKA, Jan. Smrt po česku: děláme, že neexistuje, a končíme v nedůstojných podmínkách. *iDNES.cz* [online]. 2014, [referred September 1, 2014]. <http://zpravy.idnes.cz/umirani-v-cesku-0i0-/domaci.aspx?c=A140609_091133_domaci_zt>.

TŘEŠŇÁK, P. Cena za dobrou smrt. *Respekt*, 11.–17. listopad 2013, no. 46, p. 50–56.

TŘEŠŇÁK, P. Domov máme v hlavě. *Respekt*, 2.–8. listopad 2015, no. 45, p. 45–48.

TURNER, N. C., REIS-FILHO, J. S. Genetic heterogeneity and cancer drug resistance. *The Lancet Oncology*, Apr 2012, vol. 3, no. 4, p. e178–185. DOI: https://doi.org/10.1016/S1470-2045(11)70335-7.

URBAN, T. Jak obrátit zlo naruby. *MF Dnes*, 15. října 2011, p. 34–38.

VACKOVÁ, Barbora, GALČANOVÁ, Lucie, KVAPILOVÁ BARTOŠOVÁ, Michaela, KALA, Lukáš. *Sami doma*. 1. vyd. Červený Kostelec : Pavel Mervart / Brno : Masarykova univerzita, 2014. 264 p. ISBN 978-80-7465-185-4 / 978-80-210-8123-9.

VALASTYAN, S., WEINBERG, R. A. Tumor metastasis: Molecular insights and evolving paradigms. *Cell*, Oct 2011, vol. 147, no. 2, p. 275–292. DOI: https://doi.org/10.1016/j.cell.2011.09.024.

VALENTA, Martin. Teorie rozbitého okna: Pořádkem proti zločinnosti. *PsychoLogOn*, 2015. <https://manipulatori.cz/teorie-rozbiteho-okna-poradkem-proti-zlocinnosti/>.

VAN DER BRUGGEN, P., TRAVERSARI, C., CHOMEZ, P., LURQUIN, C., DE PLAEN, E., VAN DEN EYNDE, B., KNUTH, A., BOON, T. A gene encoding an antigen recognized by cytolytic T lymphocytes on a human melanoma. *Science*, Dec 1991, vol. 254, no. 5038, p. 1643–1647. DOI: https://doi.org/10.1126/science.1840703.

VAN DEURSEN, J. M. The role of senescent cells in ageing. *Nature*, May 2014, vol. 509, no. 7501, p. 439–446. DOI: https://doi.org/10.1038/nature13193.

VAN GENT, D. C., HOEIJMAKERS, J. H. J., KANAAR, R. Chromosomal stability and the DNA double-stranded break connection. *Nature Reviews. Genetics*, Mar 2001, vol. 2, no. 3, p. 196–206. DOI: https://doi.org/10.1038/35056049.

VAVROŇ, Jiří. Lidé si stárnutí nepřipouštějí, chtějí být navždy mladí a výkonní. *Právo* [online]. 2015, [referred September 8, 2015]. <http://www.novinky.cz/domaci/379903-lide-si-starnuti-nepripousteji-chteji-byt-navzdy-mladi-a-vykonni.html>.

VEVERKA, Miroslav. *Evoluce svým vlastním tvůrcem.* 1. vyd. Praha : Prostor, 2013. 576 p. ISBN 978-80-7260-276-6.

VLASÁK, Zbyněk. Můj táta Matrix už nepochopil. Kamil Fila o filmech a (nejen) generačně rozdělené společnosti. *Salon, Právo* [online]. 2016, [referred May 11, 2016]. <https://www.novinky.cz/kultura/salon/402818-muj-tata-matrix-uz-nepochopil-kamil-fila-o-filmech-a-nejen-generacne-rozdelene-spolecnosti.html>.

VOUSDEN, K. H., LANE, D. P. p53 in health and disease. *Nature Reviews. Molecular Cell Biology*, Apr 2007, vol. 8, no. 4, p. 275–283. DOI: https://doi.org/10.1038/nrm2147.

VOUSDEN, K. H., PRIVES, C. Blinded by the light: the growing complexity of p53. *Cell*, May 2009, vol. 137, no. 3, p. 413–431. DOI: https://doi.org/10.1016/j.cell.2009.04.037.

VOUSDEN, K. H., RYAN, K. M. p53 and metabolism. *Nature Reviews. Cancer*, Oct 2009, vol. 9, no. 10, p. 691–700. DOI: https://doi.org/10.1038/nrc2715.

VRÁNKOVÁ, K., JAROŠ, M. Jak žít sám. Stále víc domácností má jen jednoho člena. *Respekt*, 31. říjen–6. listopad 2016, no. 44, p. 59–62.

VYAS, S., ZAGANJOR, E., HAIGIS, M. C. Mitochondria and cancer. *Cell*, Jul 2016, vol. 166, no. 3, p. 555–566. DOI: https://doi.org/10.1016/j.cell.2016.07.002.

VYSKOT, Boris. *Přehled vývojové biologie a genetiky.* 1. vyd. Praha : Ústav molekulární genetiky AV ČR, 1999. 241 p. ISBN 80-902588-1-6.

WANG, C., SAJI, M., JUSTINIANO, S. E., YUSOF, A. M., ZHANG, X., YU, L., FERNÁNDEZ, S., WAKELY, J. R. P., LA PERLE, K., NAKANISHI, H., POHLMAN, N., RINGEL, M. D. RCAN1-4 is a thyroid cancer growth and metastasis suppressor. *JCI Insight*, Mar 2017, vol. 2, no. 5, p. e90651. DOI: https://doi.org/10.1172/jci.insight.90651.

WANG, J. H. The role of activation-induced deaminase in antibody diversification and genomic instability. *Immunologic Research*, Mar 2013, vol. 55, no. 1–3, p. 287–297. DOI: https://doi.org/10.1007/s12026-012-8369-4.

WANG, J. Y. J. New link in a web of human genes. *Nature*, May 2000, vol. 405, no. 6785, p. 404–405. DOI: https://doi.org/10.1038/35013171.

WANG, P.-Y., ZHUANG, J., HWANG, P. M. p53: exercise capacity and metabolism. *Current Opinion in Oncology*, Jan 2012, vol. 24, no. 1, p. 76–82. DOI: https://doi.org/10.1097/CCO.0b013e32834de1d8.

WARBURG, O. On the origin of cancer cells. *Science*, Feb 1956, vol. 123, no. 3191, p. 309–314. DOI: https://doi.org/10.1126/science.123.3191.309.

WARBURG, O., WIND, F., NEGELEIN, E. The metabolism of tumors in the body. *The Journal of General Physiology*, Mar 1927, vol. 8, no. 6, p. 519–530. DOI: https://doi.org/10.1085/jgp.8.6.519.

WARD, P. S., THOMPSON, C. B. Metabolic reprogramming: A cancer hallmark even Warburg did not anticipate. *Cancer Cell*, Mar 2012, vol. 21, no. 3, p. 297–308. DOI: https://doi.org/10.1016/j.ccr.2012.02.014.

WEAVER, B. A., SILK, A. D., MONTAGNA, C. A., VERDIER-PINARD, P., CLEVELAND, D. W. Aneuploidy acts both oncogenically and as a tumor suppressor. *Cancer Cell*, Jan 2007, vol. 11, no. 1, p. 25–36. DOI: https://doi.org/10.1016/j.ccr.2006.12.003.

WEAVER, V. M., PETERSEN, O. W., WANG, F., LARABELL, C. A., BRIAND, P., DAMSKY, C., BISSELL, M. J. Reversion of the malignant phenotype of human breast cells in three-dimensional culture and in vivo by integrin blocking antibodies. *The Journal of Cell Biology*, Apr 1997, vol. 13, no. 1, p. 231–245. DOI: https://doi.org/10.1083/jcb.137.1.231.

WEIGELT, B., BISSELL, M. J. Unraveling the microenvironmental influences on the normal mammary gland and breast cancer. *Seminars in Cancer Biology*, Oct 2008, vol. 18, no. 5, p. 311–321. DOI: https://doi.org/10.1016/j.semcancer.2008.03.013.

WEINBERG, R. A. Coming full circle – From endless complexity to simplicity and back again. *Cell*, Mar 2014, vol. 157, no. 1, p. 267–271. DOI: https://doi.org/10.1016/j.cell.2014.03.004.

WEINBERG, R. A. Leaving home early: Reexamination of the canonical models of tumor progression. *Cancer Cell*, Oct 2008, vol. 14, no. 4, p. 283–284. DOI: https://doi.org/10.1016/j.ccr.2008.09.009.

WEINBERG, R. A. Oncogenes, tumor suppressor genes, and cell transformation: Trying to put it all together. p. 1–13. In: BRUGGE, Joan, CURRAN, Tom, HARLOW, Ed, McCORMICK, Frank. *Origin of human cancer: A comprehensive review*. 1st printing, Plainview, N.Y.: Cold Spring Harbor Laboratory Press, 1991. 904 p. ISBN 0-87969-404-1.

WEINBERG, Robert A. *One renegade cell : how cancer begins*. 1st printing. New York : Basic Books, 1998. 170 p. ISBN 978-0465072750.

WEINBERG, Robert A. *The Biology of Cancer*. 1st edition. New York : GS Garland Science, 2007. 864 p. ISBN 0-8153-4076-1.

WEINBERG, Robert A. *The Biology of Cancer*. 2nd edition. New York : GS Garland Science, 2014. 876 p. ISBN 978-0-8153-4220-5.

WEITEKAMP, C. A., LIBBRECHT, R., KELLER, L. Genetics and evolution of social behavior in insects. *Annual Review of Genetics*, Nov 2017, vol. 51, p. 219–239. DOI: https://doi.org/10.1146/annurev-genet-120116-024515.

WEST, S. A., FISHER, R. M., GARDNER, A., KIERS, E. T. Major evolutionary transitions in individuality. *Proceedings of the National Academy of Sciences of the USA*, Aug 2015, vol. 112, no. 33, p. 10112–10119. DOI: https://doi.org/10.1073/pnas.1421402112.

WHEELER, W. M. The ant-colony as an organism. *Journal of Morphology*, Jun 1911, vol. 22, no. 2, p. 307–325. DOI: https://doi.org/10.1002/jmor.1050220206.

WHITE, E., DIPAOLA, R. S. The double-edged sword of autophagy modulation in cancer. *Clinical Cancer Research*, Sep 2009, vol. 15, no. 17, p. 5308–5319. DOI: https://doi.org/10.1158/1078-0432.CCR-07-5023.

WILD, C. P. Complementing the genome with an "exposome": the outstanding challenge of environmental exposure measurement in molecular epidemiology. *Cancer Epidemiology, Biomarkers and Prevention*, Aug 2005, vol. 14, no. 8, p. 1847–1850. DOI: https://doi.org/10.1158/1055-9965.EPI-05-0456.

WILSON, Edward Osborne. *Smysl lidské existence*. 1. vyd. Praha : Argo : Dokořán, 2016. 184 p. ISBN 978-80-7363-766-8.

WILSON, J. Q., KELLING, G. L. Broken windows: The police and neighborhood safet. *Atlantic Monthly*, Mar 1982, vol. 249, no. 3, p. 29–38.

WOESE, C. R. A new biology for a new century. *Microbiology and Molecular Biology Reviews*, Jun 2004, vol. 68, no. 2, p. 173–186. DOI: https://doi.org/10.1128/MMBR.68.2.173-186.2004.

WOLKER, Jiří. *Zasvěcování srdce*. 1. vyd. výboru. Praha : Československý spisovatel, 1984. 256 p.

WOOD, L. D., PARSONS, D. W., JONES, S., LIN, J., SJÖBLOM, T., LEARY, R. J., SHEN, D., BOCA, S. M., BARBER, T., PTAK, J., SILLIMAN, N., SZABO, S., DEZSO, Z., USTYANKSKY, V., NIKOLSKAYA, T., NIKOLSKY, Y., KARCHIN, R., WILSON, P. A., KAMINKER, J. S., ZHANG, Z., CROSHAW, R., WILLIS, J., DAWSON, D., SHIPITSIN, M., WILLSON, J. K. V., SUKUMAR, S., POLYAK, K., PARK, B. H., PETHIYAGODA, C. L., PANT, P. V. K., BALLINGER, D. G., SPARKS, A. B., HARTIGAN, J., SMITH, D. R., SUH, E., PAPADOPOULOS, N., BUCKHAULTS, P., MARKOWITZ, S. D., PARMIGIANI, G., KINZLER, K. W., VELCULESCU, V. E., VOGELSTEIN, B. The genomic landscapes of human breast and colorectal cancers. *Science*, Nov 2007, vol. 318, no. 5853, p. 1108–1113. DOI: https://doi.org/10.1126/science.1145720.

WRIGHT, W. E., SHAY, J. W. Telomere-binding factors and general DNA repair. *Nature Genetics*, Feb 2005, vol. 37, no. 2, p. 116–118. DOI: https://doi.org/10.1038/ng0205-116.

XIN, H., LIU, D., SONGYANG, Z. The telosome/shelterin complex and its functions. *Genome Biology*, Sep 2008, vol. 9, no. 9, p. 232. DOI: https://doi.org/10.1186/gb-2008-9-9-232.

YANCOPOULOS, G. D., DAVIS, S., GALE, N. W., RUDGE, J. S., WIEGAND, S. J., HOLASH, J. Vascular-specific growth factors and blood vessel formation. *Nature*, Sep 2000, vol. 407, no. 6801, p. 242–248. DOI: https://doi.org/10.1038/35025215.

YONISH-ROUACH, E., RESNITZKY, D., LOTEM, J., SACHS, L., KIMCHI, A., OREN, M. Wild-type p53 induces apoptosis of myeloid leukaemic cells that is inhibited by interleukin-6. *Nature*, Jul 1991, vol. 352, no. 6333, p. 345–347. DOI: https://doi.org/10.1038/352345a0.

YOU, J. S., JONES, P. A. Cancer Genetics and Epigenetics: Two Sides of the Same Coin? *Cancer Cell*, Jul 2012, vol. 22, no. 1, p. 9–20. DOI: https://doi.org/10.1016/j.ccr.2012.06.008.

ZHANG, G., YANG, P., GUO, P., MIELE, L., SARKAR, F. H., WANG, Z., ZHOU, Q. Unraveling the mystery of cancer metabolism in the genesis of tumor-initiating cells and development of cancer. *Biochimica and Biophysica Acta*, Aug 2013, vol. 1836, no. 1, p. 49–59. DOI: https://doi.org/10.1016/j.bbcan.2013.03.001.

ZHANG, Y., WEINBERG, R. A. Epithelial-to-mesenchymal transition in cancer: complexity and opportunities. *Frontiers of Medicine*, Aug 2018, vol. 12, no. 4, p. 361–373. DOI: https://doi.org/10.1007/s11684-018-0656-6.

ZHOU, S., ZHAO, L., KUANG, M., ZHANG, B., LIANG, Z., YI, T., WEI, Y., ZHAO, X. Autophagy in tumorigenesis and cancer therapy: Dr. Jekyll or Mr. Hyde? *Cancer Letters*, Oct 2012, vol. 323, no. 2, p. 115–127. DOI: https://doi.org/10.1016/j.canlet.2012.02.017.

ZIMBARDO, Philip. *Luciferův efekt : Jak se z dobrých lidí stávají lidé zlí*. 1. vyd. Praha : Academia, 2014. 628 p. ISBN 978-80-200-2346-9.

ZONG, W.-X., RABINOWITZ, J. D., WHITE, E. Mitochondria and cancer. *Molecular Cell*, Mar 2016, vol. 61, no. 5, p. 667–676. DOI: https://doi.org/10.1016/j.molcel.2016.02.011.

ZRZAVÝ, J. O egoizmu všeho živého. *Vesmír*, únor 1998, vol. 77, no. 1, p. 67–70.

ZRZAVÝ, J. Počátky živočišné říše. *Vesmír*, červen 2006, vol. 85, no. 6, p. 365–368.

ZRZAVÝ, Jan, BURDA, Hynek, STORCH, David, BEGALLOVÁ, Sabine, MIHULKA, Stanislav. *Jak se dělá evoluce : labyrintem evoluční biologie*. 4. vyd. (2. v českém jazyce), Praha : Argo : Dokořán, 2017. 479 p. ISBN 978-80-7363-763-7.

ZRZAVÝ, Jan, STORCH, David, MIHULKA, Stanislav. *Jak se dělá evoluce : od sobeckého genu k rozmanitosti života*. 1. vyd. Praha : Paseka, 2004. 296 p. ISBN 80-7185-578-2.

ZWEIG, Stefan. *The World of Yesterday : An Autobiography*. New York, USA : The Viking Press, 1943, 318 p.

ŽALOUDÍK, Jan. *Vyhněte se rakovině aneb prevence zhoubných nádorů pro každého*. 1. vyd. Praha : Grada, 2008. 192 p. ISBN 978-80-247-2307-5.

ŽĎÁREK, Jan. *Hmyzí rodiny a státy*. 1. vyd. Praha : Academia, 2015. 582 p. ISBN 978-80-200-2225-7.

Name index

A

Améry, Jean 76, 77, 99
Ariely, Dan 127, 129, 238, 239, 267, 298

B

Bissell, Mina 135, 221–225, 241, 242
Byock, Ira 74, 76, 78

C

Crick, Francis 219, 234
Čapek, Karel 95, 96, 101, 102

D

Dawkins, Richard 276, 277
Drtilová, Jana 266, 267

F

Feinberg, Andrew 231
Folkman, Judah 131, 133–139, 160, 168, 230

G

Goethe, Johann Wolfgang 79
Grün, Anselm 95, 100, 102, 103, 104

H

Hanahan, Douglas 36, 37, 41, 133–137, 139,
    168, 174, 180, 205, 218, 225–228, 230,
    241, 279
Harari, Yuval Noach 60, 285, 291, 299
Hayflick, Leonard 84

H

Hofling, Charles 237
Honoré, Carl 51, 53, 127, 162, 163

J

Jakoby, Bernard 75, 78, 79, 81
Jobs, Steve 80, 81

K

Keller, Jan 156, 181, 182, 186–189
Kelling, George 236
Kessler, David 81, 96, 103
Klausner, Richard 225
Koukolík, František 157, 158, 190, 266, 267, 295
Kübler-Ross, Elisabeth 81, 95, 96, 103

L

Lacks, Henrietta 87
Lane, David 245
Lane, Nick 73–76, 78, 79, 108, 123, 125–127,
    148, 161, 206, 245, 251, 256, 257, 271
Lipton, Bruce H. 14, 15, 77, 235, 290
Lorenz, Konrad 11, 36, 39, 40, 41, 52, 53, 98,
    99, 102, 163, 181–184, 186, 190, 211, 223

M

Malaterre, Christophe 218, 221, 228, 230
Matějček, Zdeněk 128
Milgram, Stanley 236, 237, 238

N

Nordling, Carl O. 105

O

Ovidius Naso, Publius 13

P

Paget, Stephen 178
Piketty, Thomas 156, 157
Prekopová, Jiřina 184, 185, 186
Přibáň, Jiří 209, 210, 214, 215, 216, 240

R

Ringen, Stein 208, 214, 215
Roberts, Monty 184
Říčan, Pavel 76

S

Sappho 73
Sedláček, Tomáš 52, 154
Seneca 73
Sheen, Fulton John 81
Schwab, Klaus 51
Sonnenschein, Carlos 220, 221, 223, 226–229, 242
Sontag, Susan 13, 14, 15
Soto, Ana 220, 221, 223, 226–229, 242
Šiklová, Jiřina 75, 77, 81, 100

T

Tautz, Jürgen 30, 31, 270, 289
Třešňák, Petr 74, 75, 188

V

Valéry, Paul 217, 228
Veverka, Miroslav 98, 122, 125, 127, 285, 291, 299
Virchow, Rudolf 13, 14
von Bertalanffy, Ludwig 29, 30

W

Warburg, Otto 144, 146, 149, 151, 154, 249
Watson, James 69, 234
Weinberg, Robert Alan 28, 34–37, 41, 45, 86, 166, 170, 174, 176, 177, 178, 180, 200, 205, 217, 218, 221, 225, 228, 229, 241, 242, 259, 260, 279
Wilson, Edward Osborne 235, 236, 286, 288, 289, 291, 292, 294–297
Woese, Carl Richard 14, 15

Z

Zimbardo, Philip 237, 238
Zrzavý, Jan 268, 272, 277, 282, 294, 295, 297
Zweig, Stefan 129
Žďárek, Jan 286–289, 291, 292, 295

Subject index

A

adenosine triphosphate / ATP 141–146, 151, 275, 306

adhesion 135, 171–174, 178, 272

aging 84–87, 90–100, 102, 103, 104, 147, 183, 256, 276, 301, 305

agonism 215

allele 36, 254, 305, 306, 310

altruism 265, 276, 291, 293–298, 301

anaphase 43, 110

angioblast 132

angiogenesis 7, 36, 37, 131, 133–139, 151, 160, 161, 162, 164, 169, 174, 176, 179, 181, 197, 201, 216, 225, 230, 241, 249, 280, 300

angioma 54

anoikis 66, 72, 169, 171, 173, 181, 187, 188, 305

ant 263, 264, 286, 288, 290, 291, 294, 299, 309

antagonism 215

antagonistic pleiotropy 91, 94

antibody 123, 211, 305

antigen 62, 88, 123, 192, 193, 196, 197, 200, 251, 305, 308

apoptosis 7, 36, 37, 63–79, 89, 90, 96, 101, 108, 113, 115, 123, 126, 147–149, 174, 175, 200, 206, 241, 248, 251, 256, 257, 259, 264, 278, 300, 305, 306

ATM 108, 111

autophagy 69, 146–149, 162, 179, 250, 252

B

bee 30, 31, 32, 286, 294

blastocyst 20, 221, 258

blastomere 20, 23, 24

C

cadherin 171–175, 177

Caenorhabditis elegans 17, 63, 96, 259

cachexia 158, 159

cancer 7, 11, 12, 13, 15, 33–37, 39, 41, 48, 51, 53, 54, 56, 61, 74, 80, 92, 93, 95, 105, 111, 121, 124, 126, 127, 135, 158, 165, 166, 175, 185, 191, 193, 201, 210, 216–223, 225, 226, 228–230, 233, 241, 255, 256, 263, 279, 281, 300–302, 310

carcinogenesis 33–36, 53, 87, 88, 94, 105, 111, 112, 114, 128, 134, 147, 177, 218, 220, 224–227, 230, 231, 243, 245, 250, 255–257

carcinoma 87, 117, 166, 168, 173, 176, 193, 201, 310

caspase 66–69, 147

caste 286–289, 291

cell

    differentiated 23, 29, 44, 46, 89, 249, 258, 259, 261, 268, 275

    epithelial 18, 53, 117, 132, 166, 168, 169, 173–177, 222, 305

    germ 17, 20, 23, 32, 88, 193, 257, 259, 276

    malignant 34, 218

    mesenchymal 173, 175–177

    progenitor 23, 24, 72, 91, 257, 258, 260, 278, 309

stem 23, 24, 56, 88, 228, 249, 257–261, 278, 282, 309, 310
tumor 12, 32–36, 41, 48, 51, 54, 55, 60, 61, 77, 78, 83, 87–89, 98, 105, 110, 111, 113, 115, 117, 121, 127, 128, 131, 133, 134, 137–139, 144–149, 151, 154, 159–162, 165–167, 169–171, 173, 174, 176–181, 191, 193–197, 201, 202, 210, 211, 218, 220–222, 225–228, 230, 233, 241, 245, 249, 254, 256, 259, 260, 266, 267, 275, 280, 282, 283, 302, 307
centromere 22, 43
chaos 78, 122, 125, 127, 128, 162, 164, 208, 209, 217, 218
cheating 59, 127, 239, 279–283, 295–298, 300–302
choanoflagellates 257, 272
chromatid 22, 43, 109, 110
chromatin 91, 147, 306
chromosome 21, 22, 43, 63, 64, 85, 86, 88, 89, 98, 108, 110–113, 273, 305–308
cleavage 20, 24, 65–67, 135, 136, 147, 174, 306
clone 33–35, 83, 106, 113, 115, 116, 160, 306
collaboration 76, 252, 294, 298
colonization 166, 167, 170, 171, 175, 179
communication 24, 25, 30–32, 54–56, 59, 60, 92, 99, 159, 182, 184, 187, 190, 194, 202, 225, 226, 235, 271, 272, 288, 291, 292, 296, 306, 307
community 11, 17, 19, 25, 30, 36, 47, 52, 59, 60, 77, 100, 103, 122, 126, 157, 185, 187, 189, 208, 209, 211, 223, 226, 233, 234, 264, 274–276, 282, 286–289, 291, 293–296, 300, 301
cyclin 44–47, 247, 248
cytokine 25, 90, 159, 192–194, 196, 197, 199, 202, 306

D

deprived people 266, 267
differentiation 17, 19, 20, 23, 24, 26, 46, 61, 80, 87, 123, 174, 187, 192, 223, 258–260, 264, 268, 270–273, 276, 278–281, 287, 291, 306, 307

DNA
recombination 89, 306
repair 113, 118, 119, 129, 201, 251, 263, 264
double-edged sword 124, 147, 148, 197, 198, 205, 216
Drosophila 257, 259

E

ectoderm 20, 175
egoism 265, 282, 295
embryoblast 20
embryogenesis 18, 132, 175, 177
emergence / emergent 29, 34, 79, 125, 183, 267, 269, 270, 273, 274, 278, 280, 287, 289–293, 299, 300
endoderm 20, 175
endothelium / endothelial 21, 53, 54, 88, 132, 133, 135, 138, 139, 160, 161, 169, 170, 174, 199, 200, 226, 227, 230, 306, 309
epithelium / epithelial 18, 53, 54, 117, 132, 166, 168, 169, 173–177, 200, 201, 222, 228, 249, 281, 305
Escherichia coli 125, 271
eusociality 286, 287, 289, 291, 293
exosome 178, 307
exposome 124, 125
extracellular matrix 66, 90, 101, 135, 136, 138, 168, 169, 171–174, 178, 199, 200, 220, 222, 230, 271, 278, 280, 305, 307
extravasation 166, 167, 170, 171, 174, 175, 177

F

fascism 216
fibroblast 138, 175, 199, 200, 226, 227, 230, 307
fibrotization 200, 213, 307
fitness 91, 97, 125, 252, 265, 273–276, 287, 290, 293, 305, 307
fractal 285, 292

G

gamete 32, 257, 259, 276, 294, 308, 310
gastrulation 20, 175, 268

gene
    expression 20, 21, 23, 175, 231, 271, 307
    pool 277, 288, 307
genome 21, 22, 24, 32, 34, 35, 80, 106–109,
    111, 113, 115, 117, 121, 123–125,
    148, 149, 206, 219, 223, 225, 235, 245,
    250–252, 268–271, 274, 306, 307
germ-soma differentiation 276, 287
Gini coefficient 157
glucose 141, 143–146, 151, 154, 155, 162,
    249, 250, 256
glycolysis 141, 143–146, 149, 151, 162, 249,
    256, 275
gossip 59, 60, 296
grading 280, 307

H

heterogeneity 56, 113, 115, 116, 179, 194,
    196, 231, 291, 297
histogenesis 20
homeostasis 27, 28, 65, 71, 76, 83, 132, 147,
    199, 202, 207, 213, 222, 223, 278, 308
hypoxia 66, 136–138, 144–146, 160, 164,
    197, 246, 259, 263, 308

I

immunodeficiency 191, 308
immunoediting 194, 195, 197, 201
immunogenicity 194, 196, 210, 308
immunoglobulin 172, 174
inflammation
    acute 200, 201, 205, 206, 215, 250
    chronic 7, 36, 37, 199, 200–202, 205, 213,
        216, 225, 241, 250, 300
inhibitor 45–47, 69, 77, 135–139, 179,
    247–249, 260
innovation 269, 272, 274, 277–279, 282, 283,
    297, 298, 301
instability 7, 36, 37, 94, 105, 107–115,
    117–119, 121, 124, 125, 127, 128, 149,
    160, 162, 194, 196, 201, 219, 225, 231,
    241, 256, 267, 270, 300
integration 28, 67, 69, 188, 228, 265, 275,
    281, 286, 290
integrins 171–174
intravasation 166, 167, 169, 171, 174, 175, 177

K

karoshi 52
kinase 44–46, 247, 248
kinetochore 110, 308

L

lymphocyte 22, 23, 61, 62, 72, 88, 123, 124,
    191–193, 196, 197, 200, 305
lymphoma 61, 62, 72, 123, 124, 191

M

macrophage 63, 64, 90, 101, 169, 192, 193,
    197, 251, 308
membrane
    basement 132, 133, 138, 168–170,
        174–176, 279
    cytoplasmic 25, 63, 64, 171–173, 235
    mitochondrial 67, 69, 142, 143, 148
mesoderm 20, 175
metalloproteinase 90, 174, 197, 308
metaphor 13–15, 119, 206, 242, 263, 264,
    289, 302
metastasis 72, 166, 170, 174, 177–179, 197,
    225, 228, 249, 281
microenvironment 7, 28, 32, 36, 37, 75, 90,
    102, 145, 169, 170, 173, 175, 177, 179,
    180, 197, 198, 217, 220, 222–228, 231,
    241, 258, 259, 300
mitochondria 66–69, 73, 91, 106, 123, 124,
    142–145, 148, 149, 246, 248, 250, 252,
    308
mitotic spindle 43, 110, 111, 264, 308
mobilization 7, 199, 213–216, 264
morphogenesis 20, 173, 175, 223, 271, 291,
    292
morula 20, 258
mouse 131, 170, 221
multicellularity 8, 73, 267–275, 277, 278,
    280–282, 285–287, 297, 298, 300
mutation
    congenital / in the germ line 111, 254, 308
    dominant 36
    recessive 36
    somatic 218–223, 225, 226, 228, 230, 235,
        240

N

necrosis 63, 64, 69, 75, 77–79, 199, 308, 309
NF-κB 199, 251, 263, 264

O

oncogene 105, 217, 222, 245, 255, 263, 264, 309
ontogenesis 20
order 10, 19, 21, 32, 43, 44, 46, 47, 60, 69, 78, 80, 81, 122, 125, 128, 130, 145, 154, 157, 182, 188, 189, 192, 202, 208, 209, 217, 218, 237, 241, 266, 281, 287, 289, 290, 293, 296, 299
organotropism 178
overlap 7, 9, 11, 12, 13, 15, 29, 30, 41, 46, 69, 93, 163, 226, 233, 241, 242, 275, 278–280, 298
oxidative phosphorylation 141–146, 149, 151, 162, 249, 252, 256, 275
oxygen radical 90, 106, 123, 148, 149, 201, 250, 252

P

p53 8, 49, 72, 84, 89, 94, 108, 111, 113, 138, 146, 149, 245–257, 259–261, 263–270
paradigm 14, 227–230
pericyte 132, 133, 135, 138, 160, 226, 227, 309
PET / Positron Emission Tomography 154, 155, 158
phagocytosis 90, 102, 307
phosphorylation 45, 47, 141–146, 149, 151, 162, 248, 249, 252, 256, 275, 308, 309
populist 210, 216
progenitor 138
programmed cell death 20, 63, 65, 66, 69, 70, 72, 76–79, 89, 96, 115, 126, 137, 147, 148, 160, 161, 169, 171, 181, 206, 248, 250, 252, 271, 280, 282, 306
proliferation 20, 36, 47, 48, 53, 54, 61, 71, 72, 83, 87, 90, 115, 118, 123, 127, 129, 131, 135, 138, 145, 146, 151, 161, 162, 164, 192, 199–201, 220, 222, 226, 251, 264, 278–282, 306, 308, 309
protease 65, 66, 90, 174, 308, 309
proteostasis 91

R

RB 45, 47, 84, 89, 217, 248
receptor 25–27, 48, 55, 66, 68, 135, 171–174, 192, 197, 207, 235, 248, 305
replication 43, 85–89, 93–95
replication potential 7, 36, 37, 83, 84, 87, 88, 100, 113, 241, 280, 300
restriction point 45–47, 49, 50, 54, 108, 248
retinoblastoma 217, 310

S

selectin 171, 172, 174
selection 40, 52, 73, 113, 115, 116, 118, 122, 125, 195, 231, 257, 269, 273–277, 280, 283, 290, 291, 293–295, 307, 309
senescence 84, 86–95, 100, 101, 102, 108, 113, 114, 122, 137, 248–253, 263, 264, 278
shelterin 85
slippery slope 128, 129
sologamy 183
superorganism 30, 31, 288–291, 299

T

telomerase 84, 87–89, 113, 114, 258
telomere 22, 84–95, 100, 112–114, 246, 258, 264
telosome 85, 86, 113
transcription 41, 246–255
transcription factor 25, 45, 47, 145, 199, 246, 249, 251, 255, 258, 259, 263, 272, 310
transformation 41, 48, 71, 72, 88, 95, 105, 106, 110, 111, 113, 117, 121, 123, 128, 146, 149, 166, 175, 176, 194, 201, 206, 207, 217, 222, 228, 245, 254–256, 259, 260, 275
trophoblast 20
trypsin 83
tumor
    benign 165, 181
    dormant 134, 135, 148, 166, 177, 194, 195
    malignant 33, 41, 89, 110, 121, 165, 168, 181, 211, 221, 267, 310
    occult 135, 195, 222
    sporadic 112, 310

tumor suppressor 8, 35, 36, 49, 72, 84, 89,
    90, 94, 105, 108, 111–113, 123, 137, 138,
    146, 149, 162, 193, 217–219, 222, 226,
    245, 247–249, 253–257, 260, 263–265,
    267–270, 310

V

vasculature 131–133, 135, 136, 138, 139,
    144, 162, 167, 170, 176, 200
vasculogenesis 132, 138

Z

zygote 17, 18, 20–22, 289, 310

MASARYK
UNIVERSITY
MONOGRAPHS
Vol. 2

Jana Šmardová

WHAT TUMORS TEACH US
PARALLELS
IN CELL AND
HUMAN BEHAVIOR

Published as the 2nd volume in the Series "Masaryk University Monographs"
supported by Scientia est potentia fund.

Text by: Jana Šmardová
Illustrated by: Jana Koptíková
Translated by: Jan Šmarda
Language revision by: Benjamin J. Vail
Edited by: Alena Mizerová
Layout, cover design by: Jana Koptíková
Typesetting by: Radka Vyskočilová

First edition, 2023
Published by Masaryk University Press, Žerotínovo nám. 617/9, 601 77 Brno, CZ
Printed by: powerprint, Suchdolská 1018, 252 62 Horoměřice, CZ

ISBN 978-80-280-0376-0